Springer Optimization and Its Applications

VOLUME 99

Aims and Scope
Optimization has been expanding in all directions at an astonishing rate during the last few decades. New algorithmic and theoretical techniques have been developed, the diffusion into other disciplines has proceeded at a rapid pace, and our knowledge of all aspects of the field has grown even more profound. At the same time, one of the most striking trends in optimization is the constantly increasing emphasis on the interdisciplinary nature of the field. Optimization has been a basic tool in all areas of applied mathematics, engineering, medicine, economics, and other sciences.

The series *Springer Optimization and Its Applications* publishes undergraduate and graduate textbooks, monographs and state-of-the-art expository work that focus on algorithms for solving optimization problems and also study applications involving such problems. Some of the topics covered include nonlinear optimization (convex and nonconvex), network flow problems, stochastic optimization, optimal control, discrete optimization, multi-objective programming, description of software packages, approximation techniques and heuristic approaches.

More information about this series at http://www.springer.com/series/7393

Alexander J. Zaslavski

Turnpike Phenomenon and Infinite Horizon Optimal Control

 Springer

Alexander J. Zaslavski
Department of Mathematics
Technion - Israel Institute of Technology
Haifa, Israel

ISSN 1931-6828   ISSN 1931-6836 (electronic)
ISBN 978-3-319-35421-7   ISBN 978-3-319-08828-0 (eBook)
DOI 10.1007/978-3-319-08828-0
Springer Cham Heidelberg New York Dordrecht London

Mathematics Subject Classification (2010): 49J10, 49J27, 49J99, 90C31, 91A25

Printed on acid-free paper

Springer is part of Springer Science+Business Media (www.springer.com)

Preface

The monograph is devoted to the study of the structure of approximate solutions of nonconvex (nonconcave) optimal control and constrained variational problems considered on subintervals of a real line. It contains a number of recent results obtained by the author in the last 8 years. We present the results on properties of approximate solutions which are independent of the length of the interval, for all sufficiently large intervals. These results deal with the so-called turnpike property of optimal control problems. The term was first coined by P. Samuelson in 1948 when he showed that an efficient expanding economy would spend most of the time in the vicinity of a balanced equilibrium path (also called a von Neumann path). To have the turnpike property means, roughly speaking, that the approximate solutions of the problems are determined mainly by the objective function (integrand) and are essentially independent of the choice of interval and end point conditions, except in regions close to the end points. Now it is well known that the turnpike property is a general phenomenon which holds for large classes of variational problems. For these classes of problems, using the Baire category (generic) approach, it was shown that the turnpike property holds for a generic (typical) variational problem [56]. According to the generic approach we say that a property holds for a generic (typical) element of a complete metric space (or the property holds generically) if the set of all elements of the metric space possessing this property contains a G_δ everywhere dense subset of the metric space which is a countable intersection of open everywhere dense sets. In particular, in Chap. 2 of [56] we studied the turnpike property of approximate solutions of the variational problems with integrands f which belong to a complete metric space of functions $\mathcal{M}$ and showed that the turnpike property holds for a typical integrand $f \in \mathcal{M}$. In [73] we were interested in individual (nongeneric) turnpike results and in sufficient and necessary conditions for the turnpike phenomenon in the calculus of variations. In this book we are also interested in individual turnpike results but for optimal control problems.

In Chap. 1 we provide some preliminary knowledge on turnpike properties and discuss the structure of the book.

In Chap. 2 we study the turnpike phenomenon for discrete-time optimal control problems. In particular, these problems describe a general model of economic dynamics. For these problems the turnpike is a singleton. We establish the turnpike property for approximate solutions. For problems which satisfy concavity assumption common in the literature we study the structure of approximate solutions in the regions containing end points and obtain a full description of the structure of approximate solutions. We also study the stability of the turnpike phenomenon under small perturbations of objective functions and establish the existence of overtaking optimal solutions over infinite horizon.

The turnpike properties of autonomous variational problems with extended-valued integrands are studied in Chap. 3. For these integrands we establish the existence of overtaking optimal solutions over infinite horizon, compare different optimality criterions for infinite horizon problems, and establish a non-self-intersection property of overtaking optimal solutions. For convex integrands we study the structure of approximate solutions in the regions containing end points and obtain a full description of the structure of approximate solutions.

In Chap. 4 we consider large classes of infinite horizon optimal control problems without convexity (concavity) assumptions. These classes contain optimal control problems arising in economic dynamics which describe a general one-sector model of economic growth, optimal control problems which describe a general two-sector model of economic dynamics, and discrete-time and continuous-time infinite horizon optimal control problems with periodic cost functions. For these problems we establish the existence of optimal (good) solutions on infinite horizon.

In Chap. 5 we prove turnpike results for a class of dynamic discrete-time two-player zero-sum games. These results describe the structure of approximate solutions, for all sufficiently large intervals. We also show that for each initial state there exists a pair of overtaking equilibria strategies over an infinite horizon.

Rishon LeZion, Israel   Alexander J. Zaslavski
October 30, 2013

Contents

1 Introduction 1
 1.1 Convex Discrete-Time Problems 1
 1.2 The Turnpike Phenomenon 15
 1.3 Turnpike Results for Nonconcave (Nonconvex) Problems 17
 1.4 Existence Results for Nonconcave (Nonconvex) Problems 20
 1.5 Turnpike Results for Two-Player Zero-Sum Games 21

2 Turnpike Properties of Discrete-Time Problems 23
 2.1 Turnpike Results Without Convexity 23
 2.2 Auxiliary Results 28
 2.3 Completion of the Proof of Theorem 2.2 34
 2.4 Proof of Theorem 2.3 36
 2.5 Proofs of Theorems 2.4 and 2.5 39
 2.6 Proofs of Theorems 2.6 and 2.7 42
 2.7 Proof of Theorem 2.8 46
 2.8 Proof of Theorem 2.9 49
 2.9 Structure of Solutions in the Regions Containing End Points 51
 2.10 Proof of Theorem 2.22 54
 2.11 Proofs of Propositions 2.23–2.26 59
 2.12 Proof of Theorem 2.27 62
 2.13 Stability of a Turnpike Property Under Autonomous
       Perturbations 67
 2.14 Auxiliary Results for Theorems 2.29–2.32 69
 2.15 Proof of Theorem 2.29 76
 2.16 Proof of Theorem 2.30 79
 2.17 Proof of Theorems 2.31 and 2.32 80
 2.18 Stability of a Turnpike Property Under Nonautonomous
       Perturbations 85
 2.19 Proof of Theorem 2.38 88
 2.20 Proofs of Theorems 2.39 and 2.40 90

2.21   Stability for a Class of Optimal Control Systems   96
2.22   Proof of Theorem 2.41   99
2.23   Proof of Theorem 2.42   104
2.24   An Example   107
2.25   Optimal Control Systems with Discounting   108
2.26   Proof of Theorems 2.49 and 2.50   113
2.27   Proof of Theorem 2.51   120
2.28   Nonautonomous Discrete-Time Control System   123
2.29   Auxiliary Results for Theorem 2.55   125
2.30   (TP) Implies Properties (P35), (P36), and (P37)   128
2.31   A Basic Lemma for Theorem 2.55.....................................   129
2.32   Proof of Theorem 2.55   140
2.33   An Example   142

3   Variational Problems with Extended-Valued Integrands   147
3.1   Turnpike Results for Variational Problems   147
3.2   Three Propositions   151
3.3   Proof of Proposition 3.1   155
3.4   Auxiliary Results   157
3.5   Completion of the Proof of Theorem 3.2   165
3.6   Proof of Theorem 3.3   167
3.7   Proof of Theorem 3.4   170
3.8   Proof of Theorem 3.5   172
3.9   Examples   174
3.10   Behavior of Solutions in the Regions Containing End Points   178
3.11   Proof of Theorem 3.20   183
3.12   Proof of Proposition 3.21...   192
3.13   Proof of Theorem 3.22   200
3.14   Optimal Solutions for Infinite Horizon Problems   206
3.15   Proof of Theorem 3.32   207
3.16   Proofs of Theorem 3.33–3.35   211
3.17   A Property of Overtaking Optimality Functions   215

4   Infinite Horizon Problems   221
4.1   One-Dimensional Autonomous Problems   221
4.2   Auxiliary Results   225
4.3   Proof of Theorem 4.2...   240
4.4   One-Dimensional Nonautonomous Problems   245
4.5   Proof of Theorem 4.18   247
4.6   Auxiliary Results for Theorem 4.19   249
4.7   Proof of Theorem 4.19   259
4.8   Two-Dimensional Autonomous Problems...........................   260
4.9   Auxiliary Results for Theorems 4.23–4.25   265
4.10   Proof of Theorem 4.23   285
4.11   Proof of Theorem 4.24   290
4.12   Two-Dimensional Nonautonomous Problems.......................   291

4.13 Proof of Theorem 4.38 295
4.14 Auxiliary Results for Theorem 4.39 296
4.15 Proof of Theorem 4.39 312
4.16 Autonomous Discrete-Time Periodic Problems..................... 313
4.17 Variational Problems with Periodic Integrands..................... 318
4.18 Nonautonomous Discrete-Time Periodic Problems 325
4.19 Periodic Continuous-Time Problems 330

5 **Dynamic Discrete-Time Zero-Sum Games** 339
   5.1 Preliminaries 339
   5.2 Minimal Pairs of Sequences.. 341
   5.3 Main Results 344
   5.4 Auxiliary Results for Theorem 5.4................................. 345
   5.5 Proof of Theorem 5.4.. 353
   5.6 Preliminary Lemmas for Theorem 5.5............................... 354
   5.7 Proof of Theorems 5.5... 361

References... 365

Index.. 369

Chapter 1
Introduction

The study of optimal control problems and variational problems defined on infinite intervals and on sufficiently large intervals has been a rapidly growing area of research [3, 4, 8, 11–13, 18, 22, 24–27, 32, 34, 39–42, 44, 50, 51, 56, 70] which has various applications in engineering [1,29,76], in models of economic growth [2, 14–17,21,23,28,33,38,43,46–48,56], in infinite discrete models of solid-state physics related to dislocations in one-dimensional crystals [7, 49], and in the theory of thermodynamical equilibrium for materials [20, 30, 35–37].

In this chapter we discuss turnpike properties and optimality criterions over infinite horizon for a class of convex dynamic optimization problems and describe the structure of the book.

1.1 Convex Discrete-Time Problems

Let R^n be the n-dimensional Euclidean space with the inner product $\langle \cdot, \cdot \rangle$ which induces the norm

$$|x| = \left(\sum_{i=1}^{n} x_i^2 \right)^{1/2}, \quad x = (x_1, \ldots, x_n) \in R^n.$$

Let $v : R^n \times R^n \to R^1$ be a bounded from below function. We consider the minimization problem

$$\sum_{i=0}^{T-1} v(x_i, x_{i+1}) \to \min, \qquad (P_0)$$

such that $\{x_i\}_{i=0}^{T} \subset R^n$ and $x_0 = z$, $x_T = y$,

where T is a natural number and the points $y, z \in R^n$.

© Springer International Publishing Switzerland 2014
A.J. Zaslavski, *Turnpike Phenomenon and Infinite Horizon Optimal Control*,
Springer Optimization and Its Applications 99, DOI 10.1007/978-3-319-08828-0_1

The interest in discrete-time optimal problems of type (P_0) stems from the study of various optimization problems which can be reduced to it, e.g., continuous-time control systems which are represented by ordinary differential equations whose cost integrand contains a discounting factor [28], tracking problems in engineering [1, 29, 76], the study of Frenkel-Kontorova model [7, 49], and the analysis of a long slender bar of a polymeric material under tension in [20, 30, 35–37]. Optimization problems of the type (P_0) were considered in [50–52, 54].

In this section we suppose that the function $v : R^n \times R^n \to R^1$ is strictly convex and differentiable and satisfies the growth condition

$$v(y, z)/(|y| + |z|) \to \infty \text{ as } |y| + |z| \to \infty. \tag{1.1}$$

We intend to study the behavior of solutions of the problem (P_0) when the points y, z and the real number T vary and T is sufficiently large. Namely, we are interested to study a turnpike property of solutions of (P_0) which is independent of the length of the interval T, for all sufficiently large intervals. To have this property means, roughly speaking, that solutions of the optimal control problems are determined mainly by the objective function v and are essentially independent of T, y, and z. Turnpike properties are well known in mathematical economics. The term was first coined by Samuelson in 1948 (see [47]) where he showed that an efficient expanding economy would spend most of the time in the vicinity of a balanced equilibrium path (also called a von Neumann path). This property was further investigated for optimal trajectories of models of economic dynamics (see, for example, [33, 38, 46] and the references mentioned there). Many turnpike results are collected in [56, 73].

In order to meet our goal we consider the auxiliary optimization problem

$$v(x, x) \to \min, \ x \in R^n. \tag{P_1}$$

It follows from the strict convexity of v and (1.1) that problem (P_1) has a unique solution $\bar{x}$. Let

$$\nabla v(\bar{x}, \bar{x}) = (l_1, l_2), \tag{1.2}$$

where $l_1, l_2 \in R^n$. Since $\bar{x}$ is a solution of (P_1) it follows from (1.2) that for each $h \in R^n$

$$\langle l_1, h \rangle + \langle l_2, h \rangle = \langle (l_1, l_2), (h_1, h_2) \rangle$$
$$= \lim_{t \to 0^+} t^{-1}[v(\bar{x} + th, \bar{x} + th) - v(\bar{x}, \bar{x})] \geq 0.$$

Thus

$$\langle l_1 + l_2, h \rangle \geq 0 \text{ for all } h \in R^n,$$

$l_2 = -l_1$ and

$$\nabla v(\bar{x}, \bar{x}) = (l_1, -l_1),\qquad(1.3)$$

For each $(y, z) \in R^n \times R^n$ set

$$L(y, z) = v(y, z) - v(\bar{x}, \bar{x}) - \langle \nabla v(\bar{x}, \bar{x}), (y - \bar{x}, z - \bar{x}) \rangle$$
$$= v(y, z) - v(\bar{x}, \bar{x}) - \langle l_1, y - z \rangle.\qquad(1.4)$$

It is not difficult to verify that the function $L : R^n \times R^n \to R^1$ is differentiable and strictly convex. It follows from (1.1) and (1.4) that

$$L(y, z)/(|y| + |z|) \to \infty \text{ as } |y| + |z| \to \infty.\qquad(1.5)$$

Since the functions v and L are both strictly convex it follows from (1.4) that

$$L(y, z) \geq 0 \text{ for all } (y, z) \in R^n \times R^n\qquad(1.6)$$

and

$$L(y, z) = 0 \text{ if and only if } y = \bar{x}, \ z = \bar{x}.\qquad(1.7)$$

We claim that the function $L : R^n \times R^n \to R^1$ has the following property:

(C) If a sequence $\{(y_i, z_i)\}_{i=1}^{\infty} \subset R^n \times R^n$ satisfies the equality

$$\lim_{i \to \infty} L(y_i, z_i) = 0,$$

then

$$\lim_{i \to \infty} (y_i, z_i) = (\bar{x}, \bar{x}).$$

Assume that a sequence $\{(y_i, z_i)\}_{i=1}^{\infty} \subset R^n \times R^n$ satisfies $\lim_{i \to \infty} L(y_i, z_i) = 0$. In view of (1.5) the sequence $\{(y_i, z_i)\}_{i=1}^{\infty}$ is bounded. Let (y, z) be its limit point. Then it is easy to see that the equality

$$L(y, z) = \lim_{i \to \infty} L(y_i, z_i) = 0$$

holds and by (1.7) $(y, z) = (\bar{x}, \bar{x})$. This implies that $(\bar{x}, \bar{x}) = \lim_{i \to \infty} (y_i, z_i)$.

Thus the property (C) holds, as claimed.

Consider an auxiliary minimization problem

$$\sum_{i=0}^{T-1} L(x_i, x_{i+1}) \to \min,\qquad(P_2)$$

such that $\{x_i\}_{i=0}^{T} \subset R^n$ and $x_0 = z$, $x_T = y$,

where T is a natural number and the points $y, z \in R^n$.

It follows from (1.4) that for any integer $T \geq 1$ and any sequence $\{x_i\}_{i=0}^T \subset R^n$, we have

$$\sum_{i=0}^{T-1} L(x_i, x_{i+1}) = \sum_{i=0}^{T-1} v(x_i, x_{i+1}) - Tv(\bar{x}, \bar{x}) - \sum_{i=0}^{T-1} \langle l_1, x_i - x_{i+1} \rangle$$

$$= \sum_{i=0}^{T-1} v(x_i, x_{i+1}) - Tv(\bar{x}, \bar{x}) - \langle l_1, x_0 - x_T \rangle. \qquad (1.8)$$

Relation (1.8) implies that the problems (P_0) and (P_2) are equivalent. Namely, $\{x_i\}_{i=0}^T \subset R^n$ is a solution of the problem (P_0) if and only if it is a solution of the problem (P_2).

Let T be a natural number and $\Delta \geq 0$. A sequence $\{x_i\}_{i=0}^T \subset R^n$ is called (Δ)-optimal if for any sequence $\{x_i'\}_{i=0}^T \subset R^n$ satisfying $x_i = x_i'$, $i = 0, T$ the inequality

$$\sum_{i=0}^{T-1} v(x_i, x_{i+1}) \leq \sum_{i=0}^{T-1} v(x_i', x_{i+1}') + \Delta$$

holds. Clearly, if a sequence $\{x_i\}_{i=0}^T \subset R^n$ is (0)-optimal, then it is a solution of the problems (P_0) and (P_2) with $z = x_0$ and $y = x_T$.

We prove the following existence result.

Proposition 1.1. *Let $T > 1$ be an integer and $y, z \in R^n$. Then the problem (P_0) has a solution.*

Proof. It is sufficient to show that the problem (P_2) has a solution. Consider a sequence $\{x_i'\}_{i=0}^T \subset R^n$ such that $x_0' = z$, $x_T' = y$. Set

$$M_1 = \sum_{i=0}^{T-1} L(x_i', x_{i+1}')$$

and

$$M_2 = \inf \left\{ \sum_{i=0}^{T-1} L(x_i, x_{i+1}) : \{x_i\}_{i=0}^T \subset R^n, \ x_0 = z, \ x_T = y \right\}. \qquad (1.9)$$

Clearly,

$$0 \leq M_2 \leq M_1.$$

We may assume without loss of generality that

$$M_2 < M_1. \qquad (1.10)$$

There exists a sequence $\{x_i^{(k)}\}_{i=0}^{T} \subset R^n$, $k = 1, 2, \ldots$ such that for any natural number k

$$x_0^{(k)} = z, \; x_T^{(k)} = y \tag{1.11}$$

and

$$\lim_{k \to \infty} \sum_{i=0}^{T-1} L(x_i^{(k)}, x_{i+1}^{(k)}) = M_2. \tag{1.12}$$

In view of (1.10), (1.11), and (1.12) we may assume that

$$\sum_{i=0}^{T-1} L(x_i^{(k)}, x_{i+1}^{(k)}) < M_1 \text{ for all integers } k \geq 1. \tag{1.13}$$

By (1.13) and (1.5) there is $M_3 > 0$ such that

$$|x_i^{(k)}| \leq M_3 \text{ for all } i = 0, \ldots, T \text{ and all integers } k \geq 1. \tag{1.14}$$

In view of (1.14), extracting subsequences, using diagonalization process and re-indexing, if necessary, we may assume without loss of generality that for each $i \in \{0, \ldots, T\}$ there exists

$$\hat{x}_i = \lim_{k \to \infty} x_i^{(k)}. \tag{1.15}$$

By (1.15) and (1.11),

$$\hat{x}_0 = z, \; \hat{x}_T = y. \tag{1.16}$$

It follows from (1.15) and (1.12) that

$$\sum_{i=0}^{T-1} L(\hat{x}_i, \hat{x}_{i+1}) = M_2.$$

Together with (1.16) and (1.9) this implies that $\{\hat{x}_i\}_{i=0}^{T}$ is a solution of the problem (P_2). This completes the proof of Proposition 1.1. $\square$

Denote by Card(A) the cardinality of a set A.

The following result establishes a turnpike property for approximate solutions of the problem (P_0).

Proposition 1.2. *Let M_1, M_2, ϵ be positive numbers. Then there exists a natural number k_0 such that for each integer $T > 1$ and each (M_1)-optimal sequence $\{x_i\}_{i=0}^{T} \subset R^n$ satisfying*

$$|x_0| \leq M_2, \; |x_T| \leq M_2, \tag{1.17}$$

the following inequality holds:

$$Card(\{i \in \{0,\ldots,T-1\}: \ |x_i - \bar{x}| + |x_{i+1} - \bar{x}| > \epsilon\}) \leq k_0.$$

Proof. By condition (C) there is $\delta > 0$ such that for each $(y, z) \in R^n \times R^n$ satisfying

$$L(y, z) \leq \delta \tag{1.18}$$

we have

$$|y - \bar{x}| + |z - \bar{x}| \leq \epsilon. \tag{1.19}$$

Set

$$M_3 = \sup\{L(y, z): \ y, z \in R^n \text{ and } |y| + |z| \leq |\bar{x}| + M_2\}. \tag{1.20}$$

Choose a natural number:

$$k_0 > \delta^{-1}(M_1 + 2M_3). \tag{1.21}$$

Assume that an integer $T > 1$ and that an (M_1)-optimal sequence $\{x_i\}_{i=0}^T \subset R^n$ satisfies (1.17). Set

$$y_0 = x_0, \ y_T = x_T, \ y_i = \bar{x}, \ i = 1,\ldots,T-1. \tag{1.22}$$

Since the sequence $\{x_i\}_{i=0}^T$ is (M_1)-optimal it follows from (1.22) that

$$\sum_{i=0}^{T-1} v(x_i, x_{i+1}) \leq \sum_{i=0}^{T-1} v(y_i, y_{i+1}) + M_1.$$

Together with (1.7), (1.8), and (1.22) this implies that

$$\sum_{i=0}^{T-1} L(x_i, x_{i+1}) \leq \sum_{i=0}^{T-1} L(y_i, y_{i+1}) + M_1 = L(x_0, \bar{x}) + L(\bar{x}, x_T) + M_1.$$

Combined with (1.17) and (1.20) this implies that

$$\sum_{i=0}^{T-1} L(x_i, x_{i+1}) \leq M_1 + 2M_3.$$

It follows from the choice of δ (see (1.18) and (1.19)), (1.21), and the inequality above that

$$Card(\{i \in \{0, \ldots, T-1\} : |x_i - \bar{x}| + |x_{i+1} - \bar{x}| > \epsilon\})$$

$$\leq Card(\{i \in \{0, \ldots, T-1\} : L(x_i, x_{i+1}) > \delta\})$$

$$\leq \delta^{-1} \sum_{i=0}^{T-1} L(x_i, x_{i+1}) \leq \delta^{-1}(M_1 + 2M_3) \leq k_0.$$

Proposition 1.2 is proved. □

Proposition 1.2 implies the following turnpike result for exact solutions of the problem (P_0).

Proposition 1.3. *Let M, ϵ be positive numbers. Then there exists a natural number k_0 such that for each integer $T > 1$, each $y, z \in R^n$ satisfying $|y|, |z| \leq M$, and each optimal sequence $\{x_i\}_{i=0}^{T} \subset R^n$ of the problem (P_0) the following inequality holds:*

$$Card(\{i \in \{0, \ldots, T-1\} : |x_i - \bar{x}| + |x_{i+1} - \bar{x}| > \epsilon\}) \leq k_0.$$

It is easy now to see that the optimal solution $\{x_i\}_{i=0}^{T}$ of the problem (P_0) spends most of the time in an ϵ-neighborhood of $\bar{x}$. By Proposition 1.3 the number of all integers $i \in \{0, \ldots, T-1\}$ such that x_i does not belong to this ϵ-neighborhood, does not exceed the constant k_0 which depends only on M, ϵ, and does not depend on T. Following the tradition, the point $\bar{x}$ is called the turnpike. Moreover we can show that the set

$$\{i \in \{0 \ldots, T\} : |x_i - \bar{x}| > \epsilon\}$$

is contained in the union of two intervals $[0, k_1] \cup [T - k_1, T]$, where k_1 is a constant depending only on M, ϵ.

We also study the infinite horizon problem associated with the problem (P_0).

By (1.1) there is $M_* > 0$ such that

$$v(y, z) > |v(\bar{x}, \bar{x})| + 1 \tag{1.23}$$

for any $(y, z) \in R^n \times R^n$ satisfying $|y| + |z| \geq M_*$.

We suppose that the sum over empty set is zero.

Proposition 1.4. *Let $M_0 > 0$. Then there exists $M_1 > 0$ such that for each integer $T \geq 1$ and each sequence $\{x_i\}_{i=0}^{T} \subset R^n$ satisfying $|x_0| \leq M_0$,*

$$\sum_{i=0}^{T-1} v(x_i, x_{i+1}) \geq Tv(\bar{x}, \bar{x}) - M_1. \tag{1.24}$$

Proof. Put

$$M_1 = |l_1|(M_0 + M_*).$$

Assume that an integer $T \geq 1$ and a sequence $\{x_i\}_{i=0}^{T} \subset R^n$ satisfies

$$|x_0| \leq M_0. \tag{1.25}$$

If $|x_i| > M_*$, $i = 1, \ldots, T$, then by (1.23)

$$\sum_{i=0}^{T-1} v(x_i, x_{i+1}) \geq T v(\bar{x}, \bar{x})$$

and (1.24) holds. Therefore we may assume that there exists a natural number q such that

$$q \leq T, \ |x_q| \leq M_*. \tag{1.26}$$

We may assume without loss of generality that

$$|x_i| > M_* \text{ for all integers } i \text{ satisfying } q < i \leq T. \tag{1.27}$$

By (1.23) and (1.27),

$$\sum_{i=0}^{T-1}(v(x_i, x_{i+1}) - v(\bar{x}, \bar{x})) = \sum_{i=0}^{q-1}(v(x_i, x_{i+1}) - v(\bar{x}, \bar{x}))$$

$$+ \sum\{v(x_i, x_{i+1}) - v(\bar{x}, \bar{x})) : \text{ an integer } i \text{ satisfies } q \leq i < T\}$$

$$\geq \sum_{i=0}^{q-1}(v(x_i, x_{i+1}) - v(\bar{x}, \bar{x})).$$

It follows from the equation above, (1.8), (1.6), (1.25), (1.26), and the choice of M_1 that

$$\sum_{i=0}^{T-1}(v(x_i, x_{i+1}) - v(\bar{x}, \bar{x})) \geq \sum_{i=0}^{q-1}(v(x_i, x_{i+1}) - v(\bar{x}, \bar{x}))$$

$$= \sum_{i=0}^{q-1} L(x_i, x_{i+1}) + \langle l_1, x_0 - x_q \rangle \geq -|l_1|(|x_0| + |x_q|)$$

$$\geq -|l_1|(M_0 + M_*) = -M_1.$$

Proposition 1.4 is proved.   □

Fix a number $\tilde{M} > 0$ such that

$$\text{Proposition 1.4 holds with } M_0 = M_* \text{ and } M_1 = \tilde{M}. \tag{1.28}$$

Proposition 1.5. *Let $\{x_i\}_{i=0}^{\infty} \subset R^n$. Then either the sequence*

$$\left\{ \sum_{i=0}^{T-1} (v(x_i, x_{i+1}) - v(\bar{x}, \bar{x})) \right\}_{T=1}^{\infty}$$

is bounded or

$$\lim_{T \to \infty} \sum_{i=0}^{T-1} (v(x_i, x_{i+1}) - v(\bar{x}, \bar{x})) = \infty. \tag{1.29}$$

Proof. It follows from (1.23) that if for all sufficiently large natural numbers i, $|x_i| \geq M_*$, then (1.29) holds. Therefore we may assume without loss of generality that there exists a strictly increasing sequence of natural numbers $\{t_k\}_{k=1}^{\infty}$ such that

$$|x_{t_k}| < M_* \text{ for all integers } k \geq 1. \tag{1.30}$$

By Proposition 1.4 the sequence $\{\sum_{i=0}^{T-1} (v(x_i, x_{i+1}) - v(\bar{x}, \bar{x}))\}_{T=1}^{\infty}$ is bounded from below.

Assume that this sequence is not bounded from above. In order to complete the proof it is sufficient to show that (1.29) holds.

Let Q be any positive number. Then there exists a natural number T_0 such that

$$\sum_{i=0}^{T_0-1} (v(x_i, x_{i+1}) - v(\bar{x}, \bar{x})) > Q + \tilde{M}. \tag{1.31}$$

Choose a natural number k such that

$$t_k > T_0 + 4. \tag{1.32}$$

Let an integer

$$T > t_k. \tag{1.33}$$

By (1.30), (1.32), and (1.33) there exists an integer S such that

$$T > S \geq T_0, \tag{1.34}$$

$$|x_S| \leq M_*, \tag{1.35}$$

$$|x_t| > M_* \text{ for all integers } t \text{ satisfying} \tag{1.36}$$

$$S > t \geq T_0.$$

It follows from (1.31), (1.34), (1.36), (1.23), (1.35), (1.28) and Proposition 1.4 that

$$\sum_{i=0}^{T-1}(v(x_i, x_{i+1}) - v(\bar{x}, \bar{x})) = \sum_{i=0}^{T_0-1}(v(x_i, x_{i+1}) - v(\bar{x}, \bar{x}))$$

$$+ \sum \{v(x_i, x_{i+1}) - v(\bar{x}, \bar{x}) : i \text{ is an integer and } T_0 \le i < S\}$$

$$+ \sum_{i=S}^{T-1}(v(x_i, x_{i+1}) - v(\bar{x}, \bar{x}))$$

$$> Q + \tilde{M} + \sum_{i=S}^{T-1}(v(x_i, x_{i+1}) - v(\bar{x}, \bar{x})) > Q.$$

Thus for any integer $T > t_k$,

$$\sum_{i=0}^{T-1}(v(x_i, x_{i+1}) - v(\bar{x}, \bar{x})) > Q.$$

Since Q is any positive number (1.29) holds. Proposition 1.5 is proved.   □

A sequence $\{x_i\}_{i=0}^{\infty} \subset R^n$ is called good [23, 56, 73] if the sequence $\{\sum_{i=0}^{T-1}(v(x_i, x_{i+1}) - v(\bar{x}, \bar{x}))\}_{T=1}^{\infty}$ is bounded.

Proposition 1.6. *1. A sequence $\{x_i\}_{i=0}^{\infty} \subset R^n$ is good if and only if*

$$\sum_{i=0}^{\infty} L(x_i, x_{i+1}) < \infty.$$

2. If a sequence $\{x_i\}_{i=0}^{\infty} \subset R^n$ is good, then it converges to $\bar{x}$.

Proof. Assume that a sequence $\{x_i\}_{i=0}^{\infty} \subset R^n$ is good. Then there exists $M_0 > 0$ such that

$$\sum_{i=0}^{T-1}(v(x_i, x_{i+1}) - v(\bar{x}, \bar{x})) < M_0 \text{ for all integers } T \ge 1. \tag{1.37}$$

By (1.37) and (1.23) there exists a strictly increasing sequence of natural numbers $\{t_k\}_{k=1}^{\infty}$ such that

$$|x_{t_k}| < M \text{ for all natural numbers } k. \tag{1.38}$$

Let k be a natural number. By (1.8), (1.37), and (1.38),

$$M_0 > \sum_{i=0}^{t_k-1}(v(x_i, x_{i+1}) - v(\bar{x}, \bar{x})) = \sum_{i=0}^{t_k-1} L(x_i, x_{i+1}) + \langle l_1, x_0 - x_{t_k} \rangle$$

$$\geq \sum_{i=0}^{t_k-1} L(x_i, x_{i+1}) - |l_1|(|x_0| + |x_{t_k}|)$$

$$\geq \sum_{i=0}^{t_k-1} L(x_i, x_{i+1}) - |l_1|(|x_0| + M_*)$$

and

$$\sum_{i=0}^{t_k-1} L(x_i, x_{i+1}) \leq M_0 + |l_1|(|x_0| + M_*).$$

Since the inequality above holds for all natural numbers k we conclude that

$$\sum_{i=0}^{\infty} L(x_i, x_{i+1}) \leq M_0 + |l_1|(|x_0| + M_*).$$

In view of (C), the sequence $\{x_i\}_{i=0}^{\infty}$ converges to $\bar{x}$ and assertion 2 is proved.
Assume that

$$M_1 := \sum_{i=0}^{\infty} L(x_i, x_{i+1}) < \infty. \tag{1.39}$$

By (1.5) there is $M_2 > 0$ such that

$$|x_i| < M_2 \text{ for all integers } i \geq 0. \tag{1.40}$$

In view of (1.8), (1.39), and (1.40), for all natural numbers T,

$$\sum_{i=0}^{T-1} (v(x_i, x_{i+1}) - v(\bar{x}, \bar{x})) = \sum_{i=0}^{T-1} L(x_i, x_{i+1}) + \langle l_1, x_0 - x_T \rangle$$

$$\leq M_1 + 2|l_1|M_2.$$

Together with Proposition 1.5 this implies that the sequence $\{x_i\}_{i=0}^{\infty}$ is good.
Proposition 1.6 is proved. □

Proposition 1.7. *Let $x \in R^n$. Then there exists a sequence $\{x_i\}_{i=0}^{\infty} \subset R^n$ such that $x_0 = x$ and for each sequence $\{y_i\}_{i=0}^{\infty} \subset R^n$ satisfying $y_0 = x$ the inequality*

$$\sum_{i=0}^{\infty} L(x_i, x_{i+1}) \leq \sum_{i=0}^{\infty} L(y_i, y_{i+1})$$

holds.

Proof. Set

$$M_0 = \inf \left\{ \sum_{i=0}^{\infty} L(y_i, y_{i+1}) : \{y_i\}_{i=0}^{\infty} \subset R^n \text{ and } y_0 = x \right\}. \tag{1.41}$$

Clearly, M_0 is well defined and $M_0 \geq 0$. There exists a sequence $\{x_i^{(k)}\}_{i=0}^{\infty} \subset R^n$, $k = 1, 2, \ldots$ such that

$$x_0^{(k)} = x, \ k = 1, 2, \ldots, \tag{1.42}$$

$$\lim_{k \to \infty} \sum_{i=0}^{\infty} L(x_i^{(k)}, x_{i+1}^{(k)}) = M_0. \tag{1.43}$$

By (1.43) and (1.5) there exists $M_1 > 0$ such that

$$|x_i^{(k)}| < M_1 \text{ for all integers } i \geq 0 \text{ for all integers } k \geq 1. \tag{1.44}$$

In view of (1.44) using diagonalization process, extracting subsequences, and re-indexing we may assume without loss of generality that for any integer $i \geq 0$ there is

$$x_i = \lim_{k \to \infty} x_i^{(k)}. \tag{1.45}$$

By (1.42) and (1.45),

$$x_0 = x. \tag{1.46}$$

It follows from (1.6), (1.43), and (1.45) that for any natural number T

$$\sum_{i=0}^{T-1} L(x_i, x_{i+1}) = \lim_{k \to \infty} \sum_{i=0}^{T-1} L(x_i^{(k)}, x_{i+1}^{(k)}) \leq \lim_{k \to \infty} \sum_{i=0}^{\infty} L(x_i^{(k)}, x_{i+1}^{(k)}) = M_0.$$

Since T is an arbitrary natural number we conclude that

$$\sum_{i=0}^{\infty} L(x_i, x_{i+1}) \leq M_0.$$

Together with (1.41) and (1.46) this implies that

$$\sum_{i=0}^{\infty} L(x_i, x_{i+1}) = M_0.$$

This completes the proof of Proposition 1.7.   □

In our study we use the following optimality criterion introduced in the economic literature [5, 23, 48] and used in the optimal control [16, 56, 73].

A sequence $\{x_i\}_{i=0}^{\infty} \subset R^n$ is called overtaking optimal if

$$\limsup_{T \to \infty}[\sum_{i=0}^{T-1} v(x_i, x_{i+1}) - \sum_{i=0}^{T-1} v(y_i, y_{i+1})] \leq 0$$

for any sequence $\{y_i\}_{i=0}^{\infty} \subset R^n$ satisfying $y_0 = x_0$.

Proposition 1.8. *Let $\{x_i\}_{i=0}^{\infty} \subset R^n$. Then the following assertions are equivalent:*

1. the sequence $\{x_i\}_{i=0}^{\infty}$ is overtaking optimal;
2.

$$\sum_{i=0}^{\infty} L(x_i, x_{i+1}) \leq \sum_{i=0}^{\infty} L(y_i, y_{i+1})$$

for any sequence $\{y_i\}_{i=0}^{\infty} \subset R^n$ satisfying $y_0 = x_0$.

Proof. Assume that the sequence $\{x_i\}_{i=0}^{\infty}$ is overtaking optimal. Clearly, it is good. By Proposition 1.6,

$$\sum_{i=0}^{\infty} L(x_i, x_{i+1}) < \infty.$$

Let a sequence $\{y_i\}_{i=0}^{\infty} \subset R^n$ satisfies

$$y_0 = x_0. \tag{1.47}$$

We show that

$$\sum_{i=0}^{\infty} L(x_i, x_{i+1}) \leq \sum_{i=0}^{\infty} L(y_i, y_{i+1}).$$

We may assume that

$$\sum_{i=0}^{\infty} L(y_i, y_{i+1}) < \infty.$$

Then in view of (C),

$$\lim_{i \to \infty} y_i = \bar{x}, \quad \lim_{i \to \infty} x_i = \bar{x}. \tag{1.48}$$

Since the sequence $\{x_i\}_{i=0}^\infty$ is overtaking optimal it follows from (1.47), (1.8), and (1.48) that

$$0 \geq \limsup_{T \to \infty} [\sum_{i=0}^{T-1} v(x_i, x_{i+1}) - \sum_{i=0}^{T-1} v(y_i, y_{i+1})]$$

$$= \limsup_{T \to \infty} [\sum_{i=0}^{T-1} L(x_i, x_{i+1}) + \langle l_1, x_0 - x_T \rangle - \sum_{i=0}^{T-1} L(y_i, y_{i+1}) - \langle l_1, y_0 - y_T \rangle]$$

$$= \limsup_{T \to \infty} [\sum_{i=0}^{T-1} L(x_i, x_{i+1}) - \sum_{i=0}^{T-1} L(y_i, y_{i+1}) + \langle l_1, y_T - x_T \rangle]$$

$$= \sum_{i=0}^{\infty} L(x_i, x_{i+1}) - \sum_{i=0}^{\infty} L(y_i, y_{i+1}).$$

Thus assertion 2 holds.

Assume that assertion 2 holds. Let us show that the sequence $\{x_i\}_{i=0}^\infty$ is overtaking optimal. Clearly,

$$\sum_{i=0}^{\infty} L(x_i, x_{i+1}) < \infty.$$

By Proposition 1.6 the sequence $\{x_i\}_{i=0}^\infty$ is good and

$$\lim_{i \to \infty} x_i = \bar{x}. \tag{1.49}$$

Assume that a sequence $\{y_i\}_{i=0}^\infty \subset R^n$ satisfies

$$y_0 = x_0. \tag{1.50}$$

We show that

$$\limsup_{T \to \infty} [\sum_{i=0}^{T-1} v(x_i, x_{i+1}) - \sum_{i=0}^{T-1} v(y_i, y_{i+1})] \leq 0.$$

We may assume without loss of generality that the sequence $\{y_i\}_{i=0}^\infty$ is good. Then by Proposition 1.6,

$$\lim_{i \to \infty} y_i = \bar{x}, \quad \sum_{i=0}^{\infty} L(y_i, y_{i+1}) < \infty. \tag{1.51}$$

It follows from (1.8), (1.49), (1.50), (1.51) and assertion 2 that

$$\limsup_{T\to\infty}[\sum_{i=0}^{T-1} v(x_i, x_{i+1}) - \sum_{i=0}^{T-1} v(y_i, y_{i+1})]$$

$$= \limsup_{T\to\infty}[\sum_{i=0}^{T-1} L(x_i, x_{i+1}) + \langle l_1, x_0 - x_T \rangle - \sum_{i=0}^{T-1} L(y_i, y_{i+1}) - \langle l_1, y_0 - y_T \rangle]$$

$$= \sum_{i=0}^{\infty} L(x_i, x_{i+1}) - \sum_{i=0}^{\infty} L(y_i, y_{i+1}) + \langle l_1, \lim_{T\to\infty} y_T - \lim_{T\to\infty} x_T \rangle]$$

$$= \sum_{i=0}^{\infty} L(x_i, x_{i+1}) - \sum_{i=0}^{\infty} L(y_i, y_{i+1}) \leq 0.$$

Thus assertion 1 holds and Proposition 1.8 is proved.   □

Propositions 1.7 and 1.8 imply the following existence result.

Proposition 1.9. *For any* $x \in R^n$ *there exists an overtaking optimal sequence* $\{x_i\}_{i=0}^{\infty} \subset R^n$ *such that* $x_0 = x$.

1.2   The Turnpike Phenomenon

In the previous section we proved the turnpike result and the existence of overtaking optimal solutions for rather simple class of discrete-time problems. The problems of this class are unconstrained and their objective functions are convex and differentiable. In this book our goal is to establish the turnpike property and the existence of solutions over infinite horizon for several classes of constrained optimal control problems without convexity (concavity) assumptions. In particular, in Chap. 2, we study the structure of approximate solutions of an autonomous discrete-time control system with a compact metric space of states X. This control system is described by a bounded upper semicontinuous function $v : X \times X \to R^1$ which determines an optimality criterion and by a nonempty closed set $\Omega \subset X \times X$ which determines a class of admissible trajectories (programs). We study the problem

$$\sum_{i=0}^{T-1} v(x_i, x_{i+1}) \to \max, \ \{(x_i, x_{i+1})\}_{i=0}^{T-1} \subset \Omega, \ x_0 = z, \ x_T = y, \qquad (P)$$

where $T \geq 1$ is an integer and the points $y, z \in X$.

In the classical turnpike theory the objective function v possesses the turnpike property (TP) if there exists a point $\bar{x} \in X$ (a turnpike) such that the following condition holds:

For each positive number ϵ there exists an integer $L \geq 1$ such that for each integer $T \geq 2L$ and each solution $\{x_i\}_{i=0}^T \subset X$ of the problem (P) the inequality $\rho(x_i, \bar{x}) \leq \epsilon$ is true for all $i = L, \ldots, T - L$.

It should be mentioned that the constant L depends neither on T nor on y, z.

The turnpike phenomenon has the following interpretation. If one wishes to reach a point A from a point B by a car in an optimal way, then one should turn to a turnpike, spend most of time on it, and then leave the turnpike to reach the required point.

In the classical turnpike theory [23, 38, 46, 48] the space X is a compact convex subset of a finite-dimensional Euclidean space, the set Ω is convex, and the function v is strictly concave. Under these assumptions the turnpike property can be established and the turnpike $\bar{x}$ is a unique solution of the maximization problem $v(x, x) \to \max, (x, x) \in \Omega$. In this situation it is shown that for each program $\{x_t\}_{t=0}^{\infty}$ either the sequence $\{\sum_{t=0}^{T-1} v(x_t, x_{t+1}) - Tv(\bar{x}, \bar{x})\}_{T=1}^{\infty}$ is bounded (in this case the program $\{x_t\}_{t=0}^{\infty}$ is called (v)-good) or it diverges to $-\infty$. Moreover, it is also established that any (v)-good program converges to the turnpike $\bar{x}$. In the sequel this property is called as the asymptotic turnpike property.

Recently it was shown that the turnpike property is a general phenomenon which holds for large classes of variational and optimal control problems without convexity assumptions. (See, for example, [56] and the references mentioned therein.) For these classes of problems a turnpike is not necessarily a singleton but may instead be an nonstationary trajectory (in the discrete-time nonautonomous case) or an absolutely continuous function on the interval $[0, \infty)$ (in the continuous-time nonautonomous case) or a compact subset of the space X (in the autonomous case). Note that all of these results were obtained for unconstrained problems. In particular, the turnpike results for the problems of the type (P) were obtained in the case $\Omega = X \times X$.

For classes of problems considered in [56], using the Baire category approach, it was shown that the turnpike property holds for a generic (typical) problem. In this book we are interested in individual (nongeneric) turnpike results and in sufficient and necessary conditions for the turnpike phenomenon. In this book we study the problems (P) with the constraint $\{(x_i, x_{i+1})\}_{i=0}^{T-1} \subset \Omega$ where Ω is an arbitrary nonempty closed subset of $X \times X$. Clearly, these constrained problems are more difficult and less understood than their unconstrained prototypes in the previous section and in [50–52, 54]. They are also more realistic from the point of view of mathematical economics. As we have mentioned before in general a turnpike is not necessarily a singleton. Nevertheless problems of the type (P) for which the turnpike is a singleton are of great importance because of the following reasons: there are many models of economic growth for which a turnpike is a singleton; if a turnpike is a singleton, then approximate solutions of (P) have very simple structure and this is very important for applications; if a turnpike is a singleton, then it can be easily calculated as a solution of the problem $v(x, x) \to \max, (x, x) \in \Omega$.

The turnpike property is very important for applications. Suppose that our objective function v has the turnpike property and we know a finite number of "approximate" solutions of the problem (P). Then we know the turnpike $\bar{x}$, or at least its approximation, and the constant L (see the definition of (TP)) which is an estimate for the time period required to reach the turnpike. This information can be useful if we need to find an "approximate" solution of the problem (P) with a new time interval $[m_1, m_2]$ and the new values $z, y \in X$ at the end points m_1 and m_2. Namely instead of solving this new problem on the "large" interval $[m_1, m_2]$ we can find an "approximate" solution of the problem (P) on the "small" interval $[m_1, m_1 + L]$ with the values $z, \bar{x}$ at the end points and an "approximate" solution of the problem (P) on the "small" interval $[m_2 - L, m_2]$ with the values $\bar{x}, y$ at the end points. Then the concatenation of the first solution, the constant sequence $x_i = \bar{x}, i = m_1 + L, \ldots, m_2 - L$, and the second solution is an "approximate" solution of the problem (P) on the interval $[m_1, m_2]$ with the values z, y at the end points. Sometimes as an "approximate" solution of the problem (P) we can choose any admissible sequence $\{x_i\}_{i=m_1}^{m_2}$ satisfying

$$x_{m_1} = z, \; x_{m_2} = y \text{ and } x_i = \bar{x} \text{ for all } i = m_1 + L, \ldots, m_2 - L.$$

1.3 Turnpike Results for Nonconcave (Nonconvex) Problems

In Chap. 2 we study the turnpike phenomenon for discrete-time optimal control problems.

Let (X, ρ) be a compact metric space, Ω be a nonempty closed subset of $X \times X$, and $v : X \times X \to R^1$ be a bounded upper semicontinuous function.

A sequence $\{x_t\}_{t=0}^{\infty} \subset X$ is called an (Ω)-program (or just a program if the set Ω is understood) if $(x_t, x_{t+1}) \in \Omega$ for all nonnegative integers t. A sequence $\{x_t\}_{t=0}^{T}$ where $T \geq 1$ is an integer is called an (Ω)-program (or just a program if the set Ω is understood) if $(x_t, x_{t+1}) \in \Omega$ for all integers $t \in [0, T - 1]$.

In Chap. 2 we consider the problems

$$\sum_{i=0}^{T-1} v(x_i, x_{i+1}) \to \max, \; \{(x_i, x_{i+1})\}_{i=0}^{T-1} \subset \Omega, \; x_0 = y,$$

and

$$\sum_{i=0}^{T-1} v(x_i, x_{i+1}) \to \max, \; \{(x_i, x_{i+1})\}_{i=0}^{T-1} \subset \Omega, \; x_0 = y, \; x_T = z,$$

where $T \geq 1$ is an integer and the points $y, z \in X$.

We suppose that there exist a point $\bar{x} \in X$ and a positive number $\bar{c}$ such that the following assumptions hold:

(i) $(\bar{x}, \bar{x})$ is an interior point of Ω;
(ii) $\sum_{t=0}^{T-1} v(x_t, x_{t+1}) \leq Tv(\bar{x}, \bar{x}) + \bar{c}$ for any natural number T and any program $\{x_t\}_{t=0}^{T}$.

The property (ii) implies that for each program $\{x_t\}_{t=0}^{\infty}$ either the sequence

$$\{\sum_{t=0}^{T-1} v(x_t, x_{t+1}) - T v(\bar{x}, \bar{x})\}_{T=1}^{\infty}$$

is bounded or $\lim_{T\to\infty}[\sum_{t=0}^{T-1} v(x_t, x_{t+1}) - T v(\bar{x}, \bar{x})] = -\infty$.

A program $\{x_t\}_{t=0}^{\infty}$ is called (v)-good if the sequence

$$\{\sum_{t=0}^{T-1} v(x_t, x_{t+1}) - T v(\bar{x}, \bar{x})\}_{T=1}^{\infty}$$

is bounded.

In Chap. 2 we suppose that the following assumption holds:

(iii) (the asymptotic turnpike property) For any (v)-good program $\{x_t\}_{t=0}^{\infty}$, $\lim_{t\to\infty} \rho(x_t, \bar{x}) = 0$.

Note that the properties (i)–(iii) hold for models of economic dynamics considered in the classical turnpike theory.

For each positive number M denote by X_M the set of all points $x \in X$ for which there exists a program $\{x_t\}_{t=0}^{\infty}$ such that $x_0 = x$ and that for all natural numbers T the following inequality holds:

$$\sum_{t=0}^{T-1} v(x_t, x_{t+1}) - T v(\bar{x}, \bar{x}) \geq -M.$$

It is not difficult to see that $\cup\{X_M : M \in (0, \infty)\}$ is the set of all points $x \in X$ for which there exists a (v)-good program $\{x_t\}_{t=0}^{\infty}$ satisfying $x_0 = x$.

Let $T \geq 1$ be an integer and $\Delta \geq 0$. A program $\{x_i\}_{i=0}^{T} \subset R^n$ is called (Δ)-optimal if for any program $\{x_i'\}_{i=0}^{T}$ satisfying $x_0 = x_0'$, the inequality

$$\sum_{i=0}^{T-1} v(x_i, x_{i+1}) \geq \sum_{i=0}^{T-1} v(x_i', x_{i+1}') - \Delta$$

holds.

In Chap. 2 we will prove the following turnpike result for approximate solutions of our first optimization problem stated above.

Theorem 1.10. *Let ϵ, M be positive numbers. Then there exist a natural number L and a positive number δ such that for each integer $T > 2L$ and each (δ)-optimal program $\{x_t\}_{t=0}^{T}$ which satisfies $x_0 \in X_M$ there exist nonnegative integers $\tau_1, \tau_2 \leq L$ such that $\rho(x_t, \bar{x}) \leq \epsilon$ for all $t = \tau_1, \ldots, T - \tau_2$ and if $\rho(x_0, \bar{x}) \leq \delta$, then $\tau_1 = 0$.*

An analogous turnpike result for approximate solutions of our second optimization problem is also proved in Chap. 2.

A program $\{x_t\}_{t=0}^{\infty}$ is called (v)-overtaking optimal if for each program $\{y_t\}_{t=0}^{\infty}$ satisfying $y_0 = x_0$ the inequality $\limsup_{T\to\infty} \sum_{t=0}^{T-1} [v(y_t, y_{t+1}) - v(x_t, x_{t+1})] \le 0$ holds.

In Chap. 2 we prove the following result which establishes the existence of an overtaking optimal program.

Theorem 1.11. *Assume that* $x \in X$ *and that there exists a* (v)-*good program* $\{x_t\}_{t=0}^{\infty}$ *such that* $x_0 = x$. *Then there exists* (v)-*overtaking optimal program* $\{x_t^*\}_{t=0}^{\infty}$ *such that* $x_0^* = x$.

In Chap. 2 for problems which satisfy concavity assumption common in the literature we study the structure of approximate solutions in the regions containing end points and obtain a full description of the structure of approximate solutions. More precisely, we study the structure of approximate solutions of our first optimization problem stated above in the regions $[0, L]$ and $[T - L, T]$ (see the definition of the turnpike property). We will show that if $\{x_i\}_{i=0}^{T} \subset X$ is an approximate solution of our problem, then for all integers $t = 0, \dots, L$ the state x_t is closed enough to z_t where $\{z_t\}_{t=0}^{\infty} \subset X$ is a unique solution of a certain infinite horizon optimal control problem satisfying $z_0 = z$. We will also show that if $\{x_i\}_{i=0}^{T} \subset X$ is an approximate solution of our first optimization problem, then for all integers $t = 0, \dots, L$ the state x_{T-t} is closed enough to Λ_t where $\{\Lambda_t\}_{t=0}^{\infty} \subset X$ is a unique solution of a certain infinite horizon optimal control problem which does not depend on z. These results are established when the set X is a convex subset of the Euclidean space R^n, the set Ω is convex, and the function v is strictly concave. In this case we obtain the full description of the structure of approximate solutions of our first optimization problem. Note that the structure of approximate solutions in the region $[0, L]$ depends on z while their structure in the region $[T - L, T]$ does not depend on z. Actually it depends only on v and Ω.

We also study the stability of the turnpike phenomenon and show that the turnpike property is stable under perturbations of the objective function v. Note that the stability of the turnpike property is crucial in practice. One reason is that in practice we deal with a problem which consists a perturbation of the problem we wish to consider. Another reason is that the computations introduce numerical errors.

The turnpike properties of autonomous variational problems with extended-valued integrands are studied in Chap. 3. For these integrands we establish continuous-time analogs of the results of Chap. 2 obtained for discrete-time problems. We prove the existence of overtaking optimal solutions over infinite horizon, compare different optimality criterions for infinite horizon problems, and establish a non-self-intersection property of overtaking optimal solutions. For convex integrands we study the structure of approximate solutions in the regions containing end points and obtain a full description of the structure of approximate solutions.

1.4   Existence Results for Nonconcave (Nonconvex) Problems

In Chap. 4 we consider large classes of infinite horizon optimal control problems
without convexity (concavity) assumptions. These classes contain optimal control
problems arising in economic dynamics which describe a general one-sector model
of economic growth, optimal control problems which describe a general two-
sector model of economic dynamics, and discrete-time and continuous-time infinite
horizon optimal control problems with periodic cost functions. For these problems
we establish the existence of optimal (good) solutions on infinite horizon.

Let $R_+^1 = \{x \in R^1 : x \geq 0\}$, $v \in [0, 1)$, $f : R_+^1 \to R_+^1$ be an increasing
continuous function such that

$$f(0) = 0, \ f(x) > 0 \text{ for all } x > 0$$

and let $w : R_+^1 \to R_+^1$ be an increasing continuous function such that

$$w(0) = 0, \ w(x) > 0 \text{ for all } x > 0.$$

A pair of sequences $(\{x_t\}_{t=0}^\infty, \{y_t\}_{t=0}^\infty)$ is called a program if $x_t, y_t \in R_+^1$, $t =
0, 1 \ldots$ and for all nonnegative integers t

$$x_{t+1} \geq vx_t,$$

$$x_{t+1} - vx_t + y_t \leq f(x_t).$$

Let integers $T_1 \geq 0$, $T_2 > T_1$. A pair of sequences $(\{x_t\}_{t=T_1}^{T_2}, \{y_t\}_{t=T_1}^{T_2-1})$ is called
a program if $x_t \in R_+^1$, $t = T_1, \ldots, T_2$, $y_t \in R_+^1$, $t = T_1, \ldots, T_2 - 1$ and for all
integers $t = T_1, \ldots, T_2 - 1$,

$$x_{t+1} \geq vx_t, \ x_{t+1} - vx_t + y_t \leq f(x_t).$$

We study an infinite horizon optimal control problem which corresponds to a
finite horizon problem:

$$\sum_{t=0}^{T-1} w(y_t) \to \max, \ (\{x_t\}_{t=0}^T, \{y_t\}_{t=0}^{T-1}) \text{ is a program such that } x_0 = z,$$

where $T \geq 1$ is an integer and $z \in R_+^1$.

These optimal control systems describe a one-sector model of economic dynamics where x_t is funds at moment t, y_t is consumption at moment t, and $w(y_t)$
evaluates consumption at moment t. In Chap. 4 we establish the existence of good
programs.

We assume that there exists a positive number x^* such that

$$f(x) > (1-v)x \text{ for all } x \in (0, x^*),$$

$$f(x) < (1-v)x \text{ for all } x \in (x^*, \infty).$$

It is easy to see that

$$f(x^*) = (1-v)x^*.$$

In Chap. 4 we show the existence of a constant μ such that the following assertions hold:

for each program $(\{x_t\}_{t=0}^{\infty}, \{y_t\}_{t=0}^{\infty})$ either the sequence $\{\sum_{t=0}^{T-1} w(y_t) - T\mu\}_{T=1}^{\infty}$ is bounded or $\lim_{T \to \infty}[\sum_{t=0}^{T-1} w(y_t) - T\mu] = -\infty$;

let $0 < m_0 < x^* < M_0$. Then there exists $M_* > 0$ such that for each $x_0 \in [m_0, M_0]$ there is a program $(\{x_t\}_{t=0}^{\infty}, \{y_t\}_{t=0}^{\infty})$ such that for each pair of integers $T_1, T_2 \geq 0$ satisfying $T_1 < T_2$,

$$|\sum_{t=T_1}^{T_2-1} w(y_t) - (T_2 - T_1)\mu)| \leq M_*.$$

In Chap. 4 analogous results are also obtained for a class of two-dimensional optimal control problems. We also study these optimal control systems with discounting.

1.5 Turnpike Results for Two-Player Zero-Sum Games

In Chap. 5 we prove turnpike results for a class of dynamic discrete-time two-player zero-sum games. These results describe the structure of approximate solutions, for all sufficiently large intervals. We also show that for each initial state there exists a pair of overtaking equilibria strategies over an infinite horizon.

Let $X \subset R^{m_1}$ and $Y \subset R^{m_2}$ be nonempty convex compact sets. Denote by $\mathcal{M}$ the set of all continuous functions $f : X \times X \times Y \times Y \to R^1$ such that:

for each point $(y_1, y_2) \in Y \times Y$ the function $(x_1, x_2) \to f(x_1, x_2, y_1, y_2)$, $(x_1, x_2) \in X \times X$ is convex;

for each point $(x_1, x_2) \in X \times X$ the function $(y_1, y_2) \to f(x_1, x_2, y_1, y_2)$, $(y_1, y_2) \in Y \times Y$ is concave.

The set $\mathcal{M}$ is equipped with a metric $\rho : \mathcal{M} \times \mathcal{M} \to R^1$ defined by

$$\rho(f, g) = \sup\{|f(x_1, x_2, y_1, y_2) - g(x_1, x_2, y_1, y_2)| :$$

$$x_1, x_2 \in X, \quad y_1, y_2 \in Y\}, \quad f, g \in \mathcal{M}.$$

It is clearly that $(\mathcal{M}, \rho)$ is a complete metric space.

Given $f \in \mathcal{M}$ and a natural number n we consider a discrete-time two-player zero-sum game over the interval $[0, n]$. For this game $\{\{x_i\}_{i=0}^n : x_i \in X, i = 0, \ldots n\}$ is the set of strategies for the first player, $\{\{y_i\}_{i=0}^n : y_i \in Y, i = 0, \ldots n\}$ is the set of strategies for the second player, and the objective function for the first player associated with the strategies $\{x_i\}_{i=0}^n$, $\{y_i\}_{i=0}^n$ is given by $\sum_{i=0}^{n-1} f(x_i, x_{i+1}, y_i, y_{i+1})$.

Let $f \in \mathcal{M}$, n be a natural number and let $M \in [0, \infty)$. A pair of sequences $\{\bar{x}_i\}_{i=0}^n \subset X$, $\{\bar{y}_i\}_{i=0}^n \subset Y$ is called (f, M)-good if the following properties hold:

(i) for each sequence $\{x_i\}_{i=0}^n \subset X$ satisfying $x_0 = \bar{x}_0$, $x_n = \bar{x}_n$ the inequality

$$M + \sum_{i=0}^{n-1} f(x_i, x_{i+1}, \bar{y}_i, \bar{y}_{i+1}) \geq \sum_{i=0}^{n-1} f(\bar{x}_i, \bar{x}_{i+1}, \bar{y}_i, \bar{y}_{i+1})$$

holds;

(ii) for each sequence $\{y_i\}_{i=0}^n \subset Y$ satisfying $y_0 = \bar{y}_0$, $y_n = \bar{y}_n$ the inequality

$$M + \sum_{i=0}^{n-1} f(\bar{x}_i, \bar{x}_{i+1}, \bar{y}_i, \bar{y}_{i+1}) \geq \sum_{i=0}^{n-1} f(\bar{x}_i, \bar{x}_{i+1}, y_i, y_{i+1})$$

holds.

If a pair of sequences $\{x_i\}_{i=0}^n \subset X$, $\{y_i\}_{i=0}^n \subset Y$ is $(f, 0)$-good, then it is called (f)-optimal.

In Chap. 5 we study the turnpike property of good pairs of sequences.

Let $f \in \mathcal{M}$. We say that the function f possesses the *turnpike property* if there exists a unique pair $(x_f, y_f) \in X \times Y$ for which the following assertion holds:

For each positive number ϵ there exist an integer $n_0 \geq 2$ and a positive number δ such that for each integer $n \geq 2n_0$ and each (f, δ)-good pair of sequences $\{x_i\}_{i=0}^n \subset X$, $\{y_i\}_{i=0}^n \subset Y$ the inequalities $|x_i - x_f|$, $|y_i - y_f| \leq \epsilon$ hold for all integers $i \in [n_0, n - n_0]$.

In [53] we showed that the turnpike property holds for a generic $f \in \mathcal{M}$. Namely, in [53] we proved the existence of a set $\mathcal{F} \subset \mathcal{M}$ which is a countable intersection of open everywhere dense sets in $\mathcal{M}$ such that each $f \in \mathcal{F}$ has the turnpike property. Thus for most functions $f \in \mathcal{M}$ the turnpike property holds. Nevertheless it is very important to have conditions on $f \in \mathcal{M}$ which imply the turnpike property. These conditions are discussed in Chap. 5.

Chapter 2
Turnpike Properties of Discrete-Time Problems

In this chapter we study the structure of approximate solutions of an autonomous discrete-time control system with a compact metric space of states X. This control system is described by a bounded upper semicontinuous function $v : X \times X \to R^1$ which determines an optimality criterion and by a nonempty closed set $\Omega \subset X \times X$ which determines a class of admissible trajectories (programs). We are interested in turnpike properties of the approximate solutions which are independent of the length of the interval, for all sufficiently large intervals. When X is a compact convex subset of a finite-dimensional Euclidean space, the set Ω is convex, and the function v is strictly concave we obtain a full description of the structure of approximate solutions.

2.1 Turnpike Results Without Convexity

Let (X, ρ) be a compact metric space, let Ω be a nonempty closed subset of $X \times X$, and let $v : X \times X \to R^1$ be a bounded upper semicontinuous function.

A sequence $\{x_t\}_{t=0}^\infty \subset X$ is called an (Ω)-program (or just a program if the set Ω is understood) if $(x_t, x_{t+1}) \in \Omega$ for all nonnegative integers t. A sequence $\{x_t\}_{t=T_1}^{T_2}$ where integers T_1, T_2 satisfy $0 \leq T_1 < T_2$ is called an (Ω)-program (or just a program if the set Ω is understood) if $(x_t, x_{t+1}) \in \Omega$ for all integers $t \in [T_1, T_2-1]$.

In this chapter we consider the problems

$$\sum_{i=0}^{T-1} v(x_i, x_{i+1}) \to \max, \quad \{(x_i, x_{i+1})\}_{i=0}^{T-1} \subset \Omega, \; x_0 = y, \; x_T = z, \qquad (P_T^{(y,z)})$$

© Springer International Publishing Switzerland 2014

A.J. Zaslavski, *Turnpike Phenomenon and Infinite Horizon Optimal Control*,
Springer Optimization and Its Applications 99, DOI 10.1007/978-3-319-08828-0_2

and

$$\sum_{i=0}^{T-1} v(x_i, x_{i+1}) \rightarrow \max, \ \{(x_i, x_{i+1})\}_{i=0}^{T-1} \subset \Omega, \ x_0 = y, \qquad (P_T^{(y)})$$

where $T \geq 1$ is an integer and the points $y, z \in X$.

Set $\|v\| = \sup\{|v(x, y)| : \ x, y \in X\}$. For each pair of points $x, y \in X$ and each natural number T define

$$\sigma(v, T, x) = \sup \left\{ \sum_{i=0}^{T-1} v(x_i, x_{i+1}) : \ \{x_i\}_{i=0}^{T} \text{ is a program and } x_0 = x \right\}, \qquad (2.1)$$

$$\sigma(v, T, x, y) = \sup \left\{ \sum_{i=0}^{T-1} v(x_i, x_{i+1}) : \right.$$

$$\left. \{x_i\}_{i=0}^{T} \text{ is a program and } x_0 = x, \ x_T = y \right\}, \qquad (2.2)$$

$$\sigma(v, T) = \sup \left\{ \sum_{i=0}^{T-1} v(x_i, x_{i+1}) : \ \{x_i\}_{i=0}^{T} \text{ is a program} \right\}. \qquad (2.3)$$

(Here we use the convention that the supremum of an empty set is $-\infty$.)

We suppose that there exist a point $\bar{x} \in X$ and a positive constant $\bar{c}$ such that the following assumptions hold.

(A1) $(\bar{x}, \bar{x})$ is an interior point of Ω (there exists a positive number ϵ such that $\{(x, y) \in X \times X : \ \rho(x, \bar{x}), \ \rho(y, \bar{x}) \leq \epsilon\} \subset \Omega$) and the function v is continuous at the point $(\bar{x}, \bar{x})$.

(A2) $\sigma(v, T) \leq Tv(\bar{x}, \bar{x}) + \bar{c}$ for all natural numbers T.

Clearly, for each integer $T \geq 1$ and each program $\{x_t\}_{t=0}^{T}$, we have

$$\sum_{t=0}^{T-1} v(x_t, x_{t+1}) \leq \sigma(v, T) \leq Tv(\bar{x}, \bar{x}) + \bar{c}. \qquad (2.4)$$

Inequality (2.4) implies the following result.

Proposition 2.1. *For each program $\{x_t\}_{t=0}^{\infty}$ either the sequence*

$$\left\{ \sum_{t=0}^{T-1} v(x_t, x_{t+1}) - Tv(\bar{x}, \bar{x}) \right\}_{T=1}^{\infty}$$

is bounded or $\lim_{T \rightarrow \infty} [\sum_{t=0}^{T-1} v(x_t, x_{t+1}) - Tv(\bar{x}, \bar{x})] = -\infty$.

A program $\{x_t\}_{t=0}^{\infty}$ is called (v)-good if the sequence

$$\left\{\sum_{t=0}^{T-1} v(x_t, x_{t+1}) - Tv(\bar{x}, \bar{x})\right\}_{T=1}^{\infty}$$

is bounded.

We suppose that the following assumption holds.

(A3) (the asymptotic turnpike property) For every (v)-good program $\{x_t\}_{t=0}^{\infty}$,

$$\lim_{t \to \infty} \rho(x_t, \bar{x}) = 0.$$

In view of (A3) $\|v\| > 0$. For each positive number M denote by X_M the set of all points $x \in X$ for which there exists a program $\{x_t\}_{t=0}^{\infty}$ such that $x_0 = x$ and that for all natural numbers T the following inequality holds:

$$\sum_{t=0}^{T-1} v(x_t, x_{t+1}) - Tv(\bar{x}, \bar{x}) \geq -M.$$

It is not difficult to see that $\cup\{X_M : M \in (0, \infty)\}$ is the set of all points $x \in X$ such that there exists a (v)-good program $\{x_t\}_{t=0}^{\infty}$ satisfying $x_0 = x$.

Let $T \geq 1$ be an integer. Denote by Y_T the set of all points $x \in X$ such that there exists a program $\{x_t\}_{t=0}^{T}$ which satisfies $x_0 = \bar{x}$ and $x_T = x$.

In this chapter we will prove the following turnpike result for approximate solutions of problem $(P_T^{(y)})$.

Theorem 2.2. *Let ϵ, M be positive numbers. Then there exist a natural number L and a positive number δ such that for each integer $T > 2L$ and each program $\{x_t\}_{t=0}^{T}$ which satisfies*

$$x_0 \in X_M, \quad \sum_{t=0}^{T-1} v(x_t, x_{t+1}) \geq \sigma(v, T, x_0) - \delta \qquad (2.5)$$

there exist nonnegative integers $\tau_1, \tau_2 \leq L$ such that $\rho(x_t, \bar{x}) \leq \epsilon$ for all $t = \tau_1, \ldots, T - \tau_2$ and if $\rho(x_0, \bar{x}) \leq \delta$, then $\tau_1 = 0$.

In the sequel we use a notion of an overtaking optimal program introduced in [5, 23, 48].

A program $\{x_t\}_{t=0}^{\infty}$ is called (v)-overtaking optimal if for each program $\{y_t\}_{t=0}^{\infty}$ satisfying $y_0 = x_0$ the inequality

$$\limsup_{T \to \infty} \sum_{t=0}^{T-1} [v(y_t, y_{t+1}) - v(x_t, x_{t+1})] \leq 0$$

holds.

The following result establishes the existence of an overtaking optimal program.

Theorem 2.3. *Assume that $x \in X$ and that there exists a (v)-good program $\{x_t\}_{t=0}^{\infty}$ such that $x_0 = x$. Then there exists (v)-overtaking optimal program $\{x_t^*\}_{t=0}^{\infty}$ such that $x_0^* = x$.*

The next theorem is a refinement of Theorem 2.2. According to Theorem 2.2 we have $\tau_2 \leq L$ where the constant L depends on M and ϵ. The next theorem shows that $\tau_2 \leq L_0$ where the constant L_0 depends only on ϵ.

Theorem 2.4. *Let ϵ be a positive number. Then there exists a natural number L_0 such that for each positive number M there exist an integer $L > L_0$ and a positive number δ such that the following assertion holds:*

For each integer $T > 2L$ and each program $\{x_t\}_{t=0}^{T}$ which satisfies (2.5) there exist integers $\tau_1 \in [0, L]$, $\tau_2 \in [0, L_0]$ such that $\rho(x_t, \bar{x}) \leq \epsilon$ for all $t = \tau_1, \ldots, T - \tau_2$ and if $\rho(x_0, \bar{x}) \leq \delta$, then $\tau_1 = 0$.

The following result provides necessary and sufficient conditions for overtaking optimality.

Theorem 2.5. *Let $\{x_t\}_{t=0}^{\infty}$ be a program such that*

$$x_0 \in \cup\{X_M : M \in (0, \infty)\}.$$

Then the program $\{x_t\}_{t=0}^{\infty}$ is (v)-overtaking optimal if and only if the following conditions hold:

(i) $\lim_{t \to \infty} \rho(x_t, \bar{x}) = 0$;
(ii) for each natural number T and each program $\{y_t\}_{t=0}^{T}$ satisfying $y_0 = x_0$, $y_T = x_T$ the inequality $\sum_{t=0}^{T-1} v(y_t, y_{t+1}) \leq \sum_{t=0}^{T-1} v(x_t, x_{t+1})$ holds.

The next two theorems establish uniform convergence of overtaking optimal programs to $\bar{x}$.

Theorem 2.6. *Assume that the function v is continuous and let ϵ be a positive number. Then there exists a positive number δ such that for each (v)-overtaking optimal program $\{x_t\}_{t=0}^{\infty}$ satisfying $\rho(x_0, \bar{x}) \leq \delta$ the inequality $\rho(x_t, \bar{x}) \leq \epsilon$ holds for all nonnegative integers t.*

Theorem 2.7. *Assume that the function v is continuous and let M, ϵ be positive numbers. Then there exists an integer $L \geq 1$ such that for each (v)-overtaking optimal program $\{x_t\}_{t=0}^{\infty}$ satisfying $x_0 \in X_M$ the inequality $\rho(x_t, \bar{x}) \leq \epsilon$ holds for all integers $t \geq L$.*

Theorems 2.2–2.7 were obtained in [58]. The next two theorems obtained in [60] describe the structure of problem $(P_T^{(y,z)})$.

Denote by Card(A) the cardinality of the set A.

Theorem 2.8. *Let ϵ, M_0, M_1 be positive numbers and let L_0 be a natural number. Then there exist a natural number L and a natural number K such that for each integer $T > 2L$, each $z_0 \in X_{M_0}$, and each $z_1 \in Y_{L_0}$, $\sigma(v, T, z_0, z_1)$ is finite and for each program $\{x_t\}_{t=0}^{T}$ which satisfies*

$$x_0 = z_1, \ x_T = z_2, \ \sum_{t=0}^{T-1} v(x_t, x_{t+1}) \geq \sigma(v, T, z_0, z_1) - M_1,$$

the following inequality holds:

$$\text{Card}(\{t \in \{0, \ldots, T\} : \ \rho(x_t, \bar{x}) > \epsilon\}) \leq K.$$

Theorem 2.9. *Let ϵ, M_0 be positive numbers and let L_0 be a natural number. Then there exist a natural number L and a positive number δ such that for each integer $T > 2L$, each $z_0 \in X_{M_0}$, and each $z_1 \in Y_{L_0}$, $\sigma(v, T, z_0, z_1)$ is finite and for each program $\{x_t\}_{t=0}^{T}$ which satisfies*

$$x_0 = z_1, \ x_T = z_2, \ \sum_{t=0}^{T-1} v(x_t, x_{t+1}) \geq \sigma(v, T, z_0, z_1) - \delta$$

there exist integers $\tau_1, \tau_2 \in [0, L]$ such that

$$\rho(x_t, \bar{x}) \leq \epsilon, \ t = \tau_1, \ldots, T - \tau_2.$$

Moreover if $\rho(x_0, \bar{x}) \leq \delta$, then $\tau_1 = 0$, and if $\rho(x_T, \bar{x}) \leq \delta$, then $\tau_2 = 0$.

Example 2.10. Let (X, ρ) be a compact metric space, Ω be a nonempty closed subset of $X \times X$, $\bar{x} \in X$, $(\bar{x}, \bar{x})$ be an interior point of Ω, $\pi : X \to R^1$ be a continuous function, α be a real number, and $L : X \times X \to [0, \infty)$ be a continuous function such that for each $(x, y) \in X \times X$ the equality $L(x, y) = 0$ holds if and only if $(x, y) = (\bar{x}, \bar{x})$. Set

$$v(x, y) = \alpha - L(x, y) + \pi(x) - \pi(y)$$

for all $x, y \in X$. It is not difficult to see that (A1)–(A3) hold.

Example 2.11. Let X be a compact convex subset of the Euclidean space R^n with the norm $|\cdot|$ induced by the scalar product $\langle \cdot, \cdot \rangle$, let $\rho(x, y) = |x - y|$, $x, y \in R^n$, Ω be a nonempty closed subset of $X \times X$, a point $\bar{x} \in X$, $(\bar{x}, \bar{x})$ be an interior point of Ω, and let $v : X \times X \to R^1$ be a strictly concave continuous function such that

$$v(\bar{x}, \bar{x}) = \sup\{v(z, z) : \ z \in X \text{ and } (z, z) \in \Omega\}.$$

We assume that there exists a positive constant $\bar{r}$ such that

$$\{(x, y) \in R^n \times R^n : \ |x - \bar{x}|, \ |y - \bar{x}| \leq \bar{r}\} \subset \Omega.$$

It is a well-known fact of convex analysis that there exists a point $l \in R^n$ such that

$$v(x, y) \leq v(\bar{x}, \bar{x}) + \langle l, x - y \rangle$$

for any point $(x, y) \in X \times X$. Set

$$L(x, y) = v(\bar{x}, \bar{x}) + \langle l, x - y \rangle - v(x, y)$$

for all $(x, y) \in X \times X$. It is not difficult to see that this example is a particular case of Example 2.10. Therefore (A1)–(A3) hold.

2.2   Auxiliary Results

For each integer $T \geq 1$ denote by $\bar{Y}_T$ the set of all points $x \in X$ for which there exists a program $\{x_t\}_{t=0}^T$ such that $x_0 = x$ and $x_T = \bar{x}$.
   It is easy to see that

$$\bar{Y}_T \subset \bar{Y}_{T+1} \text{ for all natural numbers } T. \tag{2.6}$$

By assumption (A1), if T is a natural number and a point $x \in \bar{Y}_T$, then there exists a (v)-good program $\{x_t\}_{t=0}^{\infty}$ which satisfies $x_0 = x$. Assumptions (A1) and (A3) imply that if a program $\{x_t\}_{t=0}^{\infty}$ is (v)-good, then there exists an integer $T \geq 1$ such that $x_0 \in \bar{Y}_T$. The boundedness of v implies the following result.

Proposition 2.12. *Let T be a natural number. Then there exists a positive number M such that $\bar{Y}_T \subset X_M$.*

   In view of assumption (A1), there exists a number $\bar{r} \in (0, 1)$ such that

$$\{(x, y) \in X \times X : \rho(x, \bar{x}), \rho(y, \bar{x}) \leq \bar{r}\} \subset \Omega. \tag{2.7}$$

Lemma 2.13. *Let ϵ and M_0 be positive numbers. Then there exists an integer $T \geq 1$ such that for each program $\{x_t\}_{t=0}^T$ which satisfies*

$$\sum_{t=0}^{T-1} v(x_t, x_{t+1}) \geq Tv(\bar{x}, \bar{x}) - M_0$$

the inequality $\min\{\rho(x_i, \bar{x}) : i = 1, \ldots, T\} \leq \epsilon$ holds.

Proof. Assume the contrary. Then for each integer $k \geq 1$ there exists a program $\{x_t^{(k)}\}_{t=0}^k$ such that

$$\sum_{t=0}^{k-1} v\left(x_t^{(k)}, x_{t+1}^{(k)}\right) \geq kv(\bar{x}, \bar{x}) - M_0, \tag{2.8}$$

$$\rho\left(x_t^{(k)}, \bar{x}\right) > \epsilon \text{ for all integers } t = 1, \ldots, k. \tag{2.9}$$

Let k be a natural number. Relations (2.8) and (2.3) and assumption (A2) imply that for each natural number $j < k$,

$$\sum_{t=0}^{j-1} v\left(x_t^{(k)}, x_{t+1}^{(k)}\right)$$

$$= \sum_{t=0}^{k-1} v\left(x_t^{(k)}, x_{t+1}^{(k)}\right) - \sum_{t=j}^{k-1} v\left(x_t^{(k)}, x_{t+1}^{(k)}\right)$$

$$\geq kv(\bar{x}, \bar{x}) - M_0 - \sum_{t=j}^{k-1} v\left(x_t^{(k)}, x_{t+1}^{(k)}\right)$$

$$\geq kv(\bar{x}, \bar{x}) - M_0 - \sigma(v, k - j)$$

$$\geq kv(\bar{x}, \bar{x}) - M_0 - (k - j)v(\bar{x}, \bar{x}) - \bar{c}.$$

Combined with (2.8) this inequality implies that for each natural number k and each integer $j \in \{1, \ldots, k\}$ we have

$$\sum_{t=0}^{j-1} v(x_t^{(k)}, x_{t+1}^{(k)}) \geq jv(\bar{x}, \bar{x}) - \bar{c} - M_0. \tag{2.10}$$

There exists a strictly increasing sequence of natural numbers $\{k_i\}_{i=1}^{\infty}$ such that for each nonnegative integer t there exists a limit

$$x_t = \lim_{i \to \infty} x_t^{(k_i)}. \tag{2.11}$$

It is easy to see that the sequence $\{x_t\}_{t=0}^{\infty}$ is a program. In view of relations (2.11) and (2.9), we have

$$\rho(x_t, \bar{x}) \geq \epsilon \text{ for all natural numbers } t. \tag{2.12}$$

By relations (2.11) and (2.10), for each natural number T,

$$\sum_{t=0}^{T-1} v(x_t, x_{t+1}) \geq Tv(\bar{x}, \bar{x}) - M_0 - \bar{c}.$$

This inequality implies that the sequence $\{x_t\}_{t=0}^{\infty}$ is a (v)-good program. It follows from assumption (A3) that the equality $\lim_{t \to \infty} \rho(x_t, \bar{x}) = 0$ holds. This equality contradicts (2.12). The contradiction we have reached proves the lemma. □

Lemma 2.13 and (A1) imply the following result.

Proposition 2.14. *Let M be a positive number. Then there exists an integer $T \geq 1$ such that the inclusion $X_M \subset \bar{Y}_T$ holds.*

In view of inclusion (2.7) for each pair of points $x, y \in X$ such that $\rho(x, \bar{x}), \rho(y, \bar{x}) \leq \bar{r}$ and each natural number T, the value $\sigma(v, x, y, T)$ is finite.

Lemma 2.15. *Let ϵ be a positive number. Then there exists a number $\delta \in (0, \bar{r})$ such that for each natural number T and each program $\{x_t\}_{t=0}^T$ which satisfies*

$$\rho(x_0, \bar{x}), \ \rho(x_T, \bar{x}) \leq \delta, \tag{2.13}$$

$$\sum_{t=0}^{T-1} v(x_t, x_{t+1}) \geq \sigma(v, x_0, x_T, T) - \delta, \tag{2.14}$$

the inequality $\rho(x_t, \bar{x}) \leq \epsilon$ holds for all integers $t = 0, \ldots, T$.

Proof. Since the function v is continuous at the point $(\bar{x}, \bar{x})$ for any integer $k \geq 1$ there exists a number

$$\delta_k \in (0, 2^{-k}\bar{r}) \tag{2.15}$$

such that

$$|v(x, y) - v(\bar{x}, \bar{x})| \leq 2^{-k} \tag{2.16}$$

for each pair of points $x, y \in X$ satisfying

$$\rho(x, \bar{x}), \ \rho(y, \bar{x}) \leq \delta_k. \tag{2.17}$$

Assume that the lemma is wrong. Then for each integer $k \geq 1$ there exist a natural number T_k and a program $\{x_t^{(k)}\}_{t=0}^{T_k}$ such that

$$\rho\left(x_0^{(k)}, \bar{x}\right), \ \rho\left(x_{T_k}^{(k)}, \bar{x}\right) \leq \delta_k, \tag{2.18}$$

$$\sum_{t=0}^{T_k-1} v\left(x_t^{(k)}, x_{t+1}^{(k)}\right) \geq \sigma\left(v, x_0^{(k)}, x_{T_k}^{(k)}, T_k\right) - \delta_k, \tag{2.19}$$

$$\max\left\{\rho\left(x_t^{(k)}, \bar{x}\right) : t = 0, \ldots, T_k\right\} > \epsilon. \tag{2.20}$$

Let a natural number k be a given integer. Define a sequence $\{z_t\}_{t=0}^{T_k} \subset X$ as follows:

$$z_0 = x_0^{(k)}, \ z_{T_k} = x_{T_k}^{(k)}, \ z_t = \bar{x}, \ t \in \{0, \ldots, T_k\} \setminus \{0, T_k\}. \tag{2.21}$$

In view of relations (2.21), (2.18), (2.15), and (2.7), the sequence $\{z_t\}_{t=0}^{T_k}$ is a program. By (2.19) and (2.21), we have

$$\sum_{t=0}^{T_k-1} v(x_t^{(k)}, x_{t+1}^{(k)}) \geq \sigma(v, x_0^{(k)}, x_{T_k}^{(k)}, T_k) - \delta_k \geq \sum_{t=0}^{T_k-1} v(z_t, z_{t+1}) - \delta_k. \tag{2.22}$$

It follows from relations (2.18), (2.21) and the choice of δ_k [see (2.15)–(2.17)] that

$$\left| v(z_0, z_1) - v\left(\bar{x}, \bar{x}\right) \right| \leq 2^{-k}, \quad \left| v(z_{T_k-1}, z_{T_k}) - v(\bar{x}, \bar{x}) \right| \leq 2^{-k},$$

$$v(z_t, z_{t+1}) = v(\bar{x}, \bar{x}), \ t \in \{0, \dots, T_k - 1\} \setminus \{0, T_k - 1\}. \tag{2.23}$$

In view of (2.23) and (2.22),

$$\sum_{t=0}^{T_k-1} v(x_t^{(k)}, x_{t+1}^{(k)}) \geq T_k v(\bar{x}, \bar{x}) - 2 \cdot 2^{-k} - \delta_k. \tag{2.24}$$

Put

$$S_0 = 0, \ S_k = \sum_{i=1}^{k} (T_i + 1) - 1 \text{ for all natural numbers } k. \tag{2.25}$$

Define a sequence $\{x_t\}_{t=0}^{\infty} \subset X$ as follows:

$$x_t = x_t^{(1)}, \ t = 0, \dots, T_1, \ x_t = x_i^{(k+1)} \tag{2.26}$$

for each natural number k, each $i \in \{0, \dots, T_{k+1}\}$, and $t = S_k + i + 1$.

By relations (2.26), (2.18), (2.15), and (2.7), the sequence $\{x_t\}_{t=0}^{\infty}$ is a program. It follows from (2.25), (2.26), (2.18), and (2.15) that for each natural number k we have

$$\left| v(x_{S_k}, x_{S_k+1}) - v(\bar{x}, \bar{x}) \right| \leq 2 \cdot 2^{-k}. \tag{2.27}$$

In view of relations (2.25), (2.26), (2.24), (2.21) and the choice of δ_j, $j = 1, 2, \dots$ [see (2.15)–(2.18)] for any natural number $k \geq 2$, we have

$$\sum_{t=0}^{S_k-1} v(x_t, x_{t+1}) - S_k v(\bar{x}, \bar{x}) = \sum_{j=1}^{k} \left(\sum_{t=0}^{T_j-1} \left[v\left(x_t^{(j)}, x_{t+1}^{(j)} \right) - v(\bar{x}, \bar{x}) \right] \right)$$

$$+ \sum_{j=1}^{k-1} \left[v\left(x_{T_j}^{(j)}, x_0^{(j+1)} \right) - v(\bar{x}, \bar{x}) \right] \geq - \sum_{j=1}^{k} (2 \cdot 2^{-j} + \delta_j) - 2 \sum_{j=1}^{k-1} 2^{-j}.$$

Together with inclusion (2.15) this relation implies that for any natural number $k \geq 2$,

$$\sum_{t=0}^{S_k-1} v(x_t, x_{t+1}) - S_k v(\bar{x}, \bar{x}) \geq -5 \sum_{j=1}^{k} 2^{-j} \geq -10.$$

It follows from this inequality and Proposition 2.1 that the sequence $\{x_t\}_{t=0}^{\infty}$ is a (v)-good program. By assumption (A3), we have

$$\lim_{t \to \infty} \rho(x_t, \bar{x}) = 0.$$

On the other hand it follows from relations (2.20), (2.25), and (2.26) that $\limsup_{t \to \infty} \rho(x_t, \bar{x}) \geq \epsilon$. The contradiction we have reached proves Lemma 2.15.

$\square$

Lemma 2.16. *Let ϵ, M_0 be positive numbers. Then there exists an integer $T_0 \geq 1$ such that for each natural number $T \geq T_0$, each program $\{x_t\}_{t=0}^{T}$ which satisfies*

$$\sum_{t=0}^{T-1} v(x_t, x_{t+1}) \geq Tv(\bar{x}, \bar{x}) - M_0, \tag{2.28}$$

and each integer $s \in [0, T - T_0]$, the following inequality holds:

$$\min\{\rho(x_i, \bar{x}) : i = s+1, \ldots, s + T_0\} \leq \epsilon.$$

Proof. Lemma 2.13 implies that there exists an integer $T_0 \geq 1$ such that the following property holds:

(P1)　For each program $\{x_t\}_{t=0}^{T_0}$ which satisfies

$$\sum_{t=0}^{T_0-1} v(x_t, x_{t+1}) \geq T_0 v(\bar{x}, \bar{x}) - M_0 - 2\bar{c}$$

the inequality

$$\min\{\rho(x_i, \bar{x}) : i = 1, \ldots, T_0\} \leq \epsilon$$

holds.

Let an integer $T \geq T_0$, let a program $\{x_t\}_{t=0}^{T}$ satisfy (2.28), and let an integer $s \in [0, T - T_0]$. By relations (2.28) and (2.4), we have

$$\sum_{t=s}^{s+T_0-1} v(x_t, x_{t+1}) - T_0 v(\bar{x}, \bar{x}) \geq -M_0 - 2\bar{c}.$$

It follows from this inequality and property (P1) that

$$\min\{\rho(x_i . \bar{x}) : \ i = s + 1, \ldots, s + T_0\} \le \epsilon.$$

Lemma 2.16 is proved.   □

Lemma 2.17. *Let M_0 be a positive number and let $L_0 \ge 1$ be an integer. Then there exist an integer $T_0 \ge 1$ and a number $M_1 > 0$ such that for each natural number $T \ge T_0$, each point $z_0 \in X_{M_0}$, and each point $z_1 \in Y_{L_0}$, the inequality*

$$\sigma(v, T, z_0, z_1) \ge T v(\bar{x}, \bar{x}) - M_1$$

holds.

Proof. It follows from Lemma 2.13 that there exists an integer $L_1 \ge 1$ such that the following property holds:

(P2)   For each program $\{u_t\}_{t=0}^{L_1}$ which satisfies

$$\sum_{t=0}^{L_1-1} v(u_t, u_{t+1}) \ge L_1 v(\bar{x}, \bar{x}) - M_0$$

we have

$$\min\{\rho(u_i, \bar{x}) : \ i = 1, \ldots, L_1\} \le \bar{r}/2.$$

Put

$$T_0 = 2(L_0 + L_1 + 1), \ M_1 = M_0 + ||v||(2 + 2L_0). \tag{2.29}$$

Let a natural number $T \ge T_0$ be given and let

$$z_0 \in X_{M_0}, \ z_1 \in Y_{L_0}. \tag{2.30}$$

In view of inclusions (2.30) there exists a program $\{y_t\}_{t=0}^{\infty}$ such that

$$y_0 = z_0,$$

$$\sum_{t=0}^{T-1} v(y_t, y_{t+1}) - T v(\bar{x}, \bar{x}) \ge -M_0 \text{ for all natural numbers } T. \tag{2.31}$$

By property (P2) and relation (2.31) there exists a natural number

$$t_0 \in [1, L_1] \tag{2.32}$$

such that

$$\rho(y_{t_0}, \bar{x}) \le \bar{r}/2. \tag{2.33}$$

Since the point $z_1 \in Y_{L_0}$ there exists a program $\{u_t\}_{t=0}^{L_0}$ such that

$$u_0 = \bar{x}, \ u_{L_0} = z_1. \tag{2.34}$$

Set

$$x_t = y_t, \ t = 0, \ldots, t_0,$$
$$x_t = \bar{x}, \ t = t_0 + 1, \ldots, T - L_0,$$
$$x_t = u_{t-T+L_0}, \ t = T - L_0 + 1, \ldots, T. \tag{2.35}$$

It is easy to see that the sequence $\{x_t\}_{t=0}^{T}$ is a program. In view of relations (2.35), (2.34), (2.31), and (2.29), we have

$$\sigma(v, T, z_0, z_1) \geq \sum_{t=0}^{T-1} v(x_t, x_{t+1})$$

$$= \sum_{t=0}^{t_0-1} v(y_t, y_{t+1}) + v(y_{t_0}, \bar{x}) + (T - L_0 - t_0 - 1)v(\bar{x}, \bar{x}) + \sum_{t=0}^{L_0-1} v(u_t, u_{t+1})$$

$$\geq t_0 v(\bar{x}, \bar{x}) - M_0 + v(\bar{x}, \bar{x}) - 2||v|| + (T - L_0 - t_0 - 1)v(\bar{x}, \bar{x})$$

$$+ L_0 v(\bar{x}, \bar{x}) - 2L_0||v||$$

$$= T v(\bar{x}, \bar{x}) - M_0 - ||v||(2 + 2L_0) = T v(\bar{x}, \bar{x}) - M_1.$$

Lemma 2.17 is proved. ◻

2.3   Completion of the Proof of Theorem 2.2

Let a number $\bar{r} \in (0, 1)$ satisfy relation (2.7). We may assume without loss of generality that

$$\epsilon < \bar{r}/2, \tag{2.36}$$

$$M > 4 \sup\{|v(z_1, z_2)| : (z_1, z_2) \in X \times X\}. \tag{2.37}$$

Lemma 2.15 implies that there is a positive number $\delta < \epsilon$ such that the following property holds:

(P3)   For each natural number τ and each program $\{x_t\}_{t=0}^{\tau}$ which satisfies

$$\rho(x_0, \bar{x}), \ \rho(x_\tau, \bar{x}) \leq \delta \text{ and } \sum_{t=0}^{\tau-1} v(x_t, x_{t+1}) \geq \sigma(v, x_0, x_\tau, \tau) - \delta \tag{2.38}$$

the inequality $\rho(x_t, \bar{x}) \leq \epsilon$ holds for all integers $t = 0, \ldots, \tau$.

By Lemma 2.13 there exists an integer $L \geq 1$ such that the following property holds:

(P4)   For each program $\{x_t\}_{t=0}^{L}$ which satisfies

$$\sum_{t=0}^{L-1} v(x_t, x_{t+1}) \geq Lv(\bar{x}, \bar{x}) - M - \bar{c} - 1$$

we have $\min\{\rho(x_i, \bar{x}) : i = 1, \ldots, L\} \leq \delta$.

Assume that $T > 2L$ is a natural number and that a program $\{x_t\}_{t=0}^{T}$ satisfies

$$x_0 \in X_M, \ \sum_{t=0}^{T-1} v(x_t, x_{t+1}) \geq \sigma(v, T, x_0) - \delta. \tag{2.39}$$

In view of relations (2.39) and the definition of X_M there exists a program $\{y_t\}_{t=0}^{\infty}$ such that

$$y_0 = x_0, \ \sum_{t=0}^{\tau-1} v(y_t, y_{t+1}) - \tau v(\bar{x}, \bar{x}) \geq -M \text{ for all natural numbers } \tau. \tag{2.40}$$

It follows from relations (2.39) and (2.40) that

$$\sum_{t=0}^{T-1} v(x_t, x_{t+1}) \geq \sum_{t=0}^{T-1} v(y_t, y_{t+1}) - \delta \geq \sum_{t=0}^{T-1} v(y_t, y_{t+1}) - 1 \geq Tv(\bar{x}, \bar{x}) - M - 1.$$

$$\tag{2.41}$$

Let $S \in \{1, \ldots, T-1\}$ be given. By relations (2.41) and (2.4), we have

$$\sum_{t=S}^{T-1} v(x_t, x_{t+1}) = \sum_{t=0}^{T-1} v(x_t, x_{t+1}) - \sum_{t=0}^{S-1} v(x_t, x_{t+1})$$

$$\geq Tv(\bar{x}, \bar{x}) - M - 1 - \sum_{t=0}^{S-1} v(x_t, x_{t+1})$$

$$\geq Tv(\bar{x}, \bar{x}) - M - 1 - \sigma(v, S)$$

$$\geq Tv(\bar{x}, \bar{x}) - M - 1 - Sv(\bar{x}, \bar{x}) - \bar{c}$$

$$= (T - S)v(\bar{x}, \bar{x}) - M - \bar{c} - 1.$$

Analogously, for any $S \in \{0, \ldots, T-1\}$,

$$\sum_{t=0}^{S-1} v(x_t, x_{t+1}) = \sum_{t=0}^{T-1} v(x_t, x_{t+1}) - \sum_{t=S}^{T-1} v(x_t, x_{t+1})$$

$$\geq Tv(\bar{x}, \bar{x}) - M - 1 - \sum_{t=S}^{T-1} v(x_t, x_{t+1})$$

$$\geq Tv(\bar{x}, \bar{x}) - M - 1 - \sigma(v, T - S)$$

$$\geq Tv(\bar{x},\bar{x}) - M - 1 - (T-S)v(\bar{x},\bar{x}) - \bar{c}$$
$$= Sv(\bar{x},\bar{x}) - M - \bar{c} - 1.$$

Hence, for all $S \in \{0,\ldots,T-1\}$,

$$\sum_{t=S}^{T-1} v(x_t, x_{t+1}) \geq (T-S)v(\bar{x},\bar{x}) - M - \bar{c} - 1, \quad \sum_{t=0}^{S-1} v(x_t, x_{t+1}) \geq Sv(\bar{x},\bar{x}) - M - \bar{c} - 1.$$

$$(2.42)$$

By relation (2.42) and (P4), there exist integers $\tau_1, \tau_2 \in \{0,\ldots,L\}$ such that

$$\rho(x_{\tau_1}, \bar{x}) \leq \delta, \quad \rho(x_{T-\tau_2}, \bar{x}) \leq \delta. \tag{2.43}$$

Evidently, if $\rho(x_0, \bar{x}) \leq \delta$, then we can set $\tau_1 = 0$. By (2.39), we have

$$\sum_{t=\tau_1}^{T-\tau_2-1} v(x_t, x_{t+1}) \geq \sigma(v, x_{\tau_1}, x_{T-\tau_2}, T - \tau_1 - \tau_2) - \delta.$$

In view of this inequality, property (P3), and relation (2.43), we have $\rho(x_t, \bar{x}) \leq \epsilon$ for all integers $t = \tau_1, \ldots, T - \tau_2$. Theorem 2.2 is proved.

2.4   Proof of Theorem 2.3

There exists a (v)-good program $\{y_t\}_{t=0}^{\infty}$ such that $y_0 = x$. Since the sequence $\{y_t\}_{t=0}^{\infty}$ is a (v)-good program there exists a positive number M such that

$$\sum_{t=0}^{T-1} [v(y_t, y_{t+1}) - v(\bar{x},\bar{x})] \geq -M \text{ for all integers } T \geq 1. \tag{2.44}$$

Let $\{T_k\}_{k=1}^{\infty}$ be a strictly increasing sequence of integers with $T_1 > 4$. For each integer $k \geq 1$ there is a program $\{x_t^{(k)}\}_{t=0}^{T_k}$ such that

$$x_0^{(k)} = x, \quad \sum_{t=0}^{T_k-1} v\left(x_t^{(k)}, x_{t+1}^{(k)}\right) = \sigma(v, x, T_k). \tag{2.45}$$

Extracting a subsequence and re-indexing we may assume without loss of generality that for each nonnegative integer t there is

$$x_t = \lim_{k \to \infty} x_t^{(k)}. \tag{2.46}$$

It is easy to see that the sequence $\{x_t\}_{t=0}^{\infty}$ is a program and that $x_0 = x$. Let k be a natural number. By (2.45) we have

$$\sum_{t=0}^{T_k-1} v\left(x_t^{(k)}, x_{t+1}^{(k)}\right) \geq \sum_{t=0}^{T_k-1} v(y_t, y_{t+1}) \geq -M + T_k v(\bar{x}, \bar{x}). \tag{2.47}$$

It follows from relations (2.47) and (2.4) that for each natural number $S < T_k$

$$\sum_{t=0}^{S-1} v\left(x_t^{(k)}, x_{t+1}^{(k)}\right) = \sum_{t=0}^{T_k-1} v\left(x_t^{(k)}, x_{t+1}^{(k)}\right) - \sum_{t=S}^{T_k-1} v\left(x_t^{(k)}, x_{t+1}^{(k)}\right)$$

$$\geq -M + T_k v(\bar{x}, \bar{x}) - \sum_{t=S}^{T_k-1} v\left(x_t^{(k)}, x_{t+1}^{(k)}\right)$$

$$\geq -M + T_k v(\bar{x}, \bar{x}) - \sigma(v, T - S)$$

$$\geq -M + T_k v(\bar{x}, \bar{x}) - (T_k - S)v(\bar{x}, \bar{x}) - \bar{c}$$

$$\geq S v(\bar{x}, \bar{x}) - M - \bar{c}.$$

Therefore we have shown that for each integer $k \geq 1$ and each integer $S \in \{1, \ldots, T_k\}$ we have

$$\sum_{t=0}^{S-1} v\left(x_t^{(k)}, x_{t+1}^{(k)}\right) \geq S v(\bar{x}, \bar{x}) - M - \bar{c}. \tag{2.48}$$

By (2.48) and (2.46), for each integer $S \geq 1$, we have

$$\sum_{t=0}^{S-1} v(x_t, x_{t+1}) \geq S v(\bar{x}, \bar{x}) - M - \bar{c}. \tag{2.49}$$

In view of (2.49), the sequence $\{x_t\}_{t=0}^{\infty}$ is a (v)-good program. By assumption (A3),

$$\lim_{t \to \infty} \rho(x_t, \bar{x}) = 0. \tag{2.50}$$

We claim that the sequence $\{x_t\}_{t=0}^{\infty}$ is a (v)-overtaking optimal program. Assume the contrary. Then there exist a program $\{z_t\}_{t=0}^{\infty}$ and a number $\Delta > 0$ such that

$$z_0 = x, \ \limsup_{T \to \infty} \sum_{t=0}^{T-1} [v(z_t, z_{t+1}) - v(x_t, x_{t+1})] \geq \Delta. \tag{2.51}$$

By inequality (2.49) we have that the sequence $\{z_t\}_{t=0}^{\infty}$ is a (v)-good program. It follows from assumption (A3) that

$$\lim_{t \to \infty} \rho(z_t, \bar{x}) = 0. \tag{2.52}$$

By assumption (A1) and the continuity of the function v at the point $(\bar{x}, \bar{x})$ there exists a positive number δ such that

$$\{(x, y) \in X \times X : \rho(x, \bar{x}), \rho(y, \bar{x}) \le 2\delta\} \subset \Omega \tag{2.53}$$

and that

$$|v(x, y) - v(\bar{x}, \bar{x})| \le \Delta/16 \tag{2.54}$$

for each pair of points $x, y \in X$ which satisfy $\rho(x, \bar{x}), \rho(y, \bar{x}) \le 2\delta$. In view of (2.52) and (2.50), there exists an integer $\tau_0 \ge 4$ such that

$$\rho(z_t, \bar{x}), \; \rho(x_t, \bar{x}) \le \delta/4 \text{ for all integers } t \ge \tau_0. \tag{2.55}$$

In view of (2.51), there exists a natural number $\tau_1 \ge 4(\tau_0 + 4)$ such that

$$\sum_{t=0}^{\tau_1-1} [v(z_t, z_{t+1}) - v(x_t, x_{t+1})] \ge (3/4)\Delta. \tag{2.56}$$

It follows from equality (2.46) and the upper semicontinuity of the function v that there exists an integer $k \ge 1$ such that

$$T_k \ge 4(\tau_1 + 4), \tag{2.57}$$

$$v(x_t, x_{t+1}) - v\left(x_t^{(k)}, x_{t+1}^{(k)}\right) \ge -\Delta(16(\tau_1 + 1))^{-1} \text{ for all } t = 0, \dots, \tau_1, \tag{2.58}$$

$$\rho\left(x_t, x_t^{(k)}\right) \le \delta/4, \; t = 0, 1, \dots, 4(\tau_1 + 4). \tag{2.59}$$

In view of (2.59) and (2.55), we have

$$\rho\left(x_t^{(k)}, \bar{x}\right) \le \delta/2 \text{ for all } t = \tau_0, \dots, 4(\tau_1 + 4). \tag{2.60}$$

Define a sequence $\tilde{x}_t, t = 0, \dots, T_k$ as follows:

$$\tilde{x}_t = z_t, \; t = 0, \dots, \tau_1, \; \tilde{x}_t = x_t^{(k)}, \; t = \tau_1 + 1, \dots, T_k. \tag{2.61}$$

By (2.61), (2.55), (2.60), and (2.53), the sequence $\{\tilde{x}_t\}_{t=0}^{T_k}$ is a program. It is easy to see that $\tilde{x}_0 = x$. It follows from (2.61) and (2.56) that

$$\sum_{t=0}^{T_k-1} v(\tilde{x}_t, \tilde{x}_{t+1}) - \sum_{t=0}^{T_k-1} v\left(x_t^{(k)}, x_{t+1}^{(k)}\right)$$

$$= \sum_{t=0}^{\tau_1-1} \left[v(z_t, z_{t+1}) - v\left(x_t^{(k)}, x_{t+1}^{(k)}\right) \right] + v\left(z_{\tau_1}, x_{\tau_1+1}^{(k)}\right) - v\left(x_{\tau_1}^{(k)}, x_{\tau_1+1}^{(k)}\right)$$

$$= \sum_{t=0}^{\tau_1-1} [v(z_t, z_{t+1}) - v(x_t, x_{t+1})]$$

$$+ \sum_{t=0}^{\tau_1-1} \left[v(x_t, x_{t+1}) - v\left(x_t^{(k)}, x_{t+1}^{(k)}\right) \right] + v\left(z_{\tau_1}, x_{\tau_1+1}^{(k)}\right) - v\left(x_{\tau_1}^{(k)}, x_{\tau_1+1}^{(k)}\right)$$

$$\geq (3/4)\Delta + \sum_{t=0}^{\tau_1-1} \left[v(x_t, x_{t+1}) - v\left(x_t^{(k)}, x_{t+1}^{(k)}\right) \right] + v\left(z_{\tau_1}, x_{\tau_1+1}^{(k)}\right) - v\left(x_{\tau_1}^{(k)}, x_{\tau_1+1}^{(k)}\right).$$

$$(2.62)$$

In view of (2.55), (2.60), and (2.54), we have

$$\left| v(z_{\tau_1}, x_{\tau_1+1}^{(k)}) - v(x_{\tau_1}^{(k)}, x_{\tau_1+1}^{(k)}) \right|$$

$$= \left| v(z_{\tau_1}, x_{\tau_1+1}^{(k)}) - v(\bar{x}, \bar{x}) \right| + \left| v(\bar{x}, \bar{x}) - v(x_{\tau_1}^{(k)}, x_{\tau_1+1}^{(k)}) \right| \leq \Delta/8.$$

It follows from relations (2.62) and (2.58) that

$$\sum_{t=0}^{T_k-1} v(\tilde{x}_t, \tilde{x}_{t+1}) - \sum_{t=0}^{T_k-1} v\left(x_t^{(k)}, x_{t+1}^{(k)} \right) \geq (3/4)\Delta - \Delta/8 - \Delta(16(\tau_1+1))^{-1}\tau_1 \geq \Delta/2.$$

This inequality contradicts (2.45). The contradiction we have reached shows that the sequence $\{x_t\}_{t=0}^{\infty}$ is a (v)-overtaking optimal program. Theorem 2.3 is proved.

2.5 Proofs of Theorems 2.4 and 2.5

Proof of Theorem 2.4. It follows from assumption (A1) that there exists a positive number r such that

$$\{(x, y) \in X \times X : \rho(x, \bar{x}), \rho(y, \bar{x}) \leq r\} \subset \Omega. \qquad (2.63)$$

It is easy to see that

$$\{x \in X : \rho(x, \bar{x}) \leq r\} \subset X_{2||v||}. \qquad (2.64)$$

Theorem 2.2 implies that there are an integer $L_0 \geq 1$ and a number $\delta_0 > 0$ such that the following property holds:

(P5) For each integer $T > 2L_0$ and each program $\{x_t\}_{t=0}^{T}$ which satisfies $x_0 \in X_{2||v||+1}$ and

$$\sum_{t=0}^{T-1} v(x_t, x_{t+1}) \geq \sigma(v, T, x_0) - \delta_0$$

there exist integers $\tau_1, \tau_2 \in [0, L_0]$ such that $\rho(x_t, \bar{x}) \leq \epsilon$ for all $t = \tau_1, \ldots,$ $T - \tau_2$ and if $\rho(x_0, \bar{x}) \leq \delta_0$, then $\tau_1 = 0$.

Let $M > 0$ be given. It follows from Theorem 2.2 that there are an integer $L \geq 1$ and a number $\delta > 0$ such that the following property holds:

(P6)　For each integer $T > 2L$ and each program $\{x_t\}_{t=0}^{T}$ which satisfies

$$x_0 \in X_M, \ \sum_{t=0}^{T-1} v(x_t, x_{t+1}) \geq \sigma(v, T, x_0) - \delta \tag{2.65}$$

there exist integers $\tau_1, \tau_2 \in [0, L]$ such that $\rho(x_t, \bar{x}) \leq \min\{r, \epsilon, \delta_0\}$ for all integers $t = \tau_1, \ldots, T - \tau_2$ and if $\rho(x_0, \bar{x}) \leq \delta$, then $\tau_1 = 0$.

We may assume without loss of generality that

$$\delta < \delta_0, \ L > 2L_0. \tag{2.66}$$

Assume that an integer $T > 2L$ and that a program $\{x_t\}_{t=0}^{T}$ satisfies (2.65). It follows from (P6) that there exist integers

$$S_1, S_2 \in [0, L] \tag{2.67}$$

such that

$$\rho(x_t, \bar{x}) \leq \min\{r, \epsilon, \delta_0\} \text{ for all } t = S_1, \ldots, T - S_2; \tag{2.68}$$

$$\text{if } \rho(x_0, \bar{x}) \leq \delta, \text{ then } S_1 = 0. \tag{2.69}$$

Set

$$y_t = x_{t+S_1}, \ t = 0, \ldots, T - S_1. \tag{2.70}$$

By relations (2.66), (2.67) and the inequality $T > 2L$, we have

$$T - S_1 \geq T - L > L > 2L_0. \tag{2.71}$$

It is easy to see that $\{y_t\}_{t=0}^{T-S_1}$ is a program. In view of relations (2.65) and (2.70), we have

$$\sum_{t=0}^{T-S_1} v(y_t, y_{t+1}) \geq \sigma(v, T - S_1, y_0) - \delta. \tag{2.72}$$

It follows from (2.68) and (2.70) that

$$\rho(y_0, \bar{x}) = \rho(x_{S_1}, \bar{x}) \leq \min\{r, \epsilon, \delta_0\}. \tag{2.73}$$

In view of relations (2.73) and (2.64), the inclusion $y_0 \in X_{2\|v\|}$ is true. By this inclusion, relations (2.71)–(2.73), and property (P5) applied to the program $\{y_t\}_{t=0}^{T-S_1}$, there is $S_3 \in [0, L_0]$ such that $\rho(y_t, \bar{x}) \leq \epsilon$ for all $t = 0, \ldots, T - S_1 - S_3$. Together with relation (2.70) this inequality implies that $\rho(x_t, \bar{x}) \leq \epsilon$ for all $t = S_1, \ldots, T - S_3$. In view of this inequality, (2.67), and the inequality $S_3 \leq L_0$ the assertion of the theorem holds with $\tau_1 = S_1$, $\tau_2 = S_3$. Theorem 2.4 is proved. □

Proof of Theorem 2.5. Evidently, if the sequence $\{x_t\}_{t=0}^{\infty}$ is a (v)-overtaking optimal program, then the conditions (i) and (ii) hold. Assume that the conditions (i) and (ii) hold. We claim that the sequence $\{x_t\}_{t=0}^{\infty}$ is a (v)-overtaking optimal program.

Assume the contrary. Then there exist a program $\{y_t\}_{t=0}^{\infty}$ and a positive number Δ such that

$$y_0 = x_0, \ \limsup_{T \to \infty} \left[\sum_{t=0}^{T-1} v(y_t, y_{t+1}) - v(x_t, x_{t+1}) \right] \geq 2\Delta. \tag{2.74}$$

In view of (2.74), there exists a strictly increasing sequence of natural numbers $\{T_k\}_{k=1}^{\infty}$ such that for all natural numbers k we have

$$\sum_{t=0}^{T_k-1} v(y_t, y_{t+1}) \geq \sum_{t=0}^{T_k-1} v(x_t, x_{t+1}) + \Delta. \tag{2.75}$$

By assumption (A1) there exists a number $\delta \in (0, 1)$ such that

$$\{(z_1, z_2) \in X \times X : \rho(z_i, \bar{x}) \leq 2\delta, \ i = 1, 2\} \subset \Omega, \tag{2.76}$$

$$|v(z_1, z_2) - v(z_3, z_4)| \leq \Delta/8 \text{ for each } z_1, z_2, z_3, z_4 \in X$$

$$\text{satisfying } \rho(z_i, \bar{x}) \leq 2\delta \text{ for } i = 1, 2, 3, 4. \tag{2.77}$$

Clearly, the sequence $\{x_t\}_{t=0}^{\infty}$ is a (v)-good program. In view of (2.74) the program $\{y_t\}_{t=0}^{\infty}$ is also (v)-good. Therefore it follows from assumption (A3) that

$$\lim_{t \to \infty} \rho(x_t, \bar{x}) = \lim_{t \to \infty} \rho(y_t, \bar{x}) = 0.$$

There exists an integer $\bar{L} \geq 1$ such that

$$\rho(x_t, \bar{x}), \ \rho(y_t, \bar{x}) \leq \delta/2 \text{ for all integers } t \geq \bar{L}. \tag{2.78}$$

Fix an integer $k \geq 1$ such that $T_k > \bar{L} + 4$ and define a sequence $\{\tilde{x}_t\}_{t=0}^{T_k}$ by

$$\tilde{x}_t = y_t \text{ for all } t = 0, \ldots, T_{k-1}, \ \tilde{x}_{T_k} = x_{T_k}. \tag{2.79}$$

By relations (2.79), (2.78) and the inequality $T_k > \bar{L} + 4$, we have

$$\rho(\tilde{x}_t, \bar{x}) \le \delta/2 \text{ for all } t = \bar{L}, \dots, T_k. \tag{2.80}$$

It follows from (2.80) and (2.76) that the sequence $\{\tilde{x}_t\}_{t=0}^{T_k}$ is a program. In view of relations (2.74) and (2.79),

$$\tilde{x}_0 = x_0, \ \tilde{x}_{T_k} = x_{T_k}. \tag{2.81}$$

By (2.80), (2.78), the inequality $T_k > \bar{L} + 4$, and (2.77), we have

$$|v(\tilde{x}_{T_k-1}, \tilde{x}_{T_k}) - v(y_{T_k-1}, y_{T_k})| \le \Delta/8.$$

Combined with (2.79) this relation implies that

$$\left| \sum_{t=0}^{T_k-1} [v(\tilde{x}_t, \tilde{x}_{t+1}) - v(y_t, y_{t+1})] \right| = |v(\tilde{x}_{T_k-1}, \tilde{x}_{T_k}) - v(y_{T_k-1}, y_{T_k})| \le \Delta/8.$$

Together with inequality (2.75) this relation implies that

$$\sum_{t=0}^{T_k-1} [v(\tilde{x}_t, \tilde{x}_{t+1}) - v(x_t, x_{t+1})]$$

$$= \sum_{t=0}^{T_k-1} [v(\tilde{x}_t, \tilde{x}_{t+1}) - v(y_t, y_{t+1})] + \sum_{t=0}^{T_k-1} [v(y_t, y_{t+1}) - v(x_t, x_{t+1})] \ge 7\Delta/8.$$

This relation and (2.81) contradict the condition (ii). The contradiction we have reached proves Theorem 2.5.   □

2.6   Proofs of Theorems 2.6 and 2.7

We assume that the function v is continuous and precede the proof of Theorem 2.6 with the following auxiliary results.

Lemma 2.18. *Let M_0 be a positive number. Then there exists a positive number M_1 such that for each (v)-overtaking optimal program $\{x_t\}_{t=0}^{\infty}$ satisfying the inclusion $x_0 \in X_{M_0}$, the inequality*

$$\sum_{t=0}^{T-1} [v(x_t, x_{t+1}) - v(\bar{x}, \bar{x})] \ge -M_1$$

holds for all natural numbers T.

Proof. Put

$$M_1 = M_0 + 1 + \bar{c} \tag{2.82}$$

[see assumption (A2)]. Assume that a (v)-overtaking optimal program $\{x_t\}_{t=0}^{\infty}$ satisfies the inclusion

$$x_0 \in X_{M_0}. \tag{2.83}$$

By definition there exists a program $\{y_t\}_{t=0}^{\infty}$ such that

$$y_0 = x_0, \ \sum_{t=0}^{T-1} [v(y_t, y_{t+1}) - v(\bar{x}, \bar{x})] \geq -M_0 \ \text{for all natural numbers } T. \tag{2.84}$$

Clearly,

$$\limsup_{T \to \infty} \sum_{t=0}^{T-1} [v(y_t, y_{t+1}) - v(x_t, x_{t+1})] \leq 0. \tag{2.85}$$

Let T be a natural number. In view of inequality (2.85) there exists an integer $S > 2(T + 2)$ such that

$$\sum_{t=0}^{S-1} [v(y_t, y_{t+1}) - v(x_t, x_{t+1})] \leq 1.$$

Together with (2.84) this inequality implies that

$$\sum_{t=0}^{S-1} v(x_t, x_{t+1}) \geq Sv(\bar{x}, \bar{x}) - M_0 - 1.$$

Combined with assumption (A2) and equality (2.82) this inequality implies that

$$\sum_{t=0}^{T-1} v(x_t, x_{t+1}) = \sum_{t=0}^{S-1} v(x_t, x_{t+1}) - \sum_{t=T}^{S-1} v(x_t, x_{t+1}) \geq Sv(\bar{x}, \bar{x}) - M_0$$

$$-1 - \bar{c} - (S - T)v(\bar{x}, \bar{x}) = Tv(\bar{x}, \bar{x}) - M_0 - 1 - \bar{c} = Tv(\bar{x}, \bar{x}) - M_1.$$

Lemma 2.18 is proved.   □

Lemmas 2.13 and 2.18 imply the following result.

Lemma 2.19. *Let ϵ and M_0 be positive numbers. Then there exists an integer $L \geq 1$ such that for each (v)-overtaking optimal program $\{x_t\}_{t=0}^{\infty}$ satisfying the inclusion $x_0 \in X_{M_0}$, the inequality $\min\{\rho(x_i, \bar{x}) : i = 1, \ldots, L\} \leq \epsilon$ holds.*

In view of assumption (A1) there exists a number $\bar{r} \in (0, 1)$ such that

$$\{(x, y) \in X \times X : \rho(x, \bar{x}), \ \rho(y, \bar{x}) \leq \bar{r}\} \subset \Omega. \tag{2.86}$$

Lemma 2.20. *Let ϵ be a positive number. Then there exists a number $\delta \in (0, \bar{r}]$ such that for each natural number T and each pair of points $\xi_1, \xi_2 \in X$ satisfying $\rho(\xi_i, \bar{x}) \leq \delta$, $i = 1, 2$, the following inequality holds:*

$$|\sigma(v, \xi_1, \xi_2, T) - T v(\bar{x}, \bar{x})| \leq \epsilon. \tag{2.87}$$

Proof. It is not difficult to see that for each natural number T we have

$$\sigma(v, \bar{x}, \bar{x}, T) = T v(\bar{x}, \bar{x}). \tag{2.88}$$

Since the function v is continuous there exists a number

$$\delta \in (0, \bar{r}/4) \tag{2.89}$$

such that for each $\xi_1, \xi_2, \eta_1, \eta_2 \in X$ satisfying $\rho(\xi_i, \eta_i) \leq 2\delta, i = 1, 2$ the inequality

$$|v(\xi_1, \xi_2) - v(\eta_1, \eta_2)| \leq \epsilon/8 \tag{2.90}$$

holds.

Assume that T is a natural number and that

$$\xi_1, \xi_2 \in X, \ \rho(\xi_i, \bar{x}) \leq \delta, \ i = 1, 2. \tag{2.91}$$

We claim that inequality (2.87) holds. It is not difficult to see that (2.87) is true if $T = 1$. Therefore we may consider only the case with $T > 1$. Set

$$y_0 = \xi_1, \ y_T = \xi_2,$$
$$y_i = \bar{x} \text{ for all } i \in \{1, \ldots, T - 1\}. \tag{2.92}$$

It follows from (2.92), (2.91), (2.89), and (2.86) that $\{y_t\}_{t=0}^{T}$ is a program. It is clear that

$$\sigma(v, \xi_1, \xi_2, T) \geq \sum_{t=0}^{T-1} v(y_t, y_{t+1}) = v(\xi_1, \bar{x}) + v(\bar{x}, \xi_2) + (T - 2)v(\bar{x}, \bar{x}).$$

By this inequality, relation (2.91), and the choice of δ [see (2.90)], we have

$$\sigma(v, \xi_1, \xi_2, T) \geq T v(\bar{x}, \bar{x}) - \epsilon/4. \tag{2.93}$$

There exists a program $\{x_t\}_{t=0}^T$ such that

$$x_0 = \xi_1, \ x_T = \xi_2, \ \sum_{t=0}^{T-1} v(x_t, x_{t+1}) = \sigma(v, \xi_1, \xi_2, T). \tag{2.94}$$

Put

$$z_0 = \bar{x}, \ z_t = x_{t-1} \text{ for all } t = 1, \ldots, T+1, \ z_{T+2} = \bar{x}. \tag{2.95}$$

In view of relations (2.95), (2.94), (2.91), (2.86), and (2.89), $\{z_t\}_{t=0}^{T+2}$ is a program. It follows from (2.94), (2.95), and (2.88) that

$$(T+2)v(\bar{x}, \bar{x}) \geq \sum_{t=0}^{T+1} v(z_t, z_{t+1}) = v(\bar{x}, \xi_1) + v(\xi_2, \bar{x}) + \sigma(v, \xi_1, \xi_2, T).$$

Together with (2.91) and the choice of δ [see (2.90)] this relation implies that $(T+2)v(\bar{x}, \bar{x}) \geq \sigma(v, \xi_1, \xi_2, T) + 2v(\bar{x}, \bar{x}) - \epsilon/2$. Combined with (2.93) this inequality implies (2.87). Lemma 2.20 is proved.   □

Completion of the proof of Theorem 2.6. Lemma 2.20 implies that there exists a sequence of positive numbers $\{\delta_i\}_{i=1}^\infty$ such that

$$\delta_{i+1} < \delta_i \leq \bar{r}/4 \text{ for all natural numbers } i \geq 1 \text{ and } \delta_1 < \epsilon/4 \tag{2.96}$$

and that for each natural number i, each natural number T, and each pair of points $\xi_1, \xi_2 \in X$ satisfying $\rho(\xi_j, \bar{x}) \leq \delta_i, \ j = 1, 2$ we have

$$|\sigma(v, \xi_1, \xi_2, T) - Tv(\bar{x}, \bar{x})| \leq 2^{-i}. \tag{2.97}$$

Assume that the assertion of the theorem does not hold. Then for each natural number i there exists a (v)-overtaking optimal program $\{x_t^{(i)}\}_{t=0}^\infty$ such that

$$\rho\left(x_0^{(i)}, \bar{x}\right) \leq \delta_i, \ \sup\left\{\rho\left(x_t^{(i)}, \bar{x}\right) : \ t = 0, 1, \ldots\right\} > \epsilon. \tag{2.98}$$

It follows from (2.98) and assumption (A3) that for each natural number i there exists a pair of natural numbers $S_i, T_i > S_i + 4$ such that

$$\rho\left(x_{S_i}^{(i)}, \bar{x}\right) \geq \epsilon, \ \rho\left(x_{T_i-1}^{(i)}, \bar{x}\right) \leq \delta_{i+1}. \tag{2.99}$$

Put

$$\tau_1 = T_1, \ \tau_k = \sum_{i=1}^k T_i \text{ for all natural numbers } k, \tag{2.100}$$

$$y_t = x_t^{(1)} \text{ for all } t = 0, \ldots, T_1 - 1, \ y_t = x_{t-\tau_k}^{(k+1)} \text{ for all } t = \tau_k, \ldots, \tau_{k+1} - 1 \tag{2.101}$$

and for all natural numbers k. It is clear that the sequence $\{y_t\}_{t=0}^{\infty}$ is well defined. By relations (2.98)–(2.101), (2.96), and (2.86), the sequence $\{y_t\}_{t=0}^{\infty}$ is a program. By relations (2.98)–(2.101) and the choice of δ_{k+1} [see (2.97)], for each natural number k, we have

$$|v(y_{\tau_k-1}, y_{\tau_k}) - v(\bar{x}, \bar{x})| = \left| v\left(x_{T_k-1}^{(k)}, x_0^{(k+1)}\right) - v(\bar{x}, \bar{x})\right| \leq 2^{-k-1}. \tag{2.102}$$

It follows from relations (2.100) and (2.101) that for each natural number $k \geq 2$ we have

$$\sum_{t=0}^{\tau_k-1}[v(y_t, y_{t+1}) - v(\bar{x}, \bar{x})]$$

$$= \sum_{i=1}^{k}\sum_{t=0}^{T_i-2}\left[v\left(x_t^{(i)}, x_{t+1}^{(i)}\right) - v(\bar{x}, \bar{x})\right] + \sum_{i=1}^{k-1}[v(y_{\tau_i-1}, y_{\tau_i}) - v(\bar{x}, \bar{x})]. \tag{2.103}$$

Let a natural number i be given. Since the program $\{x_t^{(i)}\}_{t=0}^{\infty}$ is a (v)-overtaking optimal it follows from relations (2.98), (2.99), (2.96) and the choice of δ_i [see (2.97)] that

$$\left|\sum_{t=0}^{T_i-2}\left[v\left(x_t^{(i)}, x_{t+1}^{(i)}\right) - v(\bar{x}, \bar{x})\right]\right| = |\sigma(v, x_0^{(i)}, x_{T_i-1}^{(i)}, T_i-1) - (T_i-1)v(\bar{x}, \bar{x})| \leq 2^{-i}.$$

Combined with (2.103) and (2.102) this relation implies that

$$\left|\sum_{t=0}^{\tau_k-1}[v(y_t, y_{t+1}) - v(\bar{x}, \bar{x})]\right| \leq \sum_{i=1}^{\infty}2^{-i} + \sum_{i=1}^{\infty}2^{-i} = 2$$

for all natural numbers $k \geq 2$. Together with Proposition 2.1 this relation implies that the program $\{y_t\}_{t=0}^{\infty}$ is (v)-good. By assumption (A3), we have $\lim_{t\to\infty}\rho(y_t, \bar{x}) = 0$. On the other hand by relations (2.99)–(2.101), $\limsup_{t\to\infty}\rho(y_t, \bar{x}) \geq \epsilon$. The contradiction we have reached proves Theorem 2.6.

$\square$

Theorem 2.7 follows from Lemma 2.19 and Theorem 2.6.

2.7 Proof of Theorem 2.8

Lemma 2.17 implies that there exist an integer $T_0 \geq 1$ and a positive number M_2 such that for each integer $T \geq T_0$, each point $z_0 \in X_{M_0}$, and each point $z_1 \in Y_{L_0}$,

$$\sigma(v, T, z_0, z_1) \geq Tv(\bar{x}, \bar{x}) - M_2. \tag{2.104}$$

By Lemma 2.15 there exists a number $\delta \in (0, \bar{r}/2)$ such that for each natural number T and each program $\{y_t\}_{t=0}^{T}$ which satisfies

$$\rho(y_0, \bar{x}), \; \rho(y_T, \bar{x}) \le \delta,$$

$$\sum_{t=0}^{T-1} v(y_t, y_{t+1}) \ge \sigma(v, T, y_0, y_T) - \delta \tag{2.105}$$

we have

$$\rho(y_t, \bar{x}) \le \epsilon, \; t = 0, \ldots, T. \tag{2.106}$$

It follows from Lemma 2.16 that there exists an integer $T_1 \ge 1$ such that the following property holds:

For each natural number $T \ge T_1$, each program $\{y_t\}_{t=0}^{T}$ which satisfies

$$\sum_{t=0}^{T-1} v(y_t, y_{t+1}) \ge T v(\bar{x}, \bar{x}) - M_2 - M_1$$

and each integer $s \in [0, T - T_1]$ we have

$$\min\{\rho(y_i, \bar{x}) : \; i = s + 1, \ldots, s + T_1\} \le \delta.$$

Set

$$L = (2T_0 + 2T_1 + 2)8 \tag{2.107}$$

and choose a natural number

$$K > (T_1 + 2)(2 + \delta^{-1} M_1). \tag{2.108}$$

Assume that an integer $T > 2L$ and that

$$z_0 \in X_{M_0}, \; z_1 \in Y_{L_0}.$$

It follows from the choice of T_0, M_1 and relation (2.107) that inequality (2.104) is true.

Assume that a program $\{x_t\}_{t=0}^{T}$ satisfies

$$x_0 = z_1, \; x_T = z_2, \; \sum_{t=0}^{T-1} v(x_t, x_{t+1}) \ge \sigma(v, T, z_0, z_1) - M_1. \tag{2.109}$$

In view of (2.109) and (2.104), we have

$$\sum_{t=0}^{T-1} v(x_t, x_{t+1}) \ge T v(\bar{x}, \bar{x}) - M_2 - M_1. \tag{2.110}$$

It follows from (2.110), (2.107) and the choice of the integer T_1 that there exists a sequence of nonnegative integers $\{S_i\}_{i=0}^q \subset [0, T]$ such that

$$S_0 \in [0, T_1 + 1], \; S_{i+1} - S_i \in [1, T_1] \text{ for each integer } i \in [0, q-1], \; S_q + T_1 > T, \tag{2.111}$$

$$\rho(x_{S_i}, \bar{x}) \leq \delta, \; i = 0, \ldots, q. \tag{2.112}$$

Set

$$E = \{i \in \{0, \ldots, q-1\} : \sum_{t=S_i}^{S_{i+1}-1} v(x_t, x_{t+1}) < \sigma(v, S_{i+1} - S_i, x_{S_i}, x_{S_{i+1}}) - \delta\}. \tag{2.113}$$

There exists a program $\{\tilde{x}_t\}_{t=0}^T$ such that

$$\tilde{x}_t = x_t, \; t \in \{0, \ldots, S_0\} \cup \{S_q, \ldots, T\} \cup \{S_i : i = 0, \ldots, q\} \tag{2.114}$$

and that for each integer $i \in [0, \ldots, q-1]$,

$$\sum_{t=S_i}^{S_{i+1}-1} v(\tilde{x}_t, \tilde{x}_{t+1}) = \sigma(v, S_{i+1} - S_i, x_{S_i}, x_{S_{i+1}}). \tag{2.115}$$

By (2.114), (2.115), (2.109), and (2.119), we have

$$-M_1 \leq \sum_{t=0}^{T-1} v(x_t, x_{t+1}) - \sum_{t=0}^{T-1} v(\tilde{x}_t, \tilde{x}_{t+1}) \leq -\delta \mathrm{Card}(E)$$

and

$$\mathrm{Card}(E) \leq \delta^{-1} M_1. \tag{2.116}$$

In view of (2.113), (2.112) and the choice of δ [see (2.105) and (2.106)] we have for all $i \in \{0, \ldots, q-1\} \setminus E$

$$\rho(x_t, \bar{x}) \leq \epsilon, \; t \in \{S_i, \ldots, S_{i+1}\}.$$

This inequality implies that

$$\cup\{\{S_i, \ldots, S_{i+1}\} : i \in \{0, \ldots, q-1\} \setminus E\} \subset \{t \in \{0, \ldots, T\} : \rho(x_t, \bar{x}) \leq \epsilon\}$$

and

$$\{t \in \{0, \ldots, T\} : \rho(x_t, \bar{x}) > \epsilon\} \subset [0, S_0] \cup [S_q, T] \cup \{\{S_i, \ldots, S_{i+1}\} : i \in E\}.$$

Combined with (2.111), (2.116), and (2.108) this implies that

$$\text{Card}(\{t \in \{0, \ldots, T\} : \rho(x_t, \bar{x}) > \epsilon\}) \leq 2(T_1 + 2) + \text{Card}(E)(T_1 + 2)$$
$$= (T_1 + 2)(2 + \delta^{-1} M_1) < K.$$

Theorem 2.8 is proved.

2.8 Proof of Theorem 2.9

Lemma 2.17 implies that there exist an integer $T_0 \geq 1$ and a positive number M_1 such that for each natural number $T \geq T_0$, each point $z_0 \in X_{M_0}$, and each point $z_1 \in Y_{L_0}$, we have

$$\sigma(v, T, z_0, z_1) \geq T v(\bar{x}, \bar{x}) - M_1. \tag{2.117}$$

In view of Lemma 2.15 there exists a positive number

$$\delta < \min\{\bar{r}, \epsilon\}$$

such that for each natural number T and each program $\{y_t\}_{t=0}^T$ which satisfies the inequalities

$$\rho(y_0, \bar{x}), \ \rho(y_T, \bar{x}) \leq \delta,$$

$$\sum_{t=0}^{T-1} v(y_t, y_{t+1}) \geq \sigma(v, T, y_0, y_T) - \delta \tag{2.118}$$

we have

$$\rho(y_t, \bar{x}) \leq \epsilon, \ t = 0, \ldots, T. \tag{2.119}$$

It follows from Lemma 2.16 that there exists an integer $T_1 \geq 1$ such that the following property holds:

(P7) For each natural number $T \geq T_1$, each program $\{y_t\}_{t=0}^T$ which satisfies the inequality

$$\sum_{t=0}^{T-1} v(y_t, y_{t+1}) \geq T v(\bar{x}, \bar{x}) - M_1 - \bar{r} - 1,$$

and each integer $s \in [0, T - T_1]$ we have

$$\min\{\rho(y_i, \bar{x}) : i = s + 1, \ldots, s + T_1\} \leq \delta.$$

Set

$$L = (2T_0 + 2T_1 + 4)8. \tag{2.120}$$

Let $T > 2L$ be an integer and let

$$z_0 \in X_{M_0}, \; z_1 \in Y_{L_0} \tag{2.121}$$

be given. By relations (2.120), (2.121) and the choice of the numbers T_0, M_1, inequality (2.117) holds.

Assume that a program $\{x_t\}_{t=0}^T$ satisfies

$$x_0 = z_0, \; x_T = z_1, \; \sum_{t=0}^{T-1} v(x_t, x_{t+1}) \ge \sigma(v, T, z_0, z_1) - \delta. \tag{2.122}$$

In view of relations (2.122) and (2.117), we have

$$\sum_{t=0}^{T-1} v(x_t, x_{t+1}) \ge T v(\bar{x}, \bar{x}) - M_1 - \bar{r}. \tag{2.123}$$

It follows from relations (2.123), (2.120) and property (P7) that there exists a sequence of nonnegative integers $\{S_i\}_{i=0}^q \subset [0, T]$ such that

$$S_0 \in [0, T_1 + 1], \; S_{i+1} - S_i \in [1, T_1 + 1] \text{ for each integer } i \in [0, q-1], \; S_q + T_1 > T, \tag{2.124}$$

$$\rho(x_{S_i}, \bar{x}) \le \delta, \; i = 0, \ldots, q. \tag{2.125}$$

If $\rho(x_0, \bar{x}) \le \delta$, then we may assume that $S_0 = 0$, and if $\rho(x_T, \bar{x}) \le \delta$, then we may assume that $S_q = T$. In view of (2.122) for all $i = 0, \ldots, q-1$, we have

$$\sum_{t=S_i}^{S_{i+1}-1} v(x_t, x_{t+1}) \ge \sigma(v, S_{i+1} - S_i, x_{S_i}, x_{S_{i+1}}) - \delta$$

and combined with (2.125) and the choice of δ [see (2.118) and (2.119)] this implies that for all integers $i = 0, \ldots, q-1$, we have

$$\rho(x_t, \bar{x}) \le \epsilon, \; t = S_i, \ldots, S_{i+1}.$$

This implies that

$$\rho(x_t, \bar{x}) \le \epsilon, \; t \in S_0, \ldots, S_q.$$

Theorem 2.9 is proved.

2.9   Structure of Solutions in the Regions Containing End Points

Assume that X is a compact convex subset of the n-dimensional Euclidean space R^n with the norm $|\cdot|$ induced by the scalar product $\langle \cdot, \cdot \rangle$, Ω is a nonempty closed convex subset of $X \times X$, and $v : X \times X \to R^1$ is a continuous strictly concave function such that

$$v(\alpha z_1 + (1 - \alpha)z_2) > \alpha v(z_1) + (1 - \alpha)v(z_2)$$

$$\forall \alpha \in (0, 1), \forall z_1, z_2 \in X \times X \text{ such that } z_1 \neq z_2. \tag{2.126}$$

Put $\rho(x, y) = |x - y|$ for all $x, y \in X$. We assume that $\bar{x} \in X, \bar{r} \in (0, 1)$ and that

$$v(\bar{x}, \bar{x}) = \sup\{v(z, z) : z \in X \text{ and } (z, z) \in \Omega)\}, \tag{2.127}$$

$$\{(x, y) \in R^n \times R^n : |x - \bar{x}|, |y - \bar{x}| \leq 2\bar{r}\} \subset \Omega. \tag{2.128}$$

We have mentioned in Sect. 2.1 (see Example 2.11) that there is $l \in R^n$ such that

$$v(x, y) \leq v(\bar{x}, \bar{x}) + \langle l, x - y \rangle \text{ for all } (x, y) \in X \times X. \tag{2.129}$$

Set

$$L(x, y) = v(\bar{x}, \bar{x}) + \langle l, x - y \rangle - v(x, y), \ (x, y) \in X \times X. \tag{2.130}$$

It is clear that the inequality $L(x, y) \geq 0$ holds for all points $x, y \in X$. It was explained in Sect. 2.1 that assumptions (A1)–(A3) hold. Therefore Theorems 2.2–2.7 hold for the function v. Since the set Ω is convex and the function v is strictly concave Theorem 2.3 implies the following result.

Theorem 2.21. *Assume that $x \in X$ and that there exists a (v)-good program $\{x_t\}_{t=0}^\infty$ such that $x_0 = x$. Then there exists a unique (v)-overtaking program $\{x_t^*\}_{t=0}^\infty$ such that $x_0^* = x$.*

Let $z \in X$ be given and let there exist a (v)-good program $\{x_t\}_{t=0}^\infty \subset X$ such that $x_0 = z$. Denote by $\{x_t^{(v,z)}\}_{t=0}^\infty$ a unique (v)-overtaking optimal program satisfying $x_0^{(v,z)} = z$.

The following theorem which describes the structure of approximate solutions in the region containing the left end point of the interval $[0, T]$ was obtained in [58].

Theorem 2.22. *Let $M, \epsilon > 0$ be given and $L_0 \geq 1$ be an integer. Then there exist a positive number δ and an integer $L_1 > L_0$ such that for each natural number $T \geq L_1$ and each program $\{z_t\}_{t=0}^T$ which satisfies*

$$z_0 \in X_M, \ \sum_{t=0}^{T-1} v(z_t, z_{t+1}) \geq \sigma(v, T, z_0) - \delta \tag{2.131}$$

the inequality $|z_t - x_t^{(v,z_0)}| \leq \epsilon$ holds for all integers $t = 0, \ldots, L_0$.

It follows from Theorem 2.4 applied with $\epsilon = \bar{r}/4$ that there exists a natural number L_0 such that the following property holds:

(P8) For each positive number M there exist an integer $L > L_0$ and a positive number δ such that if a natural number $T > 2L$ and if a program $\{x_t\}_{t=0}^{T}$ satisfies

$$x_0 \in X_M, \ \sum_{t=0}^{T-1} v(x_t, x_{t+1}) \geq \sigma(v, T, x_0) - \delta, \tag{2.132}$$

then

$$|x_t - \bar{x}| \leq \bar{r}/4 \text{ for all integers } t = L, \ldots, T - L_0. \tag{2.133}$$

Define the functions $\bar{L}, \bar{v} : X \times X \rightarrow R^1$ and the set $\bar{\Omega}$ by

$$\bar{v}(x, y) = v(y, x), \ \bar{L}(x, y) = L(y, x), \ x, y \in X,$$

$$\bar{\Omega} = \{(x, y) \in X \times X : \ (y, x) \in \Omega\}.$$

It is easy to see that $\bar{\Omega}$ is a nonempty closed convex subset of $X \times X$, $\bar{v} : X \times X \rightarrow R^1$ is a concave function,

$$\{(\xi_1, \xi_2) \in R^n \times R^n : \ |\xi_i - \bar{x}| \leq 2\bar{r}, \ i = 1, 2\} \subset \bar{\Omega}, \tag{2.134}$$

$$\bar{v}(\bar{x}, \bar{x}) = \sup\{\bar{v}(z, z) : \ z \in X \text{ and } (z, z) \in \bar{\Omega}\}, \tag{2.135}$$

$$\bar{v}(\alpha z_1 + (1 - \alpha)z_2) > \alpha\bar{v}(z_1) + (1 - \alpha)\bar{v}(z_2) \text{ for each } z_1, z_2 \in X \times X \tag{2.136}$$

satisfying $z_1 \neq z_2$ and each $\alpha \in (0, 1)$.

Evidently, for all points $(x, y) \in X \times X$, we have

$$\bar{v}(x, y) \leq \bar{v}(\bar{x}, \bar{x}) + \langle -l, x - y \rangle, \tag{2.137}$$

$$\bar{L}(x, y) = \bar{v}(\bar{x}, \bar{x}) + \langle -l, x - y \rangle - \bar{v}(x, y). \tag{2.138}$$

It is not difficult to see that assumptions (A1)–(A3) hold for the function $\bar{v}$ and the set $\bar{\Omega}$ and that Theorems 2.2–2.7, 2.21, and 2.22 hold for $\bar{v}$ and $\bar{\Omega}$.

Denote by X_* the set of all points $x \in X$ for which there exists an $(\bar{\Omega})$-program $\{x_t\}_{t=0}^{L_0+1}$ such that

$$x_0 = x, \ x_{L_0+1} = \bar{x}. \tag{2.139}$$

It is clear that X_* is a closed and convex set. Relations (2.131), (2.133) and property (P8) imply that the following property holds:

(P9)   For each positive number M there exist a natural number $L > L_0$ and a positive number δ such that if an integer $T > 2L$ and if an (Ω)-program $\{x_t\}_{t=0}^{T}$ satisfies (2.132), then the inclusion $x_T \in X_*$ is true.

In view of Theorem 2.21 for any point $x \in X_*$ there exists a unique $(\bar{v})$-overtaking optimal $(\bar{\Omega})$-program $\{\Lambda_t(x)\}_{t=0}^{\infty}$ such that $\Lambda_0(x) = x$. For any point $x \in X_*$ put

$$\pi(x) = \lim_{T \to \infty} \sum_{t=0}^{T-1} [\bar{v}(\bar{x}, \bar{x}) - \bar{v}(\Lambda_t(x), \Lambda_{t+1}(x))]$$

$$= \lim_{T \to \infty} \left[\sum_{t=0}^{T-1} \bar{L}(\Lambda_t(x), \Lambda_{t+1}(x)) + \langle l, x - \Lambda_T(x) \rangle \right]$$

$$= \sum_{t=0}^{\infty} \bar{L}(\Lambda_t(x), \Lambda_{t+1}(x)) + \langle l, x - \bar{x} \rangle \qquad (2.140)$$

[see relation (2.138), the definitions of the functions $\bar{L}$, $\bar{v}$, and assumption (A3)]. It is easy to see that $\pi(x)$ is finite for all points $x \in X_*$.

In order to study the structure of approximate solutions of the problems $(P_T^{(y)})$ in the regions $[T - L, T]$ (see the definition of the turnpike property) we need the following auxiliary results obtained in [58].

Proposition 2.23. *An $(\bar{\Omega})$-program $\{x_t\}_{t=0}^{\infty}$ is $(\bar{v})$-good if and only if*

$$\sum_{t=0}^{\infty} \bar{L}(x_t, x_{t+1}) < \infty.$$

Proposition 2.24. *Let $x \in X_*$ and let an $(\bar{\Omega})$-program $\{x_t\}_{t=0}^{\infty}$ be $(\bar{v})$-good and satisfy $x_0 = x$. Then $\sum_{t=0}^{\infty} \bar{L}(x_t, x_{t+1}) + \langle l, x - \bar{x} \rangle \geq \pi(x)$.*

Proposition 2.25. *The function $\pi : X_* \to R^1$ is lower semicontinuous.*

Proposition 2.26. *Let points $y, z \in X_*$ satisfy $y \neq z$ and a number $\alpha \in (0, 1)$. Then $\pi(\alpha y + (1 - \alpha)z) < \alpha \pi(y) + (1 - \alpha)\pi(z)$.*

Since the function $\pi : X_* \to R^1$ is lower semicontinuous and strictly convex it possesses a unique point of minimum which will be denoted by x_*. Thus

$$\pi(x_*) < \pi(x) \text{ for all points } x \in X_* \setminus \{x_*\}. \qquad (2.141)$$

The following theorem which describes the structure of approximate solutions $\{x_t\}_{t=0}^{T}$ of the problems $(P_T^{(y)})$ in the region containing the right end point of the interval $[0, T]$ was obtained in [58]. It shows that this structure depends neither on x_0 nor T.

Theorem 2.27. *Let M, ϵ be positive numbers and let $L_1 \geq 1$ be an integer. Then there exist a positive number δ and an integer $L_2 > L_1$ such that if an integer $T > 2L_2$ and if an (Ω)-program $\{x_t\}_{t=0}^T$ satisfies*

$$x_0 \in X_M, \quad \sum_{t=0}^{T-1} v(x_t, x_{t+1}) \geq \sigma(v, T, x_0) - \delta, \tag{2.142}$$

then $|x_{T-t} - \Lambda_t(x_*)| \leq \epsilon$ for all integers $t = 0, \ldots, L_1$.

2.10 Proof of Theorem 2.22

In this section we consider only (Ω)-programs which will be called just programs. For simplicity, in this section we use the notation

$$x_t^{(z)} = x_t^{(v,z)} \text{ for all points } z \in \cup\{X_M : M > 0\} \text{ and all } t = 0, 1, \ldots. \tag{2.143}$$

Assume that the assertion of the theorem does not hold. Then for each integer $n \geq 1$ there exist an integer

$$T_n \geq L_0 + 4n \tag{2.144}$$

and a program $\{z_t^{(n)}\}_{t=0}^{T_n}$ such that

$$z_0^{(n)} \in X_M, \quad \sum_{t=0}^{T_n-1} v\left(z_t^{(n)}, z_{t+1}^{(n)}\right) \geq \sigma\left(v, T_n, z_0^{(n)}\right) - n^{-1}, \tag{2.145}$$

$$\max\left\{\left|z_t^{(n)} - x_t^{(z_0^{(n)})}\right| : t = 0, \ldots, L_0\right\} > \epsilon. \tag{2.146}$$

Extracting a subsequence and re-indexing we may assume without loss of generality that for each nonnegative integer t there exist limits

$$z_t := \lim_{n \to \infty} z_t^{(n)}, \quad x_t := \lim_{n \to \infty} x_t^{(z_0^{(n)})}. \tag{2.147}$$

It is not difficult to see that the sequences $\{z_t\}_{t=0}^\infty, \{x_t\}_{t=0}^\infty$ are programs. By (2.147) and (2.146), we have

$$\max\{|z_t - x_t| : t = 0, \ldots, L_0\} \geq \epsilon. \tag{2.148}$$

It follows from (2.147) and (2.143) that

$$z_0 = x_0. \tag{2.149}$$

We claim that $\{x_t\}_{t=0}^\infty$ and $\{y_t\}_{t=0}^\infty$ are (v)-good programs. Let a natural number n be given. In view of (2.145) there exists a program $\{\tilde{z}_t^{(n)}\}_{t=0}^\infty$ such that

$$\tilde{z}_0^{(n)} = z_0^{(n)}, \quad \sum_{t=0}^{T-1} \left[v\left(\tilde{z}_t^{(n)}, \tilde{z}_{t+1}^{(n)} \right) - v(\bar{x}, \bar{x}) \right] \geq -M \text{ for all natural numbers } T.$$

(2.150)

It is not difficult to see that the programs $\{\tilde{z}_t^{(n)}\}_{t=0}^\infty$, $\{x_t^{(z_0^{(n)})}\}_{t=0}^\infty$ (v)-are good. In view of assumption (A3),

$$\lim_{t\to\infty} \left| \tilde{z}_t^{(n)} - \bar{x} \right| = \lim_{t\to\infty} \left| x_t^{(z_0^{(n)})} - \bar{x} \right| = 0.$$

(2.151)

By relation (2.130), for all natural numbers T, we have

$$\sum_{t=0}^{T-1} \left[v\left(\tilde{z}_t^{(n)}, \tilde{z}_{t+1}^{(n)} \right) - v(\bar{x}, \bar{x}) \right] = -\sum_{t=0}^{T-1} L\left(\tilde{z}_t^{(n)}, \tilde{z}_{t+1}^{(n)} \right) + \left\langle l, \tilde{z}_0^{(n)} - \tilde{z}_T^{(n)} \right\rangle,$$

(2.152)

$$\sum_{t=0}^{T-1} \left[v\left(x_t^{(z_0^{(n)})}, x_{t+1}^{(z_0^{(n)})} \right) - v(\bar{x}, \bar{x}) \right]$$

$$= -\sum_{t=0}^{T-1} L\left(x_t^{(z_0^{(n)})}, x_{t+1}^{(z_0^{(n)})} \right) + \left\langle l, x_0^{(z_0^{(n)})} - x_T^{(z_0^{(n)})} \right\rangle.$$

(2.153)

It follows from (2.151), (2.152), and (2.150) that

$$\sum_{t=0}^{\infty} L\left(\tilde{z}_t^{(n)}, \tilde{z}_{t+1}^{(n)} \right) \leq M + \left\langle l, z_0^{(n)} - \bar{x} \right\rangle.$$

(2.154)

Since the program $\{x_t^{(z_0^{(n)})}\}_{t=0}^\infty$ is (v)-overtaking optimal relations (2.150)–(2.154) imply that

$$0 \geq \limsup_{T\to\infty} \sum_{t=0}^{T-1} \left[v\left(\tilde{z}_t^{(n)}, \tilde{z}_{t+1}^{(n)} \right) - v\left(x_t^{(z_0^{(n)})}, x_{t+1}^{(z_0^{(n)})} \right) \right]$$

$$= \limsup_{T\to\infty} \left[\sum_{t=0}^{T-1} L\left(x_t^{(z_0^{(n)})}, x_{t+1}^{(z_0^{(n)})} \right) - \left\langle l, x_0^{(z_0^{(n)})} - x_T^{(z_0^{(n)})} \right\rangle \right.$$

$$\left. - \sum_{t=0}^{T-1} L\left(\tilde{z}_t^{(n)}, \tilde{z}_{t+1}^{(n)} \right) + \left\langle l, \tilde{z}_0^{(n)} - \tilde{z}_T^{(n)} \right\rangle \right]$$

$$= \sum_{t=0}^{\infty} L\left(x_t^{(z_0^{(n)})}, x_{t+1}^{(z_0^{(n)})} \right) - \sum_{t=0}^{\infty} L\left(\tilde{z}_t^{(n)}, \tilde{z}_{t+1}^{(n)} \right).$$

(2.155)

It follows from (2.155) and (2.154) that

$$\sum_{t=0}^{\infty} L\left(x_t^{(z_0^{(n)})}, x_{t+1}^{(z_0^{(n)})}\right) \leq M + \left\langle l, z_0^{(n)} - \bar{x}\right\rangle \leq M + 2|l|\sup\{|\xi| : \xi \in X\}.$$

$$(2.156)$$

By relations (2.145), (2.150), (2.154), and (2.130), we have

$$n^{-1} \geq \sum_{t=0}^{T_n-1}\left[v\left(\tilde{z}_t^{(n)}, \tilde{z}_{t+1}^{(n)}\right) - v\left(z_t^{(n)}, z_{t+1}^{(n)}\right)\right]$$

$$= -\sum_{t=0}^{T_n-1} L\left(\tilde{z}_t^{(n)}, \tilde{z}_{t+1}^{(n)}\right) + \left\langle l, \tilde{z}_0^{(n)} - \tilde{z}_{T_n}^{(n)}\right\rangle$$

$$+ \sum_{t=0}^{T_n-1} L\left(z_t^{(n)}, z_{t+1}^{(n)}\right) - \left\langle l, z_0^{(n)} - z_{T_n}^{(n)}\right\rangle$$

$$\geq -M - \left\langle l, z_0^{(n)} - \bar{x}\right\rangle + \left\langle l, \tilde{z}_0^{(n)} - \tilde{z}_{T_n}^{(n)}\right\rangle$$

$$+ \sum_{t=0}^{T_n-1} L\left(z_t^{(n)}, z_{t+1}^{(n)}\right) - \left\langle l, z_0^{(n)} - z_{T_n}^{(n)}\right\rangle$$

$$= -M + \sum_{t=0}^{T_n-1} L\left(z_t^{(n)}, z_{t+1}^{(n)}\right) - \left\langle l, z_0^{(n)} - \bar{x} + \tilde{z}_{T_n}^{(n)} - z_{T_n}^{(n)}\right\rangle,$$

$$\sum_{t=0}^{T_n-1} L\left(z_t^{(n)}, z_{t+1}^{(n)}\right) \leq M + n^{-1} + 4|l|\sup\{|\xi| : \xi \in X\}. \qquad (2.157)$$

Since relations (2.156) and (2.157) hold for all integers $n \geq 1$ it follows from (2.147) that

$$\sum_{t=0}^{\infty} L(z_t, z_{t+1}) \leq M + 1 + 4|l|\sup\{|\xi| : \xi \in X\},$$

$$\sum_{t=0}^{\infty} L(x_t, x_{t+1}) \leq M + 2|l|\sup\{|\xi| : \xi \in X\}.$$

By these inequalities, the sequences $\{x_t\}_{t=0}^{\infty}$ and $\{z_t\}_{t=0}^{\infty}$ are (v)-good programs. In view of assumption (A3), we have

$$\lim_{t \to \infty} |x_t - \bar{x}| = \lim_{t \to \infty} |z_t - \bar{x}| = 0. \qquad (2.158)$$

It follows from relation (2.148) and the strict concavity of the function v that

$$\Delta := \sum_{t=0}^{L_0-1} [v(2^{-1}(x_t + z_t, x_{t+1} + z_{t+1})) - 2^{-1}v(x_t, x_{t+1}) - 2^{-1}v(z_t, z_{t+1})] > 0.$$

$$(2.159)$$

Since the function v is continuous and the point $(\bar{x}, \bar{x})$ belongs to the interior of the set Ω there exists $r \in (0, 1)$ such that

$$\{(\xi_1, \xi_2) \in R^n \times R^n : |\xi_i - \bar{x}| \le r, \ i = 1, 2\} \subset \Omega, \tag{2.160}$$

$$|v(\xi_1, \xi_2) - v(\eta_1, \eta_2)| \le 32^{-1}\Delta$$

for each $\xi_1, \xi_2, \eta_1, \eta_2$ satisfying $|\xi_i - \eta_i| \le 2r, \ i = 1, 2.$ $\tag{2.161}$

By (2.158) there exists an integer $L_1 \ge 1$ such that

$$|x_t - \bar{x}|, |z_t - \bar{x}| \le r/4 \text{ for all integers } t \ge L_1. \tag{2.162}$$

Put

$$L_2 = 16(L_0 + L_1 + 1). \tag{2.163}$$

Since v is a continuous function there exists $\delta \in (0, r/4)$ such that for each $\xi_1, \xi_2, \xi_3, \xi_4 \in X$ satisfying $|\xi_1 - \xi_3|, |\xi_2 - \xi_4| \le 4\delta$ we have

$$|v(\xi_1, \xi_2) - v(\xi_3, \xi_4)| \le \Delta L_2^{-1} 64^{-1}. \tag{2.164}$$

By (2.147) there exists an integer $n \ge 1$ such that

$$n > L_2, \ n^{-1} < \Delta/32, \tag{2.165}$$

$$\left| z_t^{(n)} - z_t \right| \le \delta, \ \left| x_t - x_t^{(z_0^{(n)})} \right| \le \delta \text{ for all integers } t = 0, \dots, 2L_2. \tag{2.166}$$

Put

$$y_t = 2^{-1}\left(z_t^{(n)} + x_t^{(z_0^{(n)})} \right), \ t = 0, \dots, L_2. \tag{2.167}$$

Since the function v is concave, it follows from (2.159), (2.163), (2.166) and the choice of δ [see (2.164)] that

$$\sum_{t=0}^{L_2-1} v(y_t, y_{t+1}) - 2^{-1} \sum_{t=0}^{L_2-1} v\left(z_t^{(n)}, z_{t+1}^{(n)} \right) - 2^{-1} \sum_{t=0}^{L_2-1} v\left(x_t^{(z_0^{(n)})}, x_{t+1}^{(z_0^{(n)})} \right)$$

$$\ge \sum_{t=0}^{L_2-1} v(2^{-1}(x_t + z_t, x_{t+1} + z_{t+1})) - 2^{-1} \sum_{t=0}^{L_2-1} v(x_t, x_{t+1}) - 2^{-1} \sum_{t=0}^{L_2-1} v(z_t, z_{t+1})$$

$$-L_2[\sup\{|v(y_t, y_{t+1}) - v(2^{-1}(x_t + z_t), 2^{-1}(x_{t+1}, z_{t+1}))| : t = 0, \ldots, L_2 - 1\}$$

$$+ \sup\left\{\left|v\left(z_t^{(n)}, z_{t+1}^{(n)}\right) - v(z_t, z_{t+1})\right| : t = 0, \ldots, L_2\right\}$$

$$+ \sup\left\{\left|v\left(x_t^{(z_0^{(n)})}, x_{t+1}^{(z_0^{(n)})}\right) - v(x_t, x_{t+1})\right| : t = 0, \ldots, L_2 - 1\right\}$$

$$\geq \Delta - L_2\Delta 64^{-1}L_2^{-1} \cdot 3 \geq (3/4)\Delta. \tag{2.168}$$

It is not difficult to see that $\{y_t\}_{t=0}^{L_2}$ is a program. By (2.167), we have

$$y_0 = z_0^{(n)} = x_0^{(z_0^{(n)})}. \tag{2.169}$$

Set

$$\tilde{y}_t = y_t \text{ for all } t = 0, \ldots, L_2 - 1, \ \tilde{y}_{L_2} = z_{L_2}^{(n)},$$

$$\hat{y}_t = y_t \text{ for all } t = 0, \ldots, L_2 - 1, \ \hat{y}_{L_2} = x_{L_2}^{(z_0^{(n)})}. \tag{2.170}$$

By (2.162), (2.166), (2.163), and the choice of δ for $t = L_2 - 1, L_2$, we have

$$\left|z_t^{(n)} - \bar{x}\right| \leq \left|z_t^{(n)} - z_t\right| + |z_t - \bar{x}| \leq \delta + r/4 < r/2, \tag{2.171}$$

$$\left|x_t^{(z_0^{(n)})} - \bar{x}\right| \leq \left|x_t^{(z_0^{(n)})} - x_t\right| + |x_t - \bar{x}| \leq \delta + r/4 < r/2. \tag{2.172}$$

It follows from (2.167), (2.171), and (2.172) that for $t = L_2 - 1, L_2$ we have

$$|y_t - \bar{x}| < r/2. \tag{2.173}$$

In view of (2.160) and (2.170)–(2.173), $\{\tilde{y}_t\}_{t=0}^{L_2}, \{\hat{y}_t\}_{t=0}^{L_2}$ are programs. It follows from (2.170)–(2.173) and the choice of the number r [see (2.161)] that

$$\left|\sum_{t=0}^{L_2-1} [v(y_t, y_{t+1}) - v(\tilde{y}_t, \tilde{y}_{t+1})]\right| = \left|v\left(y_{L_2-1}, z_{L_2}^{(n)}\right) - v(y_{L_2-1}, y_{L_2})\right| \leq 32^{-1}\Delta, \tag{2.174}$$

$$\left|\sum_{t=0}^{L_2-1} [v(y_t, y_{t+1}) - v(\hat{y}_t, \hat{y}_{t+1})]\right| = \left|v(y_{L_2-1}, y_{L_2}) - v\left(y_{L_2-1}, x_{L_2}^{(z_0^{(n)})}\right)\right| \leq 32^{-1}\Delta. \tag{2.175}$$

By relations (2.145), (2.165), (2.169), (2.170) and the (v)-overtaking optimality of the sequence $\{x_t^{(z_0^{(n)})}\}_{t=0}^{\infty}$ we have

$$\sum_{t=0}^{L_2-1} v(\hat{y}_t, \hat{y}_{t+1}) \le \sum_{t=0}^{L_2-1} v\left(x_t^{(z_0^{(n)})}, x_{t+1}^{(z_0^{(n)})}\right),$$

$$\sum_{t=0}^{L_2-1} v(\tilde{y}_t, \tilde{y}_{t+1}) \le \sum_{t=0}^{L_2-1} v\left(z_t^{(n)}, z_{t+1}^{(n)}\right) + n^{-1} < \sum_{t=0}^{L_2-1} v\left(z_t^{(n)}, z_{t+1}^{(n)}\right) + \Delta/32.$$

It follows from these inequalities and (2.175) that

$$\sum_{t=0}^{L_2-1} v(y_t, y_{t+1}) \le 32^{-1}\Delta + 2^{-1} \sum_{t=0}^{L_2-1} v(\tilde{y}_t, \tilde{y}_{t+1}) + 2^{-1} \sum_{t=0}^{L_2-1} v(\hat{y}_t, \hat{y}_{t+1})$$

$$\le 32^{-1}\Delta + 2^{-1} \sum_{t=0}^{L_2-1} v\left(z_t^{(n)}, z_{t+1}^{(n)}\right) + 32^{-1}\Delta + 2^{-1} \sum_{t=0}^{L_2-1} v\left(x_t^{(z_0^{(n)})}, x_{t+1}^{(z_0^{(n)})}\right).$$

This inequality contradicts (2.168). The contradiction we have reached proves Theorem 2.22.

2.11 Proofs of Propositions 2.23–2.26

Proof of Proposition 2.23. Assume that $\{x_t\}_{t=0}^{\infty}$ is an $(\bar{\Omega})$-program. It follows from relation (2.138) that for each natural number T we have

$$\sum_{t=0}^{T-1} [\bar{v}(\bar{x}, \bar{x}) - \bar{v}(x_t, x_{t+1})] = \sum_{t=0}^{T-1} \bar{L}(x_t, x_{t+1}) + \langle l, x_0 - x_T \rangle.$$

This relation implies the validity of the proposition.   □

Proof of Proposition 2.24. Proposition 2.23 implies that

$$\sum_{t=0}^{\infty} \bar{L}(x_t, x_{t+1}) < \infty.$$

Since the $(\bar{\Omega})$-program $\{\Lambda_t(x)\}_{t=0}^{\infty}$ is $(\bar{v})$-overtaking optimal and the $(\bar{\Omega})$-program $\{x_t\}_{t=0}^{\infty}$ is $(\bar{v})$-good it follows from relations (2.138) and (2.140) that

$$0 \ge \limsup_{T \to \infty} \left(\sum_{t=0}^{T-1} [\bar{v}(\bar{x}, \bar{x}) - \bar{v}(\Lambda_t(x), \Lambda_{t+1}(x))] - \sum_{t=0}^{T-1} [\bar{v}(\bar{x}, \bar{x}) - \bar{v}(x_t, x_{t+1})] \right)$$

$$= \lim_{T\to\infty} \sum_{t=0}^{T-1} [\bar{v}(\bar{x},\bar{x}) - \bar{v}(\Lambda_t(x), \Lambda_{t+1}(x))] - \lim_{T\to\infty} \sum_{t=0}^{T-1} [\bar{v}(\bar{x},\bar{x}) - \bar{v}(x_t, x_{t+1})]$$

$$= \pi(x) - \left[\sum_{t=0}^{\infty} \bar{L}(x_t, x_{t+1}) + \langle l, x - \bar{x}\rangle \right].$$

Proposition 2.24 is proved.   □

By (2.136) and (2.138), for each pair of points $z_1, z_2 \in X \times X$ satisfying $z_1 \neq z_2$ and each real number $\alpha \in (0, 1)$, we have

$$\bar{L}(\alpha z_1 + (1 - \alpha)z_2) < \alpha\bar{L}(z_1) + (1 - \alpha)\bar{L}(z_2). \tag{2.176}$$

Proof of Proposition 2.25. Assume that $\{x^{(i)}\}_{i=1}^{\infty} \subset X_*, x \in X_*$ and

$$\lim_{i\to\infty} x^{(i)} = x.$$

We claim that $\pi(x) \leq \liminf_{i\to\infty} \pi(x^{(i)})$. Extracting a subsequence and re-indexing if necessary we may assume without loss of generality that

$$\liminf_{i\to\infty} \pi(x^{(i)}) = \lim_{i\to\infty} \pi(x^{(i)}) < \infty \tag{2.177}$$

and that for each nonnegative integer t there exists a limit

$$x_t := \lim_{i\to\infty} \Lambda_t(x^{(i)}). \tag{2.178}$$

It is not difficult to see that $\{x_t\}_{t=0}^{\infty}$ is an $(\bar{\Omega})$-program. By (2.178), $x_0 = x$. Relations (2.140) and (2.178) imply that

$$\sum_{t=0}^{\infty} \bar{L}(x_t, x_{t+1}) = \lim_{T\to\infty} \sum_{t=0}^{T} \bar{L}(x_t, x_{t+1}) = \lim_{T\to\infty} \left[\lim_{i\to\infty} \sum_{t=0}^{T} \bar{L}(\Lambda_t(x^{(i)}), \Lambda_{t+1}(x^{(i)})) \right]$$

$$\leq \liminf_{i\to\infty} \sum_{t=0}^{\infty} \bar{L}(\Lambda_t(x^{(i)}), \Lambda_{t+1}(x^{(i)})) = \liminf_{i\to\infty} [\pi(x^{(i)}) - \langle l, x^{(i)} - \bar{x}\rangle]$$

$$= \liminf_{i\to\infty} \pi(x^i) - \langle l, x - \bar{x}\rangle,$$

$$\liminf_{i\to\infty} \pi(x^{(i)}) \geq \sum_{t=0}^{\infty} \bar{L}(x_t, x_{t+1}) + \langle l, x - \bar{x}\rangle.$$

By (2.177), the inequality above, the equality $x_0 = x$, and Propositions 2.23 and 2.24, we have $\pi(x) \leq \liminf_{t\to\infty} \pi(x^{(i)})$. Proposition 2.25 is proved. □

Proof of Proposition 2.26. Put

$$x_t = \alpha \Lambda_t(y) + (1 - \alpha) \Lambda_t(z) \text{ for all nonnegative integers } t.$$

It is clear that the sequence $\{x_t\}_{t=0}^{\infty}$ is an $(\bar{\Omega})$-program and $x_0 = \alpha y + (1 - \alpha)z$. By the strict convexity of the function $\bar{L}$ [see (2.176)], we have

$$\bar{L}(x_0, x_1) < \alpha \bar{L}(\Lambda_0(y), \Lambda_1(y)) + (1 - \alpha) \bar{L}(\Lambda_0(z), \Lambda_1(z)), \qquad (2.179)$$

$$\bar{L}(x_t, x_{t+1}) \leq \alpha \bar{L}(\Lambda_t(y), \Lambda_{t+1}(y))$$
$$+ (1 - \alpha) \bar{L}(\Lambda_t(z), \Lambda_{t+1}(z)) \text{ for all natural numbers } t.$$

Since the $(\bar{\Omega})$-programs

$$\{\Lambda_t(y)\}_{t=0}^{\infty}, \ \{\Lambda_t(z)\}_{t=0}^{\infty}$$

are $(\bar{v})$-overtaking optimal it follows from Proposition 2.23 that

$$\sum_{t=0}^{\infty} \bar{L}(\Lambda_t(y), \Lambda_{t+1}(y)) < \infty, \ \sum_{t=0}^{\infty} \bar{L}(\Lambda_t(z), \Lambda_{t+1}(z)) < \infty.$$

These inequalities, (2.179), and Proposition 2.23 imply that the sequence $\{x_t\}_{t=0}^{\infty}$ is a $(\bar{v})$-good $(\bar{\Omega})$-program and

$$\sum_{t=0}^{\infty} \bar{L}(x_t, x_{t+1}) < \alpha \sum_{t=0}^{\infty} \bar{L}(\Lambda_t(y), \Lambda_{t+1}(y)) + (1 - \alpha) \sum_{t=0}^{\infty} \bar{L}(\Lambda_t(z), \Lambda_{t+1}(z)).$$

In view of this inequality, the equality $x_0 = \alpha y + (1 - \alpha)z$, (2.140), and Proposition 2.24, we have

$$\pi(\alpha y + (1 - \alpha)z) \leq \langle l, \alpha y + (1 - \alpha)z - \bar{x} \rangle$$

$$+ \sum_{t=0}^{\infty} \bar{L}(x_t, x_{t+1}) < \langle l, \alpha y + (1 - \alpha)z - \bar{x} \rangle$$

$$+ \alpha \sum_{t=0}^{\infty} \bar{L}(\Lambda_t(y), \Lambda_{t+1}(y))$$

$$+ (1 - \alpha) \sum_{t=0}^{\infty} \bar{L}(\Lambda_t(z), \Lambda_{t+1}(z)) = \alpha \pi(y) + (1 - \alpha)\pi(z).$$

Proposition 2.26 is proved. $\qquad\qquad\qquad\qquad\qquad\qquad\qquad\qquad\Box$

2.12   Proof of Theorem 2.27

We precede the proof of Theorem 2.27 with the following lemma.

Lemma 2.28. *Let ϵ be a positive number and $L_1 \geq 1$ be an integer. Then there exist a positive number δ and an integer $L_2 \geq 1$ such that if an integer $T \geq L_2$ and if an $(\bar{\Omega})$-program $\{y_t\}_{t=0}^{T}$ satisfies*

$$y_0 \in X_*, \quad \sum_{t=0}^{T-1} \bar{L}(y_t, y_{t+1}) + \langle l, y_0 - \bar{x} \rangle \leq \pi(x_*) + \delta, \tag{2.180}$$

then $|y_t - \Lambda_t(x_*)| \leq \epsilon$ for all $t = 0, \ldots, L_1$.

Proof. Assume the contrary. Then for each integer $n \geq 1$ there exist an integer $T_n > n + L_1$ and an $(\bar{\Omega})$-program $\{y_t^{(n)}\}_{t=0}^{T_n}$ such that

$$y_0^{(n)} \in X_*, \quad \sum_{t=0}^{T_n-1} \bar{L}\left(y_t^{(n)}, y_{t+1}^{(n)}\right) + \left\langle l, y_0^{(n)} - \bar{x} \right\rangle \leq \pi(x_*) + n^{-1}, \tag{2.181}$$

$$\max\left\{ \left| y_t^{(n)} - \Lambda_t(x_*) \right| : t = 0, \ldots, L_1 \right\} > \epsilon. \tag{2.182}$$

Extracting a subsequence and re-indexing if necessary we may assume without loss of generality that for each nonnegative integer t there exists a limit

$$y_t = \lim_{n \to \infty} y_t^{(n)}. \tag{2.183}$$

It is easy to see that the sequence $\{y_t\}_{t=0}^{\infty}$ is an $(\bar{\Omega})$-program and that $y_0 \in X_*$. By (2.183) and (2.181), for each natural number S,

$$\sum_{t=0}^{S-1} \bar{L}(y_t, y_{t+1}) = \lim_{n \to \infty} \sum_{t=0}^{S-1} \bar{L}\left(y_t^{(n)}, y_{t+1}^{(n)}\right) \leq \limsup_{n \to \infty} \sum_{t=0}^{T_n-1} \bar{L}\left(y_t^{(n)}, y_{t+1}^{(n)}\right)$$

$$\leq \limsup_{n \to \infty} [\pi(x_*) + n^{-1} - \langle l, y_0^{(n)} - \bar{x} \rangle] = \pi(x_*) - \langle l, y_0 - \bar{x} \rangle. \tag{2.184}$$

Since (2.184) is true for any natural number S we have

$$\sum_{t=0}^{\infty} \bar{L}(y_t, y_{t+1}) \leq \pi(x_*) - \langle l, y_0 - \bar{x} \rangle. \tag{2.185}$$

By (2.185) and Proposition 2.23, the sequence $\{y_t\}_{t=0}^{\infty}$ is a $(\bar{v})$-good $(\bar{\Omega})$-program. It follows from (2.185) and Proposition 2.24 that $\pi(y_0) \leq \pi(x_*)$. Since x_* is the unique point of minimum of the function π we have

$$y_0 = x_*. \tag{2.186}$$

It follows from (2.185), (2.186) and Proposition 2.24 that

$$\pi(x_*) = \sum_{t=0}^{\infty} \bar{L}(y_t, y_{t+1}) + \langle l, y_0 - \bar{x} \rangle. \tag{2.187}$$

Since the $(\bar{\Omega})$-program $\{y_t\}_{t=0}^{\infty}$ is $(\bar{v})$-good we conclude that

$$\lim_{t \to \infty} |y_t - \bar{x}| = 0. \tag{2.188}$$

By (2.138), (2.188), (2.187), (2.140), and the equality

$$\lim_{t \to \infty} |\Lambda_t(x_*) - \bar{x}| = 0$$

we have

$$\lim_{T \to \infty} \left[\sum_{t=0}^{T-1} \bar{v}(y_t, y_{t+1}) - \bar{v}(\Lambda_t(x_*), \Lambda_{t+1}(x_*)) \right]$$

$$= \lim_{T \to \infty} \left[\sum_{t=0}^{T-1} [\bar{L}(\Lambda_t(x_*), \Lambda_{t+1}(x_*)) \right.$$

$$\left. - \bar{L}(y_t, y_{t+1})] + \langle l, y_T - y_0 \rangle - \langle l, \Lambda_T(x_*) - \Lambda_0(x_*) \rangle \right]$$

$$= \sum_{t=0}^{\infty} \bar{L}(\Lambda_t(x_*), \Lambda_{t+1}(x_*)) - \sum_{t=0}^{\infty} \bar{L}(y_t, y_{t+1})$$

$$+ \langle l, \Lambda_0(x_*) - \bar{x} \rangle - \langle l, y_0 - \bar{x} \rangle = 0.$$

Therefore the $(\bar{\Omega})$-program $\{y_t\}_{t=0}^{\infty}$ is $(\bar{v})$-overtaking optimal. In view of (2.186) and Theorem 2.21 applied with the function $\bar{v}$ and the set $\bar{\Omega}$ we have

$$y_t = \Lambda_t(x_*) \text{ for all integers } t \geq 0. \tag{2.189}$$

On the other hand by relations (2.182) and (2.183),

$$\max\{|y_t - \Lambda_t(x_*)| : t = 0, \dots, L_1\} \geq \epsilon.$$

This inequality contradicts (2.189). The contradiction we have reached proves Lemma 2.28.   □

Proof of Theorem 2.27. Lemma 2.28 implies that there exist a natural number $L_{12} > L_1$ and a real number $\delta_0 > 0$ such that the following property holds:

(P10)   If an integer $\tau \geq L_{12}$ and if an $(\bar{\Omega})$-program $\{z_t\}_{t=0}^{\tau}$ satisfies

$$z_0 \in X_*, \ \sum_{t=0}^{\tau-1} \bar{L}(z_t, z_{t+1}) + \langle l, z_0 - \bar{x} \rangle \leq \pi(x_*) + \delta_0,$$

then $|z_t - \Lambda_t(x_*)| \leq \epsilon$ for all integers $t = 0, \ldots, L_1$.

It follows from property (P9) that there exist a natural number $L_{13} > L_{12} + L_0$ and a real number $\delta_1 > 0$ such that the following property holds:

(P11)   If an integer $T > 2L_{13}$ and if an (Ω)-program $\{z_t\}_{t=0}^{T}$ satisfies $z_0 \in X_M$ and

$$\sum_{t=0}^{T-1} v(z_t, z_{t+1}) \geq \sigma(v, T, z_0) - \delta_1,$$

then the point $z_T \in X_*$.

Fix a positive number

$$\delta_2 < 64^{-1} (1 + |l|)^{-1} \min\{\delta_0, \delta_1, \epsilon, \bar{r}/4\} \tag{2.190}$$

such that

$$|v(z_1, z_2) - v(\xi_1, \xi_2)| \leq \delta_0/32 \text{ for each } z_1, z_2, \xi_1, \xi_2 \in X \tag{2.191}$$

$$\text{satisfying } |z_i - \xi_i| \leq 4\delta_2 \text{ for } i = 1, 2.$$

Theorem 2.2 implies that there exist a natural number $L_{14} > L_{13}$ and $\delta \in (0, \delta_2)$ such that the following property holds:

(P12)   If an integer $T > 2L_{14}$ and if an (Ω)-program $\{z_t\}_{t=0}^{T}$ satisfies $z_0 \in X_M$,

$$\sum_{t=0}^{T-1} v(z_t, z_{t+1}) \geq \sigma(v, T, z_0) - \delta,$$

then $|z_t - \bar{x}| \leq \delta_2$ for all integers $t = L_{14}, \ldots, T - L_{14}$.

In view of assumption (A3) applied for the function $\bar{v}$ there exists a natural number $L_2 > L_{14}$ such that

$$|\Lambda_t(x_*) - \bar{x}| \leq \delta_2 \text{ for all integers } t \geq L_2. \tag{2.192}$$

Let an integer $T > 2L_2$ be given and let an (Ω)-program $\{x_t\}_{t=0}^{T}$ satisfy

$$x_0 \in X_M, \ \sum_{t=0}^{T-1} v(x_t, x_{t+1}) \geq \sigma(v, T, x_0) - \delta. \tag{2.193}$$

We claim that

$$|x_{T-t} - \Lambda_t(x_*)| \le \epsilon \text{ for all integers } t = 0, \dots, L_1. \tag{2.194}$$

Relation (2.193) and property (P11) imply that

$$x_T \in X_*. \tag{2.195}$$

In view of relation (2.193) and property (P12) we have

$$|x_t - \bar{x}| \le \delta_2 \text{ for all } t = L_{14}, \dots, T - L_{14}. \tag{2.196}$$

Define a sequence $\{y_t\}_{t=0}^T \subset X$ as follows:

$$y_t = x_t \text{ for all } t = 0, \dots, T - L_2 - 1, \ y_t = \Lambda_{T-t}(x_*) \text{ for all } t = T - L_2, \dots, T. \tag{2.197}$$

By (2.131), (2.190), (2.192), and (2.197), the sequence $\{y_t\}_{t=0}^\infty$ is an Ω-program. In view of (2.193) and (2.197), we have

$$\delta \ge \sigma(v, T, x_0) - \sum_{t=0}^{T-1} v(x_t, x_{t+1}) \ge \sum_{t=0}^{T-1} v(y_t, y_{t+1}) - \sum_{t=0}^{T-1} v(x_t, x_{t+1})$$

$$= v(y_{T-L_2-1}, y_{T-L_2}) - v(x_{T-L_2-1}, x_{T-L_2})$$

$$+ \sum_{t=T-L_2}^{T-1} v(y_t, y_{t+1}) - \sum_{t=T-L_2}^{T-1} v(x_t, x_{t+1}). \tag{2.198}$$

Relations (2.196) and (2.197) imply that

$$|x_{T-L_2-1} - \bar{x}|, |x_{T-L_2} - \bar{x}| \le \delta_2, \ |y_{T-L_2-1} - \bar{x}| \le \delta_2. \tag{2.199}$$

It follows from (2.197) and (2.192) that

$$|y_{T-L_2} - \bar{x}| = |\Lambda_{L_2}(x_*) - \bar{x}| \le \delta_2. \tag{2.200}$$

By (2.199), (2.200) and the choice of δ_2 [see (2.190), (2.191)] we have

$$|v(x_{T-L_2-1}, x_{T-L_2}) - v(y_{T-L_2-1}, y_{T-L_2})| < \delta_0/32. \tag{2.201}$$

It follows from these inequalities and (2.198) that

$$\sum_{t=T-L_2}^{T-1} v(y_t, y_{t+1}) - \sum_{t=T-L_2}^{T-1} v(x_t, x_{t+1}) \le \delta + \delta_0/16. \tag{2.202}$$

In view of (2.197),

$$\sum_{t=T-L_2}^{T-1} v(y_t, y_{t+1}) = \sum_{t=T-L_2}^{T-1} v(\Lambda_{T-t}(x_*), \Lambda_{T-t-1}(x_*))$$

$$= \sum_{t=T-L_2}^{T-1} \bar{v}(\Lambda_{T-t-1}(x_*), \Lambda_{T-t}(x_*))$$

$$= \sum_{t=0}^{L_2-1} \bar{v}(\Lambda_t(x_*), \Lambda_{t+1}(x_*)). \tag{2.203}$$

Put

$$\tilde{x}_t = x_{T-t} \text{ for all integers } t = 0, \dots, L_2. \tag{2.204}$$

It is clear that $\{\tilde{x}_t\}_{t=0}^{L_2}$ is an $(\bar{\Omega})$-program. By relations (2.195) and (2.204),

$$\tilde{x}_0 = x_T \in X_*. \tag{2.205}$$

In view of (2.204), we have

$$\sum_{t=T-L_2}^{T-1} v(x_t, x_{t+1}) = \sum_{t=T-L_2}^{T-1} v(\tilde{x}_{T-t}, \tilde{x}_{T-t-1})$$

$$= \sum_{t=T-L_2}^{T-1} \bar{v}(\tilde{x}_{T-t-1}, \tilde{x}_{T-t}) = \sum_{t=0}^{L_2-1} \bar{v}(\tilde{x}_t, \tilde{x}_{t+1}). \tag{2.206}$$

It follows from (2.202), (2.203), (2.206), and (2.138) that

$$\delta + \delta_0/16 \geq \sum_{t=0}^{L_2-1} \bar{v}(\Lambda_t(x_*), \Lambda_{t+1}(x_*)) - \sum_{t=0}^{L_2-1} \bar{v}(\tilde{x}_t, \tilde{x}_{t+1})$$

$$= \sum_{t=0}^{L_2-1} \bar{L}(\tilde{x}_t, \tilde{x}_{t+1}) - \sum_{t=0}^{L_2-1} \bar{L}(\Lambda_t(x_*), \Lambda_{t+1}(x_*))$$

$$+ \langle l, \Lambda_{L_2}(x_*) - \Lambda_0(x_*) \rangle - \langle l, \tilde{x}_{L_2} - \tilde{x}_0 \rangle. \tag{2.207}$$

In view of (2.204) and (2.199), we have

$$|\tilde{x}_{L_2} - \bar{x}| = |x_{T-L_2} - \bar{x}| \leq \delta_2.$$

This relation, (2.190), and (2.192) imply that

$$|\langle l, \Lambda_{L_2}(x_*) - \bar{x} \rangle| \leq \delta_2 |l| < 64^{-1}\delta_0, \ |\langle l, \tilde{x}_{L_2} - \bar{x} \rangle| \leq \delta_2 |l| < 64^{-1}\delta_0.$$

It follows from these inequalities, (2.190), and (2.207) that

$$\sum_{t=0}^{L_2-1} \bar{L}(\tilde{x}_t, \tilde{x}_{t+1}) + \langle l, \tilde{x}_0 - \bar{x} \rangle$$

$$- \sum_{t=0}^{L_2-1} \bar{L}(\Lambda_t(x_*), \Lambda_{t+1}(x_*)) + \langle l, \bar{x} - x_* \rangle < 8^{-1}\delta_0.$$

In view of this inequality and (2.140), we have

$$\sum_{t=0}^{L_2-1} \bar{L}(\tilde{x}_t, \tilde{x}_{t+1}) + \langle l, \tilde{x}_0 - \bar{x} \rangle \leq \pi(x_*) + 8^{-1}\delta_0.$$

It follows from this inequality, property (P10), and (2.205) that $|\tilde{x}_t - \Lambda_t(x_*)| \leq \epsilon$ for all integers $t = 0, \ldots, L_1$. Combined with relation (2.204) this implies that for all integers $t = 0, \ldots, L_1$ we have

$$|x_{T-t} - \Lambda_t(x_*)| = |\tilde{x}_t - \Lambda_t(x_*)| \leq \epsilon.$$

Theorem 2.27 is proved.   □

2.13 Stability of a Turnpike Property Under Autonomous Perturbations

In this section based on [62, 64] we continue to study the problems $(P_T^{(y,z)})$ and $(P_T^{(y)})$. We improve the turnpike results stated in Sect. 2.1 and show that the turnpike property is stable under perturbations of the objective function. Note that the stability of the turnpike property is crucial in practice. One reason is that in practice we deal with a problem which consists a perturbation of the problem we wish to consider. Another reason is that the computations introduce numerical errors.

Let (X, ρ) be a compact metric space and Ω be a nonempty closed subset of $X \times X$. Denote by $\mathcal{M}$ the set of all bounded functions $u : \Omega \to R^1$. For each function $w \in \mathcal{M}$ we set

$$\|w\| = \sup\{|w(x, y)| : (x, y) \in \Omega\}. \tag{2.208}$$

For each pair of points $x, y \in X$, each natural number T, and each function $u \in \mathcal{M}$ set

$$\sigma(u, T, x) = \sup \left\{ \sum_{i=0}^{T-1} u(x_i, x_{i+1}) : \{x_i\}_{i=0}^{T} \text{ is a program and } x_0 = x \right\},$$

$$\tag{2.209}$$

$$\sigma(u, T, x, y) = \sup \left\{ \sum_{i=0}^{T-1} u(x_i, x_{i+1}) : \right.$$

$$\left. \{x_i\}_{i=0}^{T} \text{ is a program and } x_0 = x, \ x_T = y \right\}, \qquad (2.210)$$

$$\sigma(u, T) = \sup \left\{ \sum_{i=0}^{T-1} u(x_i, x_{i+1}) : \ \{x_i\}_{i=0}^{T} \text{ is a program} \right\}. \qquad (2.211)$$

(Here we use the convention that the supremum of an empty set is $-\infty$.)

Assume that $v \in \mathcal{M}$ is an upper semicontinuous function. Since in Sect. 2.1 we assume that objective functions are defined on the set $X \times X$ in order to apply the results of the previous sections we set

$$v(x, y) = -\|v\| - 1 \text{ for all } (x, y) \in (X \times X) \setminus \Omega.$$

We suppose that there exist a point $\bar{x} \in X$ and a positive constant $\bar{c}$ such that assumptions (A1)–(A3) introduced in Sect. 2.1 hold. In this section we use the definitions and notation of Sect. 2.1.

The following two theorems are obtained in [62].

Theorem 2.29. *Let $M_0, \epsilon > 0$ and let $L_0 \geq 1$ be an integer. Then there exist a positive number δ and an integer $L_* > L_0$ such that for each function $u \in \mathcal{M}$ satisfying $\|u - v\| \leq \delta$, each integer $T > 2L_*$, and each program $\{x_t\}_{t=0}^{T}$ which satisfies*

$$x_0 \in X_{M_0}, \ x_T \in Y_{L_0},$$

$$\sum_{t=0}^{T-1} u(x_t, x_{t+1}) \geq \sigma(u, T, x_0, x_T) - \delta$$

there exist integers $\tau_1 \in [0, L_*], \ \tau_2 \in [T - L_*, T]$ such that

$$\rho(x_t, \bar{x}) \leq \epsilon \text{ for all integers } t = \tau_1, \ldots, \tau_2.$$

Moreover if $\rho(x_0, \bar{x}) \leq \delta$, then $\tau_1 = 0$, and if $\rho(x_T, \bar{x}) \leq \delta$, then $\tau_2 = T$.

Theorem 2.30. *Let $M_0, \epsilon > 0$. Then there exist a positive number δ and an integer $L_* \geq 1$ such that for each function $u \in \mathcal{M}$ satisfying $\|u - v\| \leq \delta$, each integer $T > 2L_*$, and each program $\{x_t\}_{t=0}^{T}$ which satisfies*

$$x_0 \in X_{M_0}, \ \sum_{t=0}^{T-1} u(x_t, x_{T+1}) \geq \sigma(u, T, x_0) - \delta$$

there exist integers $\tau_1 \in [0, L_*]$, $\tau_2 \in [T - L_*, T]$ such that

$$\rho(x_t, \bar{x}) \leq \epsilon, \; t = \tau_1, \ldots, \tau_2.$$

Moreover if $\rho(x_0, \bar{x}) \leq \delta$, then $\tau_1 = 0$.

Theorems 2.29 and 2.30 establish the turnpike property for approximate solutions of the optimal control problems with an objective function u which belongs to a small neighborhood of v. They generalize the results of Sect. 2.1 which were obtained for approximate solutions of the optimal control problems with the objective function v.

Denote by $\text{Card}(A)$ the cardinality of a set A. The following two theorems were obtained in [64].

Theorem 2.31. *Let $M_0 > 0$, $M_1 > 0$, $\epsilon > 0$ and let $L_0 \geq 1$ be an integer. Then there exist a positive number δ and an integer $L_* > L_0$ such that for each function $u \in \mathcal{M}$ satisfying $\|u - v\| \leq \delta$, each integer $T > L_*$, and each program $\{x_t\}_{t=0}^{T}$ which satisfies*

$$x_0 \in X_{M_0}, \; x_T \in Y_{L_0},$$

$$\sum_{t=0}^{T-1} u(x_t, x_{t+1}) \geq \sigma(u, T, x_0, x_T) - M_1$$

the following inequality holds:

$$\text{Card}(\{t \in \{0, \ldots, T\} : \; \rho(x_t, \bar{x}) > \epsilon\}) \leq L_*.$$

Theorem 2.32. *Let M_0, M_1, ϵ be positive numbers. Then there exist a positive number δ and an integer $L_* \geq 1$ such that for each function $u \in \mathcal{M}$ satisfying $\|u - v\| \leq \delta$, each integer $T > L_*$, and each program $\{x_t\}_{t=0}^{T}$ which satisfies*

$$x_0 \in X_{M_0}, \; \sum_{t=0}^{T-1} u(x_t, x_{T+1}) \geq \sigma(u, T, x_0) - M_1$$

the following inequality holds:

$$\text{Card}(\{t \in \{0, \ldots, T\} : \; \rho(x_t, \bar{x}) > \epsilon\}) \leq L_*.$$

2.14 Auxiliary Results for Theorems 2.29–2.32

In view of assumption (A1) there exists a number $\bar{r} \in (0, 1)$ such that

$$\{(x, y) \in X \times X : \; \rho(x, \bar{x}), \; \rho(y, \bar{x}) \leq \bar{r}\} \subset \Omega. \tag{2.212}$$

It is clear that for each function $w \in \mathcal{M}$, for each pair of points $x, y \in X$ satisfying $\rho(x, \bar{x}), \rho(y, \bar{x}) \leq \bar{r}$, and any natural number T, $\sigma(w, T, x, y)$ is finite.

The next result follows immediately from Lemma 2.16.

Lemma 2.33. *Let $\epsilon > 0$, $M_0 \geq 2$, an integer $L_0 \geq 1$ be as guaranteed by Lemma 2.16, an integer $L_1 > L_0$, and a positive number $\delta < (2L_1)^{-1}$. Then the following assertions hold.*

1. *If an integer $T \in [L_0, L_1]$, a program $\{x_t\}_{t=0}^{T}$ satisfies*

$$\sum_{t=0}^{T-1} v(x_t, x_{t+1}) \geq T v(\bar{x}, \bar{x}) + 2 - M_0$$

and if a function $w \in \mathcal{M}$ satisfies $\|w - v\| \leq \delta$, then

$$\sum_{t=0}^{T-1} w(x_t, x_{t+1}) \geq T v(\bar{x}, \bar{x}) + 3/2 - M_0.$$

2. *If a function $w \in \mathcal{M}$ satisfies $\|w - v\| \leq \delta$, an integer $T \in [L_0, L_1]$, and if a program $\{x_t\}_{t=0}^{T}$ satisfies*

$$\sum_{t=0}^{T-1} w(x_t, x_{t+1}) \geq T v(\bar{x}, \bar{x}) - M_0 + 1,$$

then

$$\sum_{t=0}^{T-1} v(x_t, x_{t+1}) \geq T v(\bar{x}, \bar{x}) - M_0$$

and for each integer $s \in [0, T - L_0]$,

$$\min\{\rho(x_i, \bar{x}) : i = 1 + s, \ldots, L_0 + s]\} \leq \epsilon.$$

Lemma 2.34. *Let ϵ be a positive number. Then there exists a positive number $\delta < \bar{r}$ such that for each function $w \in \mathcal{M}$ satisfying $\|w - v\| \leq \delta$, each natural number T, and each program $\{x_t\}_{t=0}^{T}$ satisfying*

$$\rho(x_0, \bar{x}), \ \rho(x_T, \bar{x}) \leq \delta,$$

$$\sum_{t=0}^{T-1} w(x_t, x_{t+1}) \geq \sigma(w, T, x_0, x_T) - \delta,$$

the inequality $\rho(x_t, \bar{x}) \leq \epsilon$ holds for all integers $t = 0, \ldots, T$.

Proof. We have already mentioned [see (2.212)] that $\sigma(w, T, z_1, z_2)$ is finite for each natural number T, each pair of points $z_1, z_2 \in X$ satisfying $\rho(z_1, \bar{x}), \rho(z_2, \bar{x}) \leq \bar{r}$, and each function $w \in \mathcal{M}$.

Assumption (A1) implies that there exists a number

$$\delta_0 \in (0, \bar{r}/2) \tag{2.213}$$

such that

$$|v(x, y) - v(\bar{x}, \bar{x})| \leq 1 \text{ for all } x, y \in X \text{ satisfying} \tag{2.214}$$

$$\rho(x, \bar{x}), \ \rho(y, \bar{x}) \leq \delta_0.$$

It follows from Lemma 2.15 that there exists a positive number

$$\delta_1 < \min\{\bar{r}/2, \ \delta_0, \ \epsilon/4\} \tag{2.215}$$

such that the following property holds:

(P13)   for each natural number T and each program $\{x_t\}_{t=0}^{T}$ which satisfies

$$\rho(x_0, \bar{x}), \ \rho(x_T, \bar{x}) \leq \delta_1,$$

$$\sum_{t=0}^{T-1} v(x_t, x_{t+1}) \geq \sigma(v, T, x_0, x_T) - \delta_1,$$

the inequality $\rho(x_t, \bar{x}) \leq \epsilon$ holds for all $t = 0, \ldots, T$.

By Lemmas 2.16 and 2.33 there exist a natural number L_0 and a positive number $\delta_2 < \delta_1$ such that the following property holds:

(P14)   for each $w \in \mathcal{M}$ satisfying $\|w - v\| \leq \delta_2$, each integer $L \in [L_0, 8L_0]$, each program $\{x_t\}_{t=0}^{L}$ satisfying

$$\sum_{t=0}^{L-1} w(x_t, x_{t+1}) \geq Lv(\bar{x}, \bar{x}) - 1,$$

and each integer $s \in [0, L - L_0]$, we have

$$\min\{\rho(x_i, \bar{x}) : \ i = 1 + s, \ldots, L_0 + s\} \leq \delta_1.$$

Choose an integer

$$T_* \geq 32L_0 \text{ and } \delta \in (0, \min\{(16T_*)^{-1}\delta_1, \ \delta_2\}). \tag{2.216}$$

Assume that a function $w \in \mathcal{M}$ satisfies

$$\|w - v\| \leq \delta, \tag{2.217}$$

T is a natural number, and $\{x_t\}_{t=0}^T$ is a program which satisfies

$$\rho(x_0, \bar{x}), \ \rho(x_T, \bar{x}) \leq \delta, \ \sum_{t=0}^{T-1} w(x_t, x_{t+1}) \geq \sigma(w, T, x_0, x_T) - \delta. \qquad (2.218)$$

We claim that

$$\rho(x_t, \bar{x}) \leq \epsilon \text{ for all } t = 0, \dots, T. \qquad (2.219)$$

Assume the contrary. Then

$$\max\{\rho(x_t, \bar{x}) : \ t = 0, \dots, T\} > \epsilon$$

and it follows from (2.215), (2.216), and (2.218) that there exists an integer τ_0 such that

$$1 \leq \tau_0 \leq T - 1 \text{ and } \rho(x_{\tau_0}, \bar{x}) > \epsilon. \qquad (2.220)$$

By relations (2.215), (2.216), (2.218), and (2.220), there exist integers $\tau_1, \tau_2 \geq 0$ such that

$$\tau_1 < \tau_0 < \tau_2 \leq T, \ \rho(x_{\tau_i}, \bar{x}) \leq \delta_1, \ i = 1, 2, \qquad (2.221)$$

$$\rho(x_t, \bar{x}) > \delta_1 \text{ for all integers } t = \tau_1 + 1, \dots, \tau_2 - 1.$$

There are two cases:

$$\tau_2 - \tau_1 \geq T_*; \qquad (2.222)$$

$$\tau_2 - \tau_1 < T_*.$$

Assume that $\tau_2 - \tau_1 < T_*$. In view of relations (2.217) and (2.218), we have

$$\sum_{t=\tau_1}^{\tau_2-1} v(x_t, x_{t+1}) + \delta(\tau_2 - \tau_1) \geq \sum_{t=\tau_1}^{\tau_2-1} w(x_t, x_{t+1}) \geq \sigma(w, \tau_2 - \tau_1, x_{\tau_1}, x_{\tau_2}) - \delta$$

$$\geq \sigma(v, \tau_2 - \tau_1, x_{\tau_1}, x_{\tau_2}) - \delta - \delta(\tau_2 - \tau_1)$$

and by relations (2.216) and $\tau_2 - \tau_1 < T_*$,

$$\sum_{t=\tau_1}^{\tau_2-1} v(x_t, x_{t+1}) \geq \sigma(v, \tau_2 - \tau_1, x_{\tau_1}, x_{\tau_2}) - 4\delta T_*$$

$$\geq \sigma(v, \tau_2 - \tau_1, x_{\tau_1}, x_{\tau_2}) - \delta_1.$$

In view of the inequality above, property (P13), and relation (2.221), we have

$$\rho(x_t, \bar{x}) \leq \epsilon \text{ for all } t = \tau_1, \dots, \tau_2.$$

This contradicts (2.220) and (2.221). The contradiction we have reached proves (2.222). Set

$$y_{\tau_1} = x_{\tau_1}, \ y_{\tau_2} = x_{\tau_2}, \ y_t = \bar{x}, \ t = \tau_1 + 1, \ldots, \tau_2 - 1. \tag{2.223}$$

It follows from (2.14), (2.221), (2.215), and (2.212) that $\{y_t\}_{t=\tau_1}^{\tau_2}$ is a program. By (2.214), (2.215), (2.217), and (2.221), we have

$$\sum_{t=\tau_1}^{\tau_2-1} w(y_t, y_{t+1}) \geq -(\tau_2 - \tau_1)\delta + \sum_{t=\tau_1}^{\tau_2-1} v(y_t, y_{t+1})$$

$$\geq -(\tau_2 - \tau_1)\delta - 2 + (\tau_2 - \tau_1)v(\bar{x}, \bar{x}). \tag{2.224}$$

Relations (2.221), (2.218), (2.14), and (2.224) imply that

$$\sum_{t=\tau_1}^{\tau_2-1} w(x_t, x_{t+1}) \geq \sum_{t=\tau_1}^{\tau_2-1} w(y_t, y_{t+1}) - \delta \geq -\delta(\tau_2 - \tau_1 + 1) - 2 + (\tau_2 - \tau_1)v(\bar{x}, \bar{x}). \tag{2.225}$$

Define a sequence of integers $\{S_i\}_{i=0}^{q}$ such that

$$S_0 = \tau_1, \ S_{i+1} - S_i = L_0 \text{ for all integers } i \text{ satisfying } 0 \leq i \leq q - 2, \ S_q = \tau_2,$$

$$1 + L_0 \leq S_q - S_{q-1} \leq 2(L_0 + 1). \tag{2.226}$$

Assume that for all integers $i = 0, \ldots, q - 1$,

$$\sum_{t=S_i}^{S_{i+1}-1} w(x_t, x_{t+1}) < (S_{i+1} - S_i)v(\bar{x}, \bar{x}) - 1.$$

Together with relation (2.226) this implies that

$$\sum_{t=\tau_1}^{\tau_2-1} w(x_t, x_{t+1}) = \sum_{i=0}^{q-1} \sum_{t=S_i}^{S_{i+1}-1} w(x_t, x_{t+1}) < \sum_{i=0}^{q-1} (S_{i+1} - S_i)v(\bar{x}, \bar{x}) - q$$

$$= (\tau_2 - \tau_1)v(\bar{x}, \bar{x}) - q \leq (\tau_2 - \tau_1)v(\bar{x}, \bar{x}) - (\tau_2 - \tau_1)(4L_0)^{-1}.$$

Combined with (2.225), (2.222), and (2.215) this implies that

$$-\delta(\tau_2 - \tau_1 + 1) - 2 \leq -(\tau_2 - \tau_1)(4L_0)^{-1},$$

$$\delta \geq (\tau_2 - \tau_1 + 1)^{-1}[(\tau_2 - \tau_1)(4L_0)^{-1} - 2] \geq (8L_0)^{-1} - 2T_*^{-1} \geq (16L_0)^{-1}.$$

This contradicts (2.216). The contradiction we have reached proves that there exists an integer $j \in \{0, \ldots, q-1\}$ such that

$$\sum_{t=S_j}^{S_{j+1}-1} w(x_t, x_{t+1}) \geq (S_{j+1} - S_j)v(\bar{x}, \bar{x}) - 1. \tag{2.227}$$

In view of (2.227), property (P14), (2.217), (2.216), and (2.226), we have

$$\min\{\rho(x_i, \bar{x}) : i = S_j + 1, \ldots, S_j + L_0\} \leq \delta_1.$$

This contradicts (2.221) and (2.226). The contradiction we have reached proves (2.219). This completes the proof of Lemma 2.34.   □

Lemma 2.35. *Let $M_0, M_1, \epsilon > 0$ and let $L_0 \geq 1$ be an integer. Then there exist an integer $L_* > L_0 + 2$ and a positive number $\delta < \epsilon$ such that for each function $w \in \mathcal{M}$ satisfying*

$$\|w - v\| \leq \delta, \tag{2.228}$$

each integer $T \geq L_*$, each program $\{x_t\}_{t=0}^{T}$ satisfying

$$\min\{\rho(x_t, \bar{x}) : t = 1, \ldots, T - 1\} > \epsilon, \tag{2.229}$$

each point $z_0 \in X_{M_0}$, and each point $z_1 \in Y_{L_0}$, there exists a program $\{y_t\}_{t=0}^{T}$ such that

$$y_0 = z_0, \ y_T = z_1, \ \sum_{t=0}^{T-1} w(y_t, y_{t+1}) \geq \sum_{t=0}^{T-1} w(x_t, x_{t+1}) + M_1.$$

Proof. Lemma 2.16 implies that there exists an integer $L_1 \geq 2$ such that the following property holds:

(P15)   for each program $\{z_t\}_{t=0}^{L_1}$ which satisfies

$$\sum_{t=0}^{L_1-1} v(z_t, z_{t+1}) \geq L_1 v(\bar{x}, \bar{x}) - M_0$$

the inequality

$$\min\{\rho(z_t, \bar{x}) : t = 1, \ldots, L_1 - 1\} \leq \bar{r}/2$$

holds.

Lemmas 2.16 and 2.33 imply that there exist a number $\delta_1 > 0$ and an integer $L_2 \geq 1$ such that the following property holds:

(P16)   for each $w \in \mathcal{M}$ satisfying $\|w - v\| \leq \delta_1$, each integer $L \in [L_2, 8L_2]$, each program $\{x_t\}_{t=0}^{L}$ satisfying

$$\sum_{t=0}^{L-1} w(x_t, x_{t+1}) \geq Lv(\bar{x}, \bar{x}) - 1,$$

and each integer $s \in [0, L - L_2]$, we have

$$\min\{\rho(x_i, \bar{x}) : i = 1 + s, \ldots, s + L_2\} \leq \epsilon/2.$$

Choose an integer $L_* \geq 1$ such that

$$L_* > 8(L_0 + L_1 + L_2 + 2), \quad L_*(8L_2)^{-1} > 2\|v\|(L_0 + L_1 + 1) + M_1 \qquad (2.230)$$

and a number

$$\delta \in (0, \min\{\delta_1, \ (8L_*)^{-1}, \epsilon/4\}). \qquad (2.231)$$

Assume that a function $w \in \mathcal{M}$ satisfies (2.228), an integer $T \geq L_*$, a program $\{x_t\}_{t=0}^{T}$ satisfies (2.229), and points

$$z_0 \in X_{M_0}, \ z_1 \in Y_{L_0}. \qquad (2.232)$$

It follows from (2.232), the definition of the sets X_{M_0}, Y_{L_0}, property (P15), and (2.212) that there exists a program $\{y_t\}_{t=0}^{T}$ such that

$$y_0 = z_0, \ y_T = z_1, y_t = \bar{x}, \ t = L_1, \ldots, T - L_0. \qquad (2.233)$$

By (2.233), we have

$$\sum_{t=0}^{T-1} v(y_t, y_{t+1}) \geq Tv(\bar{x}, \bar{x}) - (L_0 + L_1 + 1)2\|v\|. \qquad (2.234)$$

In view of (2.234) and (2.228),

$$\sum_{t=0}^{T-1} w(y_t, y_{t+1}) \geq \sum_{t=0}^{T-1} v(y_t, y_{t+1}) - T\delta \geq Tv(\bar{x}, \bar{x}) - (L_0 + L_1 + 1)2\|v\| - T\delta. \qquad (2.235)$$

There exists a finite sequence of nonnegative integers $\{S_i\}_{i=0}^{q}$ such that

$$S_0 = 0, \ S_{i+1} = S_i + L_2, \ i = 0, \ldots, q - 2, \ S_q = T,$$
$$L_2 < S_q - S_{q-1} \leq 2L_2 + 1. \qquad (2.236)$$

By relations (2.236), (2.228), (2.229), and (2.231), for all integers $i = 0, \ldots, q - 1$, we have

$$\sum_{t=S_i}^{S_{i+1}-1} w(x_i, x_{i+1}) < (S_{i+1} - S_i)v(\bar{x}, \bar{x}) - 1. \tag{2.237}$$

Relations (2.236) and (2.237) imply that

$$\sum_{t=0}^{T-1} w(x_t, x_{t+1}) = \sum_{t=0}^{q-1} \sum_{t=S_i}^{S_{i+1}-1} w(x_t, x_{t+1}) < \sum_{i=0}^{q-1} (S_{i+1} - S_i)v(\bar{x}, \bar{x}) - q$$

$$= Tv(\bar{x}, \bar{x}) - q \leq Tv(\bar{x}, \bar{x}) - T(4L_2)^{-1}. \tag{2.238}$$

It follows from (2.235), (2.231), and (2.230) that

$$\sum_{t=0}^{T-1} w(y_t, y_{t+1}) - \sum_{t=0}^{T-1} w(x_t, x_{t+1}) \geq Tv(\bar{x}, \bar{x}) - (L_0 + L_1 + 1)2\|v\| - T\delta$$

$$-Tv(\bar{x}, \bar{x}) + T(4L_2)^{-1} \geq T(8L_2)^{-1} - (L_0 + L_1 + 1)2\|v\|$$

$$\geq L_*(8L_2)^{-1} - (L_0 + L_1 + 1)2\|v\| > M_1.$$

Lemma 2.35 is proved.   □

2.15   Proof of Theorem 2.29

Let a number $\bar{r} \in (0, 1)$ satisfy relation (2.212). We may assume that $M_0 > 2$. In view of assumption (A1) it is possible to assume that

$$|v(x, y) - v(\bar{x}, \bar{x})| \leq 4^{-1} \text{ for all points } x, y \in X \text{ satisfying } \rho(x, \bar{x}), \ \rho(\bar{x}, y) \leq \bar{r}. \tag{2.239}$$

Lemma 2.34 implies that there exists a positive number $\delta_1 < \bar{r}$ such that the following property holds:

(P17)   for each $w \in \mathcal{M}$ satisfying $\|w - v\| \leq \delta_1$, each natural number T, and each program $\{x_t\}_{t=0}^{T}$ satisfying

$$\rho(x_0, \bar{x}), \ \rho(x_T, \bar{x}) \leq \delta_1, \tag{2.240}$$

$$\sum_{t=0}^{T-1} w(x_t, x_{t+1}) \geq \sigma(w, T, x_0, x_T) - \delta_1 \tag{2.241}$$

the inequality $\rho(x_t, \bar{x}) \leq \epsilon$ holds for all integers $t = 0, \ldots, T$.

In view of Lemma 2.35 there exist a natural number $L_1 > L_0 + 2$ and a positive number $\delta < \delta_1$ such that the following property holds:

(P18)   for each function $w \in \mathcal{M}$ satisfying $\|w - v\| \le \delta$, each natural number $T \ge L_1$, each program $\{x_t\}_{t=0}^{T}$ satisfying

$$\min\{\rho(x_t, \bar{x}) : t = 1, \ldots, T - 1\} > \delta_1, \tag{2.242}$$

each point $z_0 \in X_{M_0}$, and each point $z_1 \in Y_{L_0}$ there exists a program $\{y_t\}_{t=0}^{T}$ such that

$$y_0 = z_0, \ y_T = z_1, \ \sum_{t=0}^{T-1} w(y_t, y_{t+1}) \ge \sum_{t=0}^{T-1} w(x_t, x_{t+1}) + 4.$$

Choose a natural number

$$L_* > 4(L_1 + L_0 + 2). \tag{2.243}$$

Assume that a function $u \in \mathcal{M}$ satisfies

$$\|u - v\| \le \delta, \tag{2.244}$$

an integer $T > 2L_*$, and a program $\{x_T\}_{t=0}^{T}$ satisfies

$$x_0 \in X_{M_0}, \ x_T \in Y_{L_0}, \ \sum_{t=0}^{T-1} u(x_t, x_{t+1}) \ge \sigma(u, T, x_0, x_T) - \delta. \tag{2.245}$$

In view of property (P18), (2.244), (2.243), and (2.245), we have

$$\min\{\rho(x_t, \bar{x}) : t = 1, \ldots, T - 1\} \le \delta_1. \tag{2.246}$$

Set

$$\tau_1 = \min\{t \in \{0, \ldots, T\} : \ \rho(x_t, \bar{x}) \le \delta_1\},$$

$$\tau_2 = \max\{t \in \{0, \ldots, T\} : \ \rho(x_t, \bar{x}) \le \delta_1\}. \tag{2.247}$$

If $\rho(x_0, \bar{x}) \le \delta$, then set $\tau_1 = 0$, and if $\rho(x_T, \bar{x}) \le \delta$, then set $\tau_2 = T$. It follows from (2.246), (2.247), property (P17), (2.244), and (2.245) that

$$\rho(x_t, \bar{x}) \le \epsilon \text{ for all } t = \tau_1, \ldots, \tau_2. \tag{2.248}$$

In order to complete the proof of Theorem 2.29 it is sufficient to show that

$$\tau_1 \le L_1, \ \tau_2 \ge T - L_1. \tag{2.249}$$

Assume that

$$\tau_1 > L_1. \tag{2.250}$$

By (2.245), we have

$$\sum_{t=0}^{\tau_1-1} u(x_t, x_{t+1}) \geq \sigma(u, \tau_1, x_0, x_{\tau_1}) - \delta. \tag{2.251}$$

Relations (2.247), (2.212), and (2.239) imply that

$$x_{\tau_1} \in Y_{L_0} \cap X_{M_0}.$$

Combined with property (P18) and relations (2.244), (2.247), and (2.250) this inclusion implies that there exists a program $\{y_t\}_{t=0}^{\tau_1}$ such that

$$y_0 = x_0, \ y_{\tau_1} = x_{\tau_1}, \ \sum_{t=0}^{\tau_1-1} u(y_t, y_{t+1}) \geq \sum_{t=0}^{\tau_1-1} u(x_t, x_{t+1}) + 4.$$

This contradicts (2.251). The contradiction we have reached proves that

$$\tau_1 \leq L_1. \tag{2.252}$$

Assume that

$$\tau_2 < T - L_1. \tag{2.253}$$

In view of (2.245), we have

$$\sum_{t=\tau_2}^{T-1} u(x_t, x_{t+1}) \geq \sigma(u, T - \tau_2, x_{\tau_2}, x_T) - \delta. \tag{2.254}$$

It follows from property (P18), relations (2.244), (2.253), (2.247), (2.245), and the inclusion $x_{\tau_1} \in X_{M_0}$ that there exists a program $\{y_t\}_{t=\tau_2}^{T}$ such that

$$y_{\tau_2} = x_{\tau_2}, \ y_T = x_T, \ \sum_{t=\tau_2}^{T-1} u(y_t, y_{t+1}) \geq \sum_{t=\tau_2}^{T-1} u(x_t, x_{t+1}) + 4.$$

This contradicts (2.254). The contradiction we have reached proves $\tau_2 \geq T - L_1$. This completes the proof of Theorem 2.29.

2.16   Proof of Theorem 2.30

We prove Theorem 2.30 by a modification of the proof of Theorem 2.29. Let $\bar{r} \in (0, 1)$ satisfy (2.212). We may assume that $M_0 > 2$ and that relation (2.239) holds. Lemma 2.34 implies that there exists a positive number $\delta_1 < \bar{r}$ such that property (P17) holds. Set $L_0 = 4$ and in view of Lemma 2.35 there exist a natural number $L_1 > L_0 + 2$ and a positive number $\delta < \delta_1$ such that property (P18) holds. Fix an integer

$$L_* > 4(L_1 + L_0 + 2). \tag{2.255}$$

Assume that a function $u \in \mathcal{M}$ satisfies

$$\|u - v\| \leq \delta, \tag{2.256}$$

an integer $T > 2L_*$, and a program $\{x_t\}_{t=0}^{T}$ satisfies

$$x_0 \in X_{M_0}, \ \sum_{t=0}^{T-1} u(x_t, x_{t+1}) \geq \sigma(u, T, x_0) - \delta. \tag{2.257}$$

Property (P18), (2.256), and (2.257) imply that (2.246) holds. Define τ_1, τ_2 by (2.247). If $\rho(x_0, \bar{x}) \leq \delta$, then $\tau_1 = 0$, and if $\rho(x_T, \bar{x}) \leq \delta$, then $\tau_2 = T$. Relations (2.246), (2.247), (2.257) and property (P17) imply that (2.248) holds. In order to complete the proof of Theorem 2.30 it is sufficient to show that $\tau_1 \leq L_1$ and $\tau_2 \geq T - L_1$. Arguing as in the proof of Theorem 2.29 we show that $\tau_1 \leq L_1$.
    Assume that

$$\tau_2 < T - L_1. \tag{2.258}$$

By (2.257), we have

$$\sum_{t=\tau_2}^{T-1} u(x_t, x_{t+1}) \geq \sigma(u, T - \tau_2, x_{\tau_2}) - \delta. \tag{2.259}$$

It follows from property (P18), (2.256), (2.258), (2.246), and (2.247) that there exists a program $\{y_t\}_{t=\tau_2}^{T}$ such that

$$y_{\tau_2} = x_{\tau_2}, \ y_T = \bar{x},$$

$$\sum_{t=\tau_2}^{T-1} w(y_t, y_{t+1}) \geq \sum_{t=\tau_2}^{T-1} w(x_t, x_{t+1}) + 4.$$

This contradicts (2.259). The contradiction we have reached proves $\tau_2 \geq T - L_1$. This completes the proof of Theorem 2.30.

2.17   Proof of Theorems 2.31 and 2.32

We prove Theorems 2.31 and 2.32 simultaneously. Let $\bar{r} \in (0, 1)$ satisfy (2.212).
We may assume that $M_0 > 2$ and that

$$|v(x, y) - v(\bar{x}, \bar{x})| \leq 1/4 \text{ for all } x, y \in X \text{ satisfying } \rho(x, \bar{x}), \rho(y, \bar{x}) \leq \bar{r}.$$
$$(2.260)$$

Lemma 2.34 implies that there exists

$$\delta_1 \in (0, \min\{\epsilon, \bar{r}\}) \tag{2.261}$$

such that the following property holds:

(P19)   for each function $w \in \mathcal{M}$ satisfying $\|w - v\| \leq \delta_1$, each natural number T,
and each program $\{x_t\}_{t=0}^{T}$ satisfying

$$\rho(x_0, \bar{x}), \ \rho(x_T, \bar{x}) \leq \delta_1, \ \sum_{t=0}^{T-1} w(x_t, x_{t+1}) \geq \sigma(w, T, x_0, x_T) - \delta_1$$

the inequality $\rho(x_t, \bar{x}) \leq \epsilon$ holds for all $t = 0, \ldots, T$.

In the case of Theorem 2.31 the natural number L_0 is given. In the case of
Theorem 2.32 put $L_0 = 4$.

Lemma 2.35 implies that there exist an integer $L_1 > L_0 + 2$ and a positive
number $\delta < \delta_1$ such that the following property holds:

(P20)   for each function $w \in \mathcal{M}$ satisfying $\|w - v\| \leq \delta$, each integer $T \geq L_1$,
each program $\{x_t\}_{t=0}^{T}$ satisfying

$$\min\{\rho(x_t, \bar{x}) : \ t = 1, \ldots, T - 1\} > \delta_1,$$

each point $z_0 \in X_{M_0}$ and each point $z_1 \in Y_{L_0}$ there exists a program $\{y_t\}_{t=0}^{T}$ such
that

$$y_0 = z_0, \ y_T = z_1, \ \sum_{t=0}^{T-1} w(y_t, y_{t+1}) \geq \sum_{t=0}^{T-1} w(x_t, x_{t+1}) + M_1 + 4.$$

In view of (2.212), the choice of $\bar{r}$, and (2.260), we have

$$\{z \in X : \ \rho(x, \bar{x}) \leq \bar{r}\} \subset X_1 \cap Y_1 \subset X_{M_0} \cap Y_{L_0}. \tag{2.262}$$

Fix an integer

$$L_2 > 4 + L_1 \tag{2.263}$$

and an integer

$$L_* > 8(L_0 + L_1 + L_2 + 2) + L_2(2 + M_1 \delta_1^{-1}). \tag{2.264}$$

Assume that a function $u \in \mathcal{M}$ satisfies

$$\|u - v\| \le \delta, \tag{2.265}$$

an integer $T > L_*$, and a program $\{x_t\}_{t=0}^T$ satisfies

$$x_0 \in X_{M_0}, \ x_T \in Y_{L_0},$$

$$\sum_{t=0}^{T-1} u(x_t, x_{t+1}) \ge \sigma(u, T, x_0, x_T) - M_1 \tag{2.266}$$

in the case of Theorem 2.31 and

$$x_0 \in X_{M_0}, \ \sum_{t=0}^{T-1} u(x_t, x_{t+1}) \ge \sigma(u, T, x_0) - M_1 \tag{2.267}$$

in the case of Theorem 2.32.
   Let an integer

$$\tau \in [0, T - L_2] \tag{2.268}$$

be given. We claim that

$$\min\{\rho(x_t, \bar{x}) : \ t = \tau + 1, \dots, \tau + L_2\} \le \delta_1. \tag{2.269}$$

Assume the contrary. Then

$$\rho(x_t, \bar{x}) > \delta_1, \ t = \tau + 1, \dots, \tau + L_2. \tag{2.270}$$

Relations (2.266) and (2.267) imply that there exists an integer S_1 such that

$$0 \le S_1 \le \tau, \ x_{S_1} \in X_{M_0},$$

$$x_t \notin X_{M_0} \text{ for all integers } t \text{ satisfying } S_1 < t \le \tau. \tag{2.271}$$

In view of (2.261), (2.262), and (2.271), for all integers t satisfying $S_1 < t \le \tau$, we have

$$\rho(x_t, \bar{x}) > \bar{r} > \delta_1. \tag{2.272}$$

We claim that there exists an integer S_2 such that

$$\tau + L_2 \le S_2 \le T, \ x_{S_2} \in Y_{L_0}. \tag{2.273}$$

In the case of Theorem 2.31 the existence of an integer S_2 satisfying (2.273) follows from (2.266). Consider the case of Theorem 2.32 and show that in this case an integer S_2 satisfying (2.273) also exists.

Assume the contrary. Then

$$x_t \notin Y_{L_0}, \ t = \tau + L_2, \ldots, T$$

and by (2.261) and (2.262), we have

$$\rho(x_t, \bar{x}) > \bar{r} > \delta_1, \ t = \tau + L_2, \ldots, T.$$

Together with (2.272) and (2.270) this relation implies that

$$\rho(x_t, \bar{x}) > \delta_1, \ t = S_1 + 1, \ldots, T. \tag{2.274}$$

In view of (2.263), (2.268), and (2.271), we have

$$T - S_1 \ge T - \tau \ge L_2 > L_1. \tag{2.275}$$

It follows from (2.265), (2.271), (2.274), (2.275), and (P20) that there exists a program $\{y_t\}_{t=S_1}^{T}$ such that

$$y_{S_1} = x_{S_1}, \ y_T = \bar{x}, \ \sum_{t=S_1}^{T-1} u(y_t, y_{t+1}) \ge \sum_{t=S_1}^{T-1} u(x_t, x_{t+1}) + M_1 + 4. \tag{2.276}$$

Put

$$y_t = x_t, \ t = 0, \ldots, S_1.$$

It is clear that $\{y_t\}_{t=0}^{T}$ is a program and by (2.276) and the equation above we have

$$y_0 = x_0,$$

$$\sum_{t=0}^{T-1} u(y_t, y_{t+1}) - \sum_{t=0}^{T-1} u(x_t, x_{t+1}) = \sum_{t=S_1}^{T-1} u(y_t, y_{t+1}) - \sum_{t=S_1}^{T-1} u(x_t, x_{t+1}) \ge M_1 + 4.$$

This contradicts (2.267). The contradiction we have reached proves that there exists an integer S_2 satisfying (2.273). Thus in the case of Theorem 2.31 and in the case of Theorem 2.32 there exists an integer S_2 such that (2.273) holds.

We may assume without loss of generality that for all integers t satisfying $\tau + L_2 < t < S_2$ the inclusion

$$x_t \notin Y_{L_0} \tag{2.277}$$

is true. In view of (2.277), (2.261), and (2.262), for all integers t satisfying $\tau + L_2 < t < S_2$, we have

$$\rho(x_t, \bar{x}) > \bar{r} > \delta_1. \tag{2.278}$$

Relations (2.273), (2.271), (2.270), (2.272), and (2.278) imply that

$$S_2 - S_1 \geq L_2, \; x_{S_1} \in X_{M_0}, \; x_{S_2} \in Y_{L_0},$$

$$\rho(x_t, \bar{x}) > \delta_1, \; t = S_1 + 1, \ldots, S_2 - 1. \tag{2.279}$$

It follows from (2.263), (2.265), (2.279) and property (P20) that there exists a program $\{y_t\}_{t=S_1}^{S_2}$ such that

$$y_{S_1} = x_{S_1}, \; y_{S_2} = x_{S_2},$$

$$\sum_{t=S_1}^{S_2-1} u(y_t, y_{t+1}) \geq \sum_{t=S_1}^{S_2-1} u(x_t, x_{t+1}) + M_1 + 4. \tag{2.280}$$

Set

$$y_t = x_t \text{ for all integers } t \text{ satisfying } 0 \leq t < S_1$$

$$\text{and for all integers } t \text{ satisfying } S_2 < t \leq T. \tag{2.281}$$

It is clear that $\{y_t\}_{t=0}^T$ is a program and that

$$y_0 = x_0, \; y_T = x_T. \tag{2.282}$$

In view of (2.280) and (2.281), we have

$$\sum_{t=0}^{T-1} u(y_t, y_{t+1}) - \sum_{t=0}^{T-1} u(x_t, x_{t+1}) = \sum_{t=S_1}^{S_2-1} u(y_t, y_{t+1}) - \sum_{t=S_1}^{S_2-1} u(x_t, x_{t+1}) \geq M_1 + 4.$$

Combined with (2.282) this contradicts (2.266). The contradiction we have reached proves (2.269).

Thus we have shown that the following property holds:

(P21)   for each integer $\tau \in [0, \ldots, T - L_2]$ we have

$$\min\{\rho(x_t, \bar{x}) : t = \tau + 1, \ldots, \tau + L_2\} \leq \delta_1.$$

Using (P21) by induction we construct a sequence of natural numbers $\{S_i\}_{i=1}^q$ such that

$$S_1 \in [1, L_2], \quad \text{for each integer } i \text{ satisfying } 1 \leq i \leq q-1, \tag{2.283}$$

$$S_{i+1} - S_i \in [1, L_2[, \ 0 \leq T - S_q < L_2,$$

$$\rho(x_{S_i}, \bar{x}) < \delta_1, \ i = 1, \ldots, q. \tag{2.284}$$

Relations (2.264) and (2.283) imply that $q \geq 6$. Put

$$E_1 = \{i \in \{1, \ldots, q-1\} : \sum_{t=S_i}^{S_{i+1}-1} u(x_i, x_{i+1}) \geq \sigma(u, S_{i+1} - S_i, x_{S_i}, x_{S_{i+1}}) - \delta_1\},$$
$$\tag{2.285}$$

$$E_2 = \{1, \ldots, q-1\} \setminus E_1. \tag{2.286}$$

In view of (2.265), (2.284), (2.285) and property (P19), for each $i \in E_1$, we have

$$\rho(x_t, \bar{x}) \leq \epsilon, \ t = S_i, \ldots, S_{i+1}.$$

Combined with (2.261), (2.283), and (2.286) this inequality implies that

$$\{t \in \{0, \ldots, T\} : \rho(x_t, \bar{x}) > \epsilon\}$$
$$\subset \{0, \ldots, S_1 - 1\} \cup \{t : t \text{ is an integer such that } S_q < t \leq T\}$$
$$\cup_{i \in E_2} \{t : t \text{ is an integer such that } S_i < t < S_{i+1}\}.$$

Together with (2.283) this implies that

$$\text{Card}(\{t \in \{0, \ldots, T\} : \rho(x_t, \bar{x}) > \epsilon\}) \leq 2L_2 + L_2 \text{Card}(E_2). \tag{2.287}$$

In view of (2.266), (2.267), (2.283), (2.285), and (2.286), we have

$$M_1 \geq \sigma(u, T, x_0, x_T) - \sum_{t=0}^{T-1} u(x_i, x_{i+1})$$

$$\geq \sum_{i \in E_2} \left[\sigma(u, S_{i+1} - S_i, x_{S_i}, x_{S_{i+1}}) - \sum_{t=S_i}^{S_{i+1}-1} u(x_i, x_{i+1}) \right] \geq \delta_1 \text{Card}(E_2)$$

and

$$\text{Card}(E_2) \leq \delta_1^{-1} M_1.$$

Combined with (2.264) and (2.287) this implies that

$$\text{Card}(\{t \in \{0, \dots, T\} : \rho(x_t, \bar{x}) > \epsilon\}) \leq 2L_2 + L_2 M_1 \delta^{-1} < L_*.$$

This completes the proof of Theorems 2.31 and 2.32.

2.18   Stability of a Turnpike Property Under Nonautonomous Perturbations

Let (X, ρ) be a compact metric space and Ω be a nonempty closed subset of $X \times X$. In this section we use the notation and definitions of Sects. 2.1 and 2.13.

Recall that $\mathcal{M}$ is the set of all bounded functions $u : \Omega \to R^1$ and that for each function $w \in \mathcal{M}$,

$$\|w\| = \sup\{|w(x, y)| : (x, y) \in \Omega\}. \tag{2.288}$$

For each pair of points $x, y \in X$, each pair of integers T_1, T_2 satisfying $0 \leq T_1 < T_2$, and each finite sequence $\{u_t\}_{t=T_1}^{T_2-1} \subset \mathcal{M}$ set

$$\sigma\left(\{u_t\}_{t=T_1}^{T_2-1}, T_1, T_2, x\right) = \sup\left\{\sum_{t=T_1}^{T_2-1} u_t(x_t, x_{t+1}) : \{x_t\}_{t=T_1}^{T_2} \text{ is a program and } x_{T_1} = x\right\}, \tag{2.289}$$

$$\sigma\left(\{u_t\}_{t=T_1}^{T_2-1}, T_1, T_2, x, y\right)$$
$$= \sup\left\{\sum_{t=T_1}^{T_2-1} u_t(x_t, x_{t+1}) : \{x_t\}_{t=T_1}^{T_2} \text{ is a program and } x_{T_1} = x, \ x_{T_2} = y\right\}, \tag{2.290}$$

$$\sigma\left(\{u_t\}_{t=T_1}^{T_2-1}, T_1, T_2\right) = \sup\left\{\sum_{t=T_1}^{T_2-1} u_t(x_t, x_{t+1}) : \{x_t\}_{t=T_1}^{T_2} \text{ is a program}\right\}. \tag{2.291}$$

Assume that $v \in \mathcal{M}$ is an upper semicontinuous function. We set $v(x, y) = -\|v\| - 1$ for all $(x, y) \in (X \times X) \setminus \Omega$.

We suppose that there exist a point $\bar{x} \in X$ and a positive constant $\bar{c}$ such that assumptions (A1)–(A3) introduced in Sect. 2.1 hold.

Assumption (A1) implies that there exists a positive number $\bar{r} < 1$ such that

$$\{(x, y) \in X \times X : \rho(x, \bar{x}), \rho(y, \bar{x}) \leq \bar{r}\} \subset \Omega \tag{2.292}$$

and

$$|v(x, y) - v(\bar{x}, \bar{x})| \leq 1/8 \text{ for all } x, y \in X \text{ satisfying } \rho(x, \bar{x}), \rho(y, \bar{x}) \leq \bar{r}.$$
(2.293)

It is clear that for each pair of integers T_1, T_2 satisfying $0 \leq T_1 < T_2$, each sequence $\{w_t\}_{t=T_1}^{T_2-1} \subset \mathcal{M}$, and each pair of points $x, y \in X$ satisfying $\rho(x, \bar{x}), \rho(y, \bar{x}) \leq \bar{r}$ the value $\sigma(\{w_t\}_{t=T_1}^{T_2-1}, T_1, T_2, x, y)$ is finite.

Let $T \geq 1$ be an integer. Recall that we denote by Y_T the set of all points $x \in X$ for which there exists a program $\{x_t\}_{t=0}^{T}$ satisfying $x_0 = \bar{x}$ and $x_T = x$ and that we denote by $\bar{Y}_T$ the set of all points $x \in X$ for which there exists a program $\{x_t\}_{t=0}^{T}$ such that $x_0 = x$ and $x_T = \bar{x}$.

It is easy to see that the following result holds.

Proposition 2.36. *Let $L \geq 1$ be an integer. Then $\bar{Y}_L \subset X_{L\|v\|}$.*

Proposition 2.37. *Let M be a positive number. Then there exists an integer $L \geq 1$ such that $X_M \subset \bar{Y}_L$.*

For the proof of Proposition 2.37 see Lemma 2.16.

The following three theorems which describe the structure of approximate solutions of our discrete-time control system were obtained in [65].

Theorem 2.38. *Let $\epsilon \in (0, 1)$ and M be a positive number. Then there exist an integer $L_0 \geq 1$ and a positive number $\delta_0 < \min\{\epsilon, \bar{r}\}$ such that for each integer $L_1 \geq L_0$, the following assertion holds with $\delta = \delta_0(4L_1)^{-1}$.*

Assume that an integer $T > 2L_1$, $\{u_t\}_{t=0}^{T-1} \subset \mathcal{M}$, a program $\{x_t\}_{t=0}^{T}$, and a finite sequence of integers $\{S_i\}_{i=0}^{q}$ satisfy

$$\|u_t - v\| \leq \delta, \ t = 0, \ldots, T - 1,$$

$$S_0 = 0, \ S_{i+1} - S_i \in [L_0, L_1], i = 0, \ldots, q - 1, \ S_q > T - L_1,$$

$$\sum_{t=S_i}^{S_{i+1}-1} u_t(x_t, x_{t+1}) \geq \sum_{t=S_i}^{S_{i+1}-1} u_t(\bar{x}, \bar{x}) - M$$

for each integer $i \in [0, q - 1]$,

$$\sum_{t=S_i}^{S_{i+2}-1} u_t(x_t, x_{t+1}) \geq \sigma\left(\{u_t\}_{t=S_i}^{S_{i+2}-1}, S_i, S_{i+2}, x_{S_i}, x_{S_{i+2}}\right) - \delta_0$$
(2.294)

for each integer $i \in [0, q - 2]$, and

$$\sum_{t=S_{q-2}}^{T-1} u_t(x_t, x_{t+1}) \geq \sigma(\{u_t\}_{t=S_{q-2}}^{T-1}, S_{q-2}, T, x_{S_{q-2}}, x_T) - \delta_0.$$
(2.295)

Then there exist integers $\tau_1, \tau_2 \in [0, T]$ such that $\tau_1 \leq 2L_0$, $\tau_2 > T - 2L_1$, and

$$\rho(x_t, \bar{x}) \leq \epsilon, \ t = \tau_1, \ldots, \tau_2.$$

Moreover, if $\rho(x_0, \bar{x}) \leq \delta_0$, then $\tau_1 = 0$, and if $\rho(x_T, \bar{x}) \leq \delta_0$, then $\tau_2 = T$.

Theorem 2.39. *Let a positive number $\epsilon < \bar{r}$, $L_0 \geq 1$ be an integer and M_0 be a positive number. Then there exist an integer $L \geq 1$ and a positive number $\delta < \epsilon$ such that for each integer $T > 2L$, each $\{u_t\}_{t=0}^{T-1} \subset \mathcal{M}$ satisfying*

$$\|u_t - v\| \leq \delta, \ t = 0, \ldots T - 1,$$

and each program $\{x_t\}_{t=0}^{T}$ which satisfies

$$x_0 \in \bar{Y}_{L_0}, \ x_T \in Y_{L_0},$$

$$\sum_{t=0}^{T-1} u_t(x_t, x_{t+1}) \geq \sigma(\{u_t\}_{t=0}^{T-1}, 0, T, x_0, x_T) - M_0$$

and

$$\sum_{t=\tau}^{\tau+L-1} u_t(x_t, x_{t+1}) \geq \sigma(\{u_t\}_{t=\tau}^{\tau+L-1}, \tau, \tau + L, x_\tau, x_{\tau+L}) - \delta \qquad (2.296)$$

for each integer $\tau \in [0, T - L]$ there exist integers $\tau_1 \in [0, L]$, $\tau_2 \in [T - L, T]$ such that

$$\rho(x_t, \bar{x}) \leq \epsilon, \ t = \tau_1, \ldots, \tau_2.$$

Moreover, if $\rho(x_0, \bar{x}) \leq \delta$, then $\tau_1 = 0$, and if $\rho(x_T, \bar{x}) \leq \delta$, then $\tau_2 = T$.

Theorem 2.40. *Let a positive number $\epsilon < \bar{r}$, $L_0 \geq 1$ be an integer and M_0 be a positive number. Then there exist an integer $L \geq 1$ and a positive number $\delta < \epsilon$ such that for each integer $T > 2L$, each $\{u_t\}_{t=0}^{T-1} \subset \mathcal{M}$ satisfying*

$$\|u_t - v\| \leq \delta, \ t = 0, \ldots, T - 1,$$

and each program $\{x_t\}_{t=0}^{T}$ which satisfies

$$x_0 \in \bar{Y}_{L_0}, \ \sum_{t=0}^{T-1} u_t(x_t, x_{t+1}) \geq \sigma(\{u_t\}_{t=0}^{T-1}, 0, T, x_0) - M_0$$

and

$$\sum_{t=\tau}^{\tau+L-1} u_t(x_t, x_{t+1}) \geq \sigma(\{u_t\}_{t=\tau}^{\tau+L-1}, \tau, \tau + L, x_\tau, x_{\tau+L}) - \delta \qquad (2.297)$$

for each integer $\tau \in [0, T - L]$ there exist integers $\tau_1 \in [0, L]$, $\tau_2 \in [T - L, T]$ such that

$$\rho(x_t, \bar{x}) \leq \epsilon, \ t = \tau_1, \ldots, \tau_2.$$

Moreover if $\rho(x_0, \bar{x}) \leq \delta$, then $\tau_1 = 0$.

Theorems 2.38–2.40 establish the turnpike property for approximate solutions of the optimal control problems with objective functions u_t, $t = 0, \ldots, T - 1$ which belong to a small neighborhood of v.

2.19   Proof of Theorem 2.38

Lemma 2.15 implies that there exists

$$\delta_0 \in (0, \min\{\epsilon, \bar{r}\}) \tag{2.298}$$

such that the following property holds:

(P21)   for each natural number T and each program $\{x_t\}_{t=0}^T$ satisfying

$$\rho(x_0, \bar{x}), \ \rho(x_T, \bar{x}) \leq \delta_0, \ \sum_{t=0}^{T-1} v(x_t, x_{t+1}) \geq \sigma(v, T, x_0, x_T) - 2\delta_0$$

the inequality $\rho(x_t, \bar{x}) \leq \epsilon$ holds for all integers $t = 0, \ldots, T$.
In view of Lemma 2.16 there exists an integer $L_0 \geq 1$ such that the following property holds:

(P22)   for each natural number $T \geq L_0$, each program $\{x_t\}_{t=0}^T$ which satisfies

$$\sum_{t=0}^{T-1} v(x_t, x_{t+1}) \geq T v(\bar{x}, \bar{x}) - M - 2$$

and each integer $S \in [0, T - L_0]$ we have

$$\min\{\rho(x_t, \bar{x}) : \ t = S + 1, \ldots, S + L_0\} \leq \delta_0.$$

Let an integer

$$L_1 \geq L_0 \tag{2.299}$$

be given and set

$$\delta = \delta_0 (4L_1)^{-1}. \tag{2.300}$$

Assume that an integer $T > 2L_1$, $\{u_t\}_{t=0}^{T-1} \subset \mathcal{M}$ satisfies

$$\|u_t - v\| \leq \delta, \ t = 0, \ldots, T-1, \tag{2.301}$$

$\{x_t\}_{t=0}^{T}$ is a program, and $\{S_i\}_{i=0}^{q}$ is a sequence of integers such that

$$S_0 = 0, \ S_{i+1} - S_i \in [L_0, L_1], \ i = 0, \ldots, q-1, \ S_q > T - L_1, \tag{2.302}$$

for each integer $i \in [0, q-1]$

$$\sum_{t=S_i}^{S_{i+1}-1} u_t(x_t, x_{t+1}) \geq \sum_{t=S_i}^{S_{i+1}-1} u_t(\bar{x}, \bar{x}) - M, \tag{2.303}$$

for each integer $i \in [0, q-2]$

$$\sum_{t=S_i}^{S_{i+2}-1} u_t(x_t, x_{t+1}) \geq \sigma\left(\{u_t\}_{t=S_i}^{S_{i+2}-1}, S_i, S_{i+2}, x_{S_i}, x_{S_{i+2}}\right) - \delta_0, \tag{2.304}$$

$$\sum_{t=S_{q-2}}^{T-1} u_t(x_t, x_{t+1}) \geq \sigma\left(\{u_t\}_{t=S_{q-2}}^{T-1}, S_{q-2}, T, x_{S_{q-2}}, x_T\right) - \delta_0. \tag{2.305}$$

Let an integer $i \in [0, q-1]$ be given. In view of (2.300)–(2.303), we have

$$\sum_{t=S_i}^{S_{i+1}-1} v(x_t, x_{t+1}) \geq \sum_{t=S_i}^{S_{i+1}-1} u_t(x_t, x_{t+1}) - \delta(S_{i+1} - S_i)$$

$$\geq \sum_{t=S_i}^{S_{i+1}-1} u_t(\bar{x}, \bar{x}) - M - \delta L_1$$

$$\geq v(\bar{x}, \bar{x})(S_{i+1} - S_i) - \delta L_1 - M - \delta L_1$$

$$= v(\bar{x}, \bar{x})(S_{i+1} - S_i) - M - 1.$$

It follows from the equation above, property (P22), and (2.302) that there exists an integer τ_i such that

$$\tau_i \in [S_i + 1, S_i + L_0], \ \rho(x_{\tau_i}, \bar{x}) \leq \delta_0. \tag{2.306}$$

Thus for any integer $i \in [0, q-1]$ there exists an integer τ_i satisfying (2.306). In view of (2.306) and (2.302), we have

$$\tau_0 \leq 2L_0, \ \tau_{q-1} > T - 2L_1. \tag{2.307}$$

For any integer $i \in [0, q - 2]$, we have

$$0 < \tau_{i+1} - \tau_i \le 2L_1, \quad \tau_i, \tau_{i+1} \in [S_i, S_{i+2}]. \tag{2.308}$$

In view of inequality (2.304), for any integer $i \in [0, q - 2]$, we have

$$\sum_{t=\tau_i}^{\tau_{i+1}-1} u_t(x_t, x_{t+1}) \ge \sigma(\{u_t\}_{t=\tau_i}^{\tau_{i+1}-1}, \tau_i, \tau_{i+1}, x_{\tau_i}, x_{\tau_{i+1}}) - \delta_0. \tag{2.309}$$

Thus we have shown that there is a finite sequence of integers $\{\tau_i\}_{i=0}^{p}$ such that

$$0 \le \tau_0 \le 2L_0, \quad T \ge \tau_p > T - 2L_1,$$

for each integer i satisfying $0 \le i < p$

$$1 \le \tau_{i+1} - \tau_i \le 2L_1 \tag{2.310}$$

and that inequality (2.309) is valid.

It is easy to see that we may assume without loss of generality that if $\rho(x_0, \bar{x}) \le \delta_0$, then $\tau_0 = 0$ and if $\rho(x_T, \bar{x}) \le \delta_0$, then $\tau_p = 0$.

Let an integer $i \in \{0, \ldots, p - 1\}$ be given. Relations (2.301), (2.309), (2.310), and (2.300) imply that

$$\sum_{t=\tau_i}^{\tau_{i+1}-1} v(x_t, x_{t+1}) \ge \sum_{t=\tau_i}^{\tau_{i+1}-1} u_t(x_t, x_{t+1}) - \delta(\tau_{i+1} - \tau_i)$$

$$\ge \sigma(\{u_t\}_{t=\tau_i}^{\tau_{i+1}-1}, \tau_i, \tau_{i+1}, x_{\tau_i}, x_{\tau_{i+1}}) - \delta_0 - \delta 2L_1$$

$$\ge \sigma(v, \tau_{i+1} - \tau_i, x_{\tau_i}, x_{\tau_{i+1}}) - \delta_0 - \delta 4L_1$$

$$\ge \sigma(v, \tau_{i+1} - \tau_i, x_{\tau_i}, x_{\tau_{i+1}}) - 2\delta_0.$$

In view of the equation above, (2.308), and property (P21), we have

$$\rho(x_t, \bar{x}) \le \epsilon, \quad t = \tau_i, \ldots, \tau_{i+1}, \quad i = 0, \ldots, p - 1.$$

Theorem 2.38 is proved.

2.20 Proofs of Theorems 2.39 and 2.40

We prove Theorems 2.39 and 2.40 simultaneously. Choose a real number

$$M_1 > 4. \tag{2.311}$$

Lemma 2.15 implies that there exists a number $\delta_0 \in (0, \epsilon)$ such that the following property holds:

(P23)   for each natural number T and each program $\{x_t\}_{t=0}^T$ satisfying

$$\rho(x_0, \bar{x}), \; \rho(x_T, \bar{x}) \leq \delta_0, \; \sum_{t=0}^{T-1} v(x_t, x_{t+1}) \geq \sigma(v, T, x_0, x_T) - 2\delta_0$$

the inequality $\rho(x_t, \bar{x}) \leq \epsilon$ holds for all $t = 0, \dots, T$.
In view of Lemma 2.16 there exists a natural number $L_1 > L_0 + 4$ such that the following property holds:
(P24)   for each natural number $T \geq L_1$, each program $\{x_t\}_{t=0}^T$ which satisfies

$$\sum_{t=0}^{T-1} v(x_t, x_{t+1}) \geq Tv(\bar{x}, \bar{x}) - M_1 - 2,$$

and each integer $S \in [0, T - L_1]$, we have

$$\min\{\rho(x_t, \bar{x}) : \; t = S + 1, \dots, S + L_1\} \leq \delta_0.$$

Fix an integer $k \geq 1$ such that

$$k > 8L_1(\|v\| + 1) + M_0 + 4, \tag{2.312}$$

put

$$L_2 = kL_1, \tag{2.313}$$

and choose an integer

$$L > 2L_2 \tag{2.314}$$

and a number $\delta > 0$ such that

$$8L_2\delta < \delta_0. \tag{2.315}$$

Assume that an integer $T > 2L$, $\{u_t\}_{t=0}^{T-1} \subset \mathcal{M}$ satisfies

$$\|u_t - v\| \leq \delta, \; t = 0, \dots, T - 1 \tag{2.316}$$

and that a program $\{x_t\}_{t=0}^T$ satisfies the inclusion

$$x_0 \in \bar{Y}_{L_0}, \tag{2.317}$$

for each integer $\tau \in [0, T - L]$

$$\sum_{t=\tau}^{\tau+L-1} u_t(x_t, x_{t+1}) \geq \sigma\left(\{u_t\}_{t=\tau}^{\tau+L-1}, \tau, \tau + L, x_\tau, x_{\tau+L}\right) - \delta \tag{2.318}$$

and that

$$x_T \in Y_{L_0}, \ \sum_{t=0}^{T-1} u_t(x_t, x_{t+1}) \geq \sigma\left(\{u_t\}_{t=0}^{T-1}, 0, T, x_0, x_T\right) - M_0 \qquad (2.319)$$

in the case of Theorem 2.39 and

$$\sum_{t=0}^{T-1} u_t(x_t, x_{t+1}) \geq \sigma\left(\{u_t\}_{t=0}^{T-1}, 0, T, x_0\right) - M_0 \qquad (2.320)$$

in the case of Theorem 2.40.

Assume that an integer S satisfies

$$S \in [0, T - L_2], \ x_S \in \bar{Y}_{L_0}. \qquad (2.321)$$

We claim that there exists an integer $t \in [S + 1, S + L_2]$ which satisfies $\rho(x_t, \bar{x}) \leq \delta_0$.

Assume the contrary. Then

$$\rho(x_t, \bar{x}) > \delta_0, \ t = S + 1, \ldots, S + L_2. \qquad (2.322)$$

There are two cases:

$$\rho(x_t, \bar{x}) > \delta_0 \text{ for all integers } t = S + 1, \ldots, T; \qquad (2.323)$$

$$\rho(x_t, \bar{x}) \leq \delta_0 \text{ for some integer } t \text{ satisfying } S + L_2 < t \leq T. \qquad (2.324)$$

Assume that (2.323) holds. In the case of Theorem 2.40 in view of (2.321) there exists a program $\{y_t\}_{t=0}^T$ such that

$$y_t = x_t, \ t = 0, \ldots, S, \ y_t = \bar{x} \text{ for all integers } t \in [S + L_0, T]. \qquad (2.325)$$

In the case of Theorem 2.39 by relations (2.321) and (2.319) there exists a program $\{y_t\}_{t=0}^T$ such that

$$y_t = x_t, \ t = 0, \ldots, S, \ y_t = \bar{x} \text{ for all integers } t \in [S + L_0, T - L_0], \ y_T = x_T. \qquad (2.326)$$

In view of (2.319), (2.320), (2.325), and (2.326), we have

$$- M_0 \leq \sum_{t=0}^{T-1} u_t(x_t, x_{t+1}) - \sum_{t=0}^{T-1} u_t(y_t, y_{t+1}) = \sum_{t=S}^{T-1} u_t(x_t, x_{t+1}) - \sum_{t=S}^{T-1} u_t(y_t, y_{t+1}). \qquad (2.327)$$

By (2.321) there is an integer $p \geq 0$ such that

$$T - S \in [pL_1, (p + 1)L_1). \qquad (2.328)$$

In view of (2.313), (2.321), and (2.328), we have

$$p \geq k. \tag{2.329}$$

It follows from (2.323), (2.328) and property (P24) that for each integer $i \in [0, p-1]$ we have

$$\sum_{t=S+iL_1}^{S+(i+1)L_1-1} v(x_t, x_{t+1}) \leq L_1 v(\bar{x}, \bar{x}) - M_1 - 2.$$

By this inequality, (2.315), and (2.316), for each integer $i \in [0, p-1]$,

$$\sum_{t=S+iL_1}^{S+(i+1)L_1-1} u_t(x_t, x_{t+1}) \leq \delta L_1 + L_1 v(\bar{x}, \bar{x}) - M_1 - 2$$

$$\leq 2\delta L_1 + \sum_{t=S+iL_1}^{S+(i+1)L_1-1} u_t(\bar{x}, \bar{x}) - M_1 - 2$$

$$\leq \sum_{t=S+iL_1}^{S+(i+1)L_1-1} u_t(\bar{x}, \bar{x}) - M_1 - 1. \tag{2.330}$$

In view of (2.312), (2.313), (2.316), (2.326), (2.328), (2.329), and (2.330), we have

$$\sum_{t=S}^{T-1} u_t(x_t, x_{t+1}) - \sum_{t=S}^{T-1} u_t(y_t, y_{t+1})$$

$$\leq \sum_{t=S}^{S+pL_1-1} u_t(\bar{x}, \bar{x}) - (M_1 + 1)p$$

$$+ \sum \{\|u_t\| : t \text{ is an integer and } S + pL_1 \leq t \leq T - 1\}$$

$$- \sum_{t=S+L_0}^{T-L_0-1} u_t(\bar{x}, \bar{x}) + 2L_0(\|v\| + 1)$$

$$\leq \sum_{t=S}^{S+pL_1-1} u_t(\bar{x}, \bar{x}) - p(M_1 + 1) + L_1(\|v\| + 1) + 2L_0(\|v\| + 1)$$

$$- \sum_{t=S}^{T-1} u_t(\bar{x}, \bar{x}) + 2L_0(\|v\| + 1)$$

$$\leq -p(M_1 + 1) + 6L_1(\|v\| + 1) \leq -k(M_1 + 1) + 6L_1(\|v\| + 1) \leq -M_0 - 4.$$

This contradicts (2.327). The contradiction we have reached proves that (2.323) does not hold. Thus (2.324) holds.

We may assume without loss of generality that there is an integer $\tilde{S}$ such that

$$S + L_2 < \tilde{S} \le T, \ \rho(x_{\tilde{s}}, \bar{x}) \le \delta_0, \tag{2.331}$$

$$\rho(x_t, \bar{x}) > \delta_0 \text{ for all integers } t \text{ satisfying } S < t < \tilde{S}. \tag{2.332}$$

In view of (2.321) and (2.331) there exists a program $\{y_t\}_{t=0}^T$ such that

$$y_t = x_t, \ t = 0, \dots, S, \ y_t = \bar{x} \text{ for all integers } t \in [S + L_0, \tilde{S} - 1],$$

$$y_t = x_t \text{ for all integers satisfying } \tilde{S} \le t \le T. \tag{2.333}$$

Relations (2.333), (2.320), (2.319) imply that

$$- M_0 \le \sum_{t=0}^{T-1} u_t(x_t, x_{t+1}) - \sum_{t=0}^{T-1} u_t(y_t, y_{t+1}) = \sum_{t=S}^{T-1} u_t(x_t, x_{t+1}) - \sum_{t=S}^{T-1} u_t(y_t, y_{t+1}). \tag{2.334}$$

In view of relation (2.331) there exists a nonnegative integer p for which

$$\tilde{S} - S - 1 \in [pL_1, (p+1)L_1). \tag{2.335}$$

In view of (2.313), (2.331), and (2.335), we have

$$p \ge k.$$

It follows from (2.335), (2.332) and property (P24) that for each integer $i \in [0, p-1]$,

$$\sum_{t=S+iL_1}^{S+(i+1)L_1-1} v(x_t, x_{t+1}) \le L_1 v(\bar{x}, \bar{x}) - M_1 - 2.$$

Combined with (2.315) and (2.316) this inequality implies that for each integer $i \in [0, p-1]$ we have

$$\sum_{t=S+iL_1}^{S+(i+1)L_1-1} u_t(x_t, x_{t+1}) \le \delta L_1 + L_1 v(\bar{x}, \bar{x}) - M_1 - 2$$

$$\le 2\delta L_1 + \sum_{t=S+iL_1}^{S+(i+1)L_1-1} u_t(\bar{x}, \bar{x}) - M_1 - 2$$

$$\le \sum_{t=S+iL_1}^{S+(i+1)L_1-1} u_t(\bar{x}, \bar{x}) - M_1 - 1. \tag{2.336}$$

It follows from relations (2.333), (2.331), (2.316), (2.336), (2.335) and the inequality $p \geq k$ that

$$\sum_{t=S}^{T-1} u_t(x_t, x_{t+1}) - \sum_{t=S}^{T-1} u_t(y_t, y_{t+1})$$

$$= \sum_{t=S}^{\tilde{S}-1} u_t(x_t, x_{t+1}) - \sum_{t=S}^{\tilde{S}-1} u_t(y_t, y_{t+1})$$

$$\leq \sum_{t=S}^{S+pL_1-1} u_t(x_t, x_{t+1}) + 2L_1(\|v\| + 1) - \sum_{t=S}^{\tilde{S}-1} u_t(\bar{x}, \bar{x}) + 4L_0(\|v\| + 1)$$

$$\leq -p(M_1 + 1) + 8L_1(\|v\| + 1) < -k + 8L_1(\|v\| + 1) < -M_0 - 4.$$

This contradicts (2.334).

The contradiction we have reached proves that there is an integer $t \in [S + 1, S + L_2]$ for which $\rho(x_t, \bar{x}) \leq \delta_0$.

Thus we have shown that the following property holds:

(P25) for each integer S satisfying $S \in [0, T - L_2]$ and $x_S \in \bar{Y}_{L_0}$, there exists an integer $t \in [S + 1, S + L_2]$ such that $\rho(x_t, \bar{x}) \leq \delta_0$.

Using (2.317) and property (P25) by induction we construct an increasing sequence of integers $\{S_i\}_{i=1}^q$ such that

$$S_1 \in [0, L_2], \ S_q \in (T - L_2, T], \ S_{i+1} - S_i \in [1, L_2]. \ i = 1, \ldots, q - 1,$$

$$\rho(x_{S_i}, \bar{x}) \leq \delta_0, \ i = 1, \ldots, q. \tag{2.337}$$

Evidently, we may assume that if $\rho(x_0, \bar{x}) \leq \delta_0$, then $S_1 = 0$ and if $\rho(x_T, \bar{x}) \leq \delta_0$, then $S_q = T$.

Let an integer $i \in \{0, \ldots, q - 1\}$ be given. Relations (2.318), (2.319), (2.337), and (2.314) imply that

$$\sum_{t=S_i}^{S_{i+1}-1} u_t(x_t, x_{t+1}) \geq \sigma\left(\{u_t\}_{t=S_i}^{S_{i+1}-1}, S_i, S_{i+1}, x_{S_i}, x_{S_{i+1}}\right) - \delta.$$

Combined with (2.315), (2.316), and (2.337) this inequality implies that

$$\sum_{t=S_i}^{S_{i+1}-1} v(x_t, x_{t+1}) \geq \sigma(v, S_i, S_{i+1}, x_{S_i}, x_{S_{i+1}}) - \delta_0.$$

It follows from the equation above, (2.337), and property (P23) that

$$\rho(x_t, \bar{x}) \leq \epsilon, \ t = S_i, \ldots, S_{i+1}, \ i = 0, \ldots, q - 1.$$

Theorems 2.39 and 2.40 are proved.

2.21　Stability for a Class of Optimal Control Systems

Let (X, ρ) be a compact metric space and Ω be a nonempty closed subset of $X \times X$. In this section which is based on [66] we use the notation and definitions of Sects. 2.1, 2.13, and 2.18. Recall that we denote by $\mathcal{M}$ the set of all bounded functions $u : \Omega \to R^1$ and that for each function $w \in \mathcal{M}$,

$$\|w\| = \sup\{|w(x, y)| : (x, y) \in \Omega\}. \tag{2.338}$$

Assume that $v \in \mathcal{M}$ is an upper semicontinuous function and set $v(x, y) = -\|v\| - 1$ for all $(x, y) \in (X \times X) \setminus \Omega$.

We suppose that there exist a point $\bar{x} \in X$ and a positive constant $\bar{c}$ such that assumptions (A1)–(A3) introduced in Sect. 2.1 hold.

Assumption (A1) implies that there exists a positive number $\bar{r} < 1$ such that

$$\{(x, y) \in X \times X : \rho(x, \bar{x}), \rho(y, \bar{x}) \leq \bar{r}\} \subset \Omega \tag{2.339}$$

and

$$|v(x, y) - v(\bar{x}, \bar{x})| \leq 1/8 \text{ for all } x, y \in X \text{ satisfying } \rho(x, \bar{x}), \rho(y, \bar{x}) \leq \bar{r}. \tag{2.340}$$

It is clear that for each pair of integers T_1, T_2 satisfying $0 \leq T_1 < T_2$, each sequence of functions $\{w_t\}_{t=T_1}^{T_2-1} \subset \mathcal{M}$, and each pair of points $x, y \in X$ satisfying $\rho(x, \bar{x}), \rho(y, \bar{x}) \leq \bar{r}$, the value $\sigma(\{w_t\}_{t=T_1}^{T_2-1}, T_1, T_2, x, y)$ is finite.

In this section we suppose that the following assumption holds.

(A4)　There exist integers $\tilde{L}_1 \geq 1$, $\tilde{L}_2 \geq 1$, and $\tilde{L}_3 \geq 1$ such that if a program $\{x_t\}_{t=0}^{\tilde{L}_2}$ satisfies $x_0 \notin \bar{Y}_{\tilde{L}_1}$, then $x_{\tilde{L}_2} \in Y_{\tilde{L}_3}$.

Note that many control systems satisfy this assumption (see Sect. 2.24).

The following four theorems which describe the structure of approximate solutions of the discrete-time control system were obtained in [66]. In all these theorems we suppose that assumptions (A1)–(A4) hold and that integers $\tilde{L}_1$, $\tilde{L}_2$, and $\tilde{L}_3$ are as guaranteed by assumption (A4).

Theorem 2.41. *Let $\epsilon \in (0, 1)$ and $L_0 \geq 1$ be an integer. Then there exist an integer $L \geq 1$ and a positive number $\delta < \min\{\epsilon, \bar{r}\}$ such that for each integer $T > 2L$ and each program $\{x_t\}_{t=0}^{T}$ which satisfies the inclusions*

$$x_0 \in \bar{Y}_{L_0}, \quad x_T \in Y_{L_0}$$

and the inequality

$$\sum_{t=\tau}^{\tau+L-1} v(x_t, x_{t+1}) \geq \sigma(v, L, x_\tau, x_{\tau+L}) - \delta \tag{2.341}$$

for each integer $\tau \in [0, T - L]$ *there exist integers* $\tau_1 \in [0, L]$ *and* $\tau_2 \in [T - L, T]$ *such that*

$$\rho(x_t, \bar{x}) \leq \epsilon, \ t = \tau_1, \ldots, \tau_2.$$

Moreover, if $\rho(x_0, \bar{x}) \leq \delta$, *then* $\tau_1 = 0$, *and if* $\rho(x_T, \bar{x}) \leq \delta$, *then* $\tau_2 = T$.

Theorem 2.42. *Let* $\epsilon \in (0, 1)$ *and* $L_0 \geq 1$ *be an integer. Then there exist an integer* $L \geq 1$ *and* $\delta \in (0, \min\{\epsilon, \bar{r}\})$ *such that for each integer* $T > 2L$ *and each program* $\{x_t\}_{t=0}^{T}$ *which satisfies the inclusion*

$$x_0 \in \bar{Y}_{L_0},$$

(2.341) for each integer $\tau \in [0, T - L]$, *and the inequality*

$$\sum_{t=T-L}^{T-1} v(x_t, x_{t+1}) \geq \sigma(v, L, x_{T-L}) - \delta$$

there exist integers $\tau_1 \in [0, L]$, $\tau_2 \in [T - L, T]$ *such that*

$$\rho(x_t, \bar{x}) \leq \epsilon, \ t = \tau_1, \ldots, \tau_2.$$

Moreover, if $\rho(x_0, \bar{x}) \leq \delta$, *then* $\tau_1 = 0$.

It is not difficult to see that Theorem 2.41 implies the following result.

Theorem 2.43. *Let* $\epsilon \in (0, 1)$, *let* L_0 *be a natural number, let a natural number* L *and* $\delta \in (0, \min\{\epsilon, \bar{r}\})$ *be as guaranteed by Theorem 2.41, and let*

$$\delta_1 = \delta/2, \ \delta_2 = \delta(4L)^{-1}.$$

Then for each integer $T > 2L$, *each* $\{u_t\}_{t=0}^{T-1} \subset \mathcal{M}$ *satisfying*

$$\|u_t - v\| \leq \delta_2, \ t = 0, \ldots T - 1,$$

and each program $\{x_t\}_{t=0}^{T}$ *which satisfies*

$$x_0 \in \bar{Y}_{L_0}, \ x_T \in Y_{L_0}$$

and

$$\sum_{t=\tau}^{\tau+L-1} u_t(x_t, x_{t+1}) \geq \sigma\left(\{u_t\}_{t=\tau}^{\tau+L-1}, \tau, \tau + L, x_\tau, x_{\tau+L}\right) - \delta_1 \qquad (2.342)$$

for each integer $\tau \in [0, T - L]$ there exist integers $\tau_1 \in [0, L]$ and $\tau_2 \in [T - L, T]$
such that

$$\rho(x_t, \bar{x}) \le \epsilon, \ t = \tau_1, \ldots, \tau_2.$$

Moreover, if $\rho(x_0, \bar{x}) \le \delta$, then $\tau_1 = 0$, and if $\rho(x_T, \bar{x}) \le \delta$, then $\tau_2 = T$.

It is not difficult to see that Theorem 2.42 implies the following result.

Theorem 2.44. *Let $\epsilon \in (0, 1)$, $L_0 \ge 1$ be an integer, let an integer $L \ge 1$ and*
$\delta \in (0, \min\{\epsilon, \bar{r}\})$ be as guaranteed by Theorem 2.42, and let

$$\delta_1 = \delta/2, \ \delta_2 = \delta(4L)^{-1}.$$

Then for each integer $T > 2L$, each sequence of functions $\{u_t\}_{t=0}^{T-1} \subset \mathcal{M}$ satisfying

$$\|u_t - v\| \le \delta_2, \ t = 0, \ldots, T - 1,$$

and each program $\{x_t\}_{t=0}^{T}$ which satisfies the inclusion

$$x_0 \in \bar{Y}_{L_0},$$

(2.342) for each integer $\tau \in [0, T - L]$, and the inequality

$$\sum_{t=T-L}^{T-1} u_t(x_t, x_{t+1}) \ge \sigma\big(\{u_t\}_{t=T-L}^{T-1}, T - L, T, x_{T-L}\big) - \delta_1, \qquad (2.343)$$

there exist integers $\tau_1 \in [0, L]$ and $\tau_2 \in [T - L, T]$ such that

$$\rho(x_t, \bar{x}) \le \epsilon, \ t = \tau_1, \ldots, \tau_2.$$

Moreover, if $\rho(x_0, \bar{x}) \le \delta$, then $\tau_1 = 0$.

Theorems 2.41–2.44 establish the stability of the turnpike phenomenon for approximate solutions of the optimal control problems with objective functions u_t, $t = 0, \ldots, T - 1$ which belong to a small neighborhood of v. Note that in Sects. 2.13–2.17 the stability of the turnpike phenomenon was established for programs which are approximately optimal on the whole interval $[0, T]$ while in this section we show that the turnpike phenomenon is stable for those programs which are approximately optimal on all subintervals of $[0, T]$ that have a fixed length L which does not depend on T.

In Sect. 2.24 we present optimal control systems which satisfy (A1)–(A4).

2.22   Proof of Theorem 2.41

We may assume without loss of generality that

$$\epsilon < \bar{r}, \ L_0 > \tilde{L}_1 + \tilde{L}_2 + \tilde{L}_3 + 4. \tag{2.344}$$

Lemma 2.15 implies that there exists a number

$$\delta \in (0, \epsilon) \tag{2.345}$$

such that the following property holds:

(P26)   for each natural number T and each program $\{x_t\}_{t=0}^{T}$ satisfying

$$\rho(x_0, \bar{x}), \ \rho(x_T, \bar{x}) \le \delta, \ \sum_{t=0}^{T-1} v(x_t, x_{t+1}) \ge \sigma(v, T, x_0, x_T) - 2\delta$$

the inequality $\rho(x_t, \bar{x}) \le \epsilon$ holds for all $t = 0, \dots, T$.

Choose a positive number

$$M_0 > 4 + 4(\|v\| + 1)(L_0 + \tilde{L}_2 + \tilde{L}_3 + 2\bar{c} + 4). \tag{2.346}$$

In view of Lemma 2.16 there exists an integer $L_1 > 2L_0 + 2$ such that the following property holds:

(P27)   for each integer $T \ge L_1$, each program $\{x_t\}_{t=0}^{T}$ which satisfies

$$\sum_{t=0}^{T-1} v(x_t, x_{t+1}) \ge T v(\bar{x}, \bar{x}) - M_0 - 2,$$

and each integer $S \in [0, T - L_1]$, we have

$$\min\{\rho(x_t, \bar{x}) : \ t = S + 1, \dots, S + L_1\} \le \delta.$$

Let integers

$$L_2 > 2L_1 + \tilde{L}_2 + 4, \ L > 4(1 + L_1 + L_2 + \tilde{L}_2 + \tilde{L}_3). \tag{2.347}$$

Assume that an integer $T > 2L$ and a program $\{x_t\}_{t=0}^{T}$ satisfies the inclusions

$$x_0 \in \bar{Y}_{L_0}, \ x_T \in Y_{L_0} \tag{2.348}$$

and the inequality

$$\sum_{t=\tau}^{\tau+L-1} v(x_t, x_{t+1}) \geq \sigma(v, L, x_\tau, x_{\tau+L}) - \delta \qquad (2.349)$$

for each integer $\tau \in [0, T - L]$.

Let an integer

$$\tau \in [L_2, T - L_2] \qquad (2.350)$$

be given. We claim that

$$\sum_{t=\tau}^{\tau+L_1-1} v(x_t, x_{t+1}) \geq L_1 v(\bar{x}, \bar{x}) - M_0 - 2. \qquad (2.351)$$

In view of relations (2.348) and (2.350) there exist integers

$$S_1 \in [0, \tau], \ S_2 \in [\tau + L_1, T] \qquad (2.352)$$

such that

$$x_{S_1} \in \bar{Y}_{L_0}, \ x_t \notin \bar{Y}_{L_0} \ \text{for each integer } t \text{ satisfying } S_1 < t \leq \tau, \qquad (2.353)$$

$$x_{S_2} \in Y_{L_0}, \ x_t \notin Y_{L_0} \ \text{for each integer } t \text{ satisfying } \tau + L_1 \leq t < S_2. \qquad (2.354)$$

We claim that

$$S_1 \geq \tau - \tilde{L}_2 - \tilde{L}_3 - L_1. \qquad (2.355)$$

Assume the contrary. Then

$$S_1 < \tau - \tilde{L}_2 - \tilde{L}_3 - L_1. \qquad (2.356)$$

It follows from relations (2.344), (2.353), and (2.356) that

$$x_{S_1+L_1+\tilde{L}_3} \notin \bar{Y}_{L_0} \supset \bar{Y}_{\tilde{L}_1}. \qquad (2.357)$$

In view of (2.353) there exists a program $\{y_t^{(1)}\}_{t=S_1}^{S_1+L_0}$ such that

$$y_{S_1}^{(1)} = x_{S_1}, \ y_{S_1+L_0}^{(1)} = \bar{x}. \qquad (2.358)$$

It follows from assumption (A4), the choice of $\tilde{L}_1, \tilde{L}_2, \tilde{L}_3$, (2.356), and (2.357) that there exists a program

$$\{y_t^{(2)}\}_{t=S_1+L_1+\tilde{L}_2}^{S_1+L_1+\tilde{L}_2+\tilde{L}_3}$$

such that

$$y^{(2)}_{S_1+L_1+\tilde{L}_2+\tilde{L}_3} = x_{S_1+L_1+\tilde{L}_2+\tilde{L}_3}, \ y^{(2)}_{S_1+L_1+\tilde{L}_2} = \bar{x}. \qquad (2.359)$$

Put

$$y_t = y^{(1)}_t, \ t = S_1, \ldots, S_1 + L_0,$$

$$y_t = y^{(2)}_t, \ t = S_1 + L_1 + \tilde{L}_2, \ldots, S_1 + L_1 + \tilde{L}_2 + \tilde{L}_3,$$

$$y_t = \bar{x}, \ t = S_1 + L_0 + 1, \ldots, S_1 + L_1 + \tilde{L}_2 - 1. \qquad (2.360)$$

It is clear that by (2.358)–(2.360), $\{y_t\}^{S_1+L_1+\tilde{L}_2+\tilde{L}_3}_{t=S_1}$ is a program. In view of (2.347), (2.349), and (2.358)–(2.360), we have

$$-\delta + \sum_{t=S_1}^{S_1+L_1+\tilde{L}_2+\tilde{L}_3-1} v(y_t, y_{t+1}) \le \sum_{t=S_1}^{S_1+L_1+\tilde{L}_2+\tilde{L}_3-1} v(x_t, x_{t+1}). \qquad (2.361)$$

By (2.358)–(2.360),

$$\sum_{t=S_1}^{S_1+L_1+\tilde{L}_2+\tilde{L}_3-1} v(y_t, y_{t+1}) \ge -L_0\|v\| - \tilde{L}_3\|v\| + (L_1 + \tilde{L}_2 - L_0 - 2)v(\bar{x}, \bar{x}).$$

$$(2.362)$$

Relations (2.346), (2.361), and (2.362) imply that

$$\sum_{t=S_1}^{S_1+L_1+\tilde{L}_2+\tilde{L}_3-1} v(x_t, x_{t+1}) \ge (L_1 + \tilde{L}_2 - L_0 - 2)v(\bar{x}, \bar{x}) - 1 - \|v\|(L_0 + \tilde{L}_3)$$

$$\ge (L_1 + \tilde{L}_2 + \tilde{L}_3)v(\bar{x}, \bar{x}) - 1 - \|v\|(2L_0 + 2\tilde{L}_3 + 2\tilde{L}_2 + 2)$$

$$\ge (L_1 + \tilde{L}_2 + \tilde{L}_3)v(\bar{x}, \bar{x}) - M_0.$$

It follows from property (P27) and the equation above that there exists an integer

$$t_0 \in [S_1 + 1, S_1 + L_1] \qquad (2.363)$$

such that $\rho(x_{t_0}, \bar{x}) \le \delta$. This implies the inclusion $x_{t_0} \in \bar{Y}_{L_0}$. Combined with relations (2.356) and (2.363) this contradicts (2.353). The contradiction we have reached proves that (2.355) holds, as claimed.

We claim that

$$S_2 \le \tau + 2L_1 + \tilde{L}_2 + \tilde{L}_3. \qquad (2.364)$$

Assume the contrary. Then

$$S_2 > \tau + 2L_1 + \tilde{L}_2 + \tilde{L}_3. \qquad (2.365)$$

In view of (2.354), (2.365), and (2.344), we have

$$x_{S_2-L_1-\tilde{L}_3} \notin Y_{L_0} \supset Y_{\tilde{L}_3}. \tag{2.366}$$

It follows from relations (2.365), (2.366), assumption (A4), and the choice of $\tilde{L}_1, \tilde{L}_2, \tilde{L}_3$ that

$$x_{S_2-L_1-\tilde{L}_3-\tilde{L}_2} \in \bar{Y}_{\tilde{L}_1}. \tag{2.367}$$

In view of inclusion (2.367), there exists a program $\{z_t^{(1)}\}_{t=S_2-L_1-\tilde{L}_3-\tilde{L}_2}^{S_2-L_1-\tilde{L}_3-\tilde{L}_2+\tilde{L}_1}$ such that

$$z_{S_2-L_1-\tilde{L}_3-\tilde{L}_2}^{(1)} = x_{S_2-L_1-\tilde{L}_3-\tilde{L}_2}, \; z_{S_2-L_1-\tilde{L}_3-\tilde{L}_2+\tilde{L}_1}^{(1)} = \bar{x}. \tag{2.368}$$

It follows from relations (2.345) and (2.354) that there exists a program $\{z_t^{(2)}\}_{t=S_2-L_0}^{S_2}$ such that

$$z_{S_2-L_0}^{(2)} = \bar{x}, \; z_{S_2}^{(2)} = x_{S_2}. \tag{2.369}$$

Define

$$z_t = z_t^{(1)}, \; t = S_2 - L_1 - \tilde{L}_3 - \tilde{L}_2, \ldots, S_2 - L_1 - \tilde{L}_3 - \tilde{L}_2 + \tilde{L}_1, \tag{2.370}$$

$$z_t = z_t^{(2)}, \; t = S_2 - L_0, \ldots, S_2,$$

$$z_t = \bar{x}, \; t = S_2 - L_1 - \tilde{L}_2 - \tilde{L}_3 + \tilde{L}_1 + 1, \ldots, S_2 - L_0 - 1.$$

It is clear that $\{z_t\}_{t=S_2-L_1-\tilde{L}_2-\tilde{L}_3}^{S_2}$ is a well-defined program by (2.344) and (2.368)–(2.370). In view of (2.346), (2.347), (2.349), and (2.368)–(2.370), we have

$$\sum_{t=S_2-L_1-\tilde{L}_2-\tilde{L}_3}^{S_2-1} v(x_t, x_{t+1}) \geq -\delta + \sum_{t=S_2-L_1-\tilde{L}_2-\tilde{L}_3}^{S_2-1} v(y_t, y_{t+1})$$

$$\geq -(L_0 + \tilde{L}_1 + 2)\|v\| + (L_1 + \tilde{L}_2 + \tilde{L}_3 - \tilde{L}_1 - L_0 - 2)v(\bar{x}, \bar{x})$$

$$\geq (L_1 + \tilde{L}_2 + \tilde{L}_3)v(\bar{x}, \bar{x}) - \|v\|(2L_0 + 2\tilde{L}_1 + 4)$$

$$\geq (L_1 + \tilde{L}_2 + \tilde{L}_3)v(\bar{x}, \bar{x}) - M_0.$$

It follows from the equation above and property (P27) that there exists an integer t_2 such that

$$t_2 \in [S_2 - L_1 - \tilde{L}_2 + 1, S_2 - \tilde{L}_2], \; \rho(x_{t_2}, \bar{x}) \leq \delta. \tag{2.371}$$

This implies the inclusion $x_{t_2} \in Y_{L_0}$. Combined with (2.371) and (2.365) this inclusion contradicts (2.354). The contradiction we have reached proves that (2.364) holds, as claimed.

In view of relations (2.52)–(2.354), (2.355), and (2.364) there exists a program $\{\xi_t\}_{t=S_1}^{S_2}$ such that

$$\xi_{S_1} = x_{S_1}, \; \xi_{S_1+L_0} = \bar{x}, \; \xi_t = \bar{x}, \; t = S_1 + L_0, \dots, S_2 - L_0,$$

$$\xi_{S_2} = x_{S_2}, \; \xi_{S_2-L_0} = \bar{x}. \tag{2.372}$$

By relation (2.372), we have

$$\sum_{t=S_1}^{S_2-1} v(\xi_t, \xi_{t+1}) \geq (S_2 - S_1 - 2L_0)v(\bar{x}, \bar{x}) - 2L_0\|v\|$$

$$\geq (S_2 - S_1)v(\bar{x}, \bar{x}) - 4L_0\|v\|. \tag{2.373}$$

It follows from (2.355), (2.364), and (2.347) that

$$S_2 - S_1 \leq 4L_1 + \tilde{L}_2 + \tilde{L}_3 < L. \tag{2.374}$$

By (2.374), (2.372), (2.349), and (2.373), we have

$$\sum_{t=S_1}^{S_2-1} v(x_t, x_{t+1}) \geq \sum_{t=S_1}^{S_2-1} v(\xi_t, \xi_{t+1}) - \delta$$

$$\geq -(S_2 - S_1)v(\bar{x}, \bar{x}) - 4L_0\|v\| - 1. \tag{2.375}$$

Relations (2.352), (2.375), (A2), and (2.346) imply that

$$\sum_{t=\tau}^{\tau+L_1-1} v(x_t, x_{t+1}) \geq \sum_{t=S_1}^{S_2-1} v(x_t, x_{t+1})$$

$$- \sum\{v(x_t, x_{t+1}) : \; t \text{ is an integer such that } S_1 \leq t < \tau\}$$

$$- \sum\{v(x_t, x_{t+1}) : \; t \text{ is an integer such that } \tau + L_1 \leq t < S_2\}$$

$$\geq (S_2 - S_1)v(\bar{x}, \bar{x}) - 4L_0\|v\| - 4$$

$$-(S_2 - \tau - L_1)v(\bar{x}, \bar{x}) - \bar{c} - (\tau - S_1)v(\bar{x}, \bar{x}) - \bar{c}$$

$$\geq L_1 v(\bar{x}, \bar{x}) - 4L_0\|v\| - 4 - 2\bar{c} \leq L_1 v(\bar{x}, \bar{x}) - M_0.$$

Thus (2.351) holds, as claimed.

Therefore we have shown that the following property holds:
for each integer $\tau \in [L_2, T - L_2]$ we have

$$\sum_{t=\tau}^{\tau+L_1-1} v(x_t, x_{t+1}) \geq L_1 v(\bar{x}, \bar{x}) - M_0 - 2.$$

Together with (P27) this property implies that the following property holds:

(P28)   for each integer $\tau \in [L_2, T - L_2]$ there exists an integer $t \in [\tau + 1, \tau + L_1]$
such that $\rho(x_t, \bar{x}) \leq \delta$.

Applying property (P28) by induction we construct a sequence of integers $\tau_i \in$
$[0, T], i = 1, \ldots, q$ such that

$$\tau_1 \leq L_1 + L_2, \ \tau_q > T - L_2, \tag{2.376}$$

for each integer i satisfying $1 \leq i \leq q - 1$

$$\tau_{i+1} - \tau_i \in [1, 2L_1] \tag{2.377}$$

and that for all integers $i = 1, \ldots, q$ we have

$$\rho(x_{\tau_i}, \bar{x}) \leq \delta. \tag{2.378}$$

It follows from (2.347), (2.349), (2.376)–(2.378) and property (P26) that for all
integers $i = 1, \ldots, q - 1$, we have

$$\rho(x_t, \bar{x}) \leq \epsilon, \ t = \tau_i, \ldots, \tau_{i+1}$$

and

$$\rho(x_t, \bar{x}) \leq \epsilon, \ t = \tau_1, \ldots, \tau_q.$$

This implies that

$$\rho(x_t, \bar{x}) \leq \epsilon, \ t = L_1 + L_2, \ldots, T - L_2.$$

Evidently, if $\rho(x_0, \bar{x}) \leq \delta$, then in view of (2.347), (2.349), (2.376)–(2.378) and
property (P26),

$$\rho(x_t, \bar{x}) \leq \epsilon, \ t = 0, \ldots, \tau_1,$$

and if $\rho(x_T, \bar{x}) \leq \delta$, then

$$\rho(x_t, \bar{x}) \leq \epsilon, \ t = \tau_q, \ldots, T.$$

This completes the proof of Theorem 2.41.

2.23 Proof of Theorem 2.42

We may assume without loss of generality that

$$\epsilon < \bar{r}, \ L_0 > \tilde{L}_1 + \tilde{L}_3. \tag{2.379}$$

Let an integer $L \geq 1$ and a number $\delta \in (0, \epsilon)$ be as guaranteed by Theorem 2.41. Fix a number

$$M > 1 + (\|v\| + 1)(2L_0 + 2\tilde{L}_3). \tag{2.380}$$

In view of Lemma 2.16 there exists a natural number $L_1 > L_0 + 2$ such that the following property holds:

(P29)   for each integer $T \geq L_1$, each program $\{x_t\}_{t=0}^T$ which satisfies the inequality

$$\sum_{t=0}^{T-1} v(x_t, x_{t+1}) \geq Tv(\bar{x}, \bar{x}) - M,$$

and each integer $S \in [0, T - L_1]$, we have

$$\min\{\rho(x_t, \bar{x}) : t = S + 1, \ldots, S + L_1\} \leq \delta.$$

Put

$$Q = 1 + L_1 + \tilde{L}_1 + \tilde{L}_2 + \tilde{L}_3. \tag{2.381}$$

Choose a natural number

$$\tilde{L} > L + Q + L_1. \tag{2.382}$$

Assume that an integer $T > 2\tilde{L}$ and that a program $\{x_t\}_{t=0}^T$ satisfies the inclusion

$$x_0 \in \bar{Y}_{L_0}, \tag{2.383}$$

for each integer $\tau \in [0, T - \tilde{L}]$ we have

$$\sum_{t=\tau}^{\tau+\tilde{L}-1} v(x_t, x_{t+1}) \geq \sigma(v, \tilde{L}, x_\tau, x_{\tau+\tilde{L}}) - \delta \tag{2.384}$$

and that

$$\sum_{t=T-\tilde{L}}^{T-1} v(x_t, x_{t+1}) \geq \sigma(v, \tilde{L}, x_{T-\tilde{L}}) - \delta. \tag{2.385}$$

Assume that an integer $S \geq 0$ satisfies

$$S + Q \leq T, \; x_S \in \bar{Y}_{L_0}. \tag{2.386}$$

We claim that there is an integer $j \in \{S + 1, \ldots, S + Q\}$ such that $x_j \in \bar{Y}_{L_0}$.

Assume the contrary. Then

$$x_t \notin \bar{Y}_{L_0} \supset \bar{Y}_{\tilde{L}_1}, \ t = S + 1, \ldots, S + Q. \tag{2.387}$$

In view of (2.339) and (2.387), we have

$$\rho(x_t, \bar{x}) > \bar{r} > \epsilon > \delta, \ t = S + 1, \ldots, S + Q. \tag{2.388}$$

Property (P29) and relations (2.381), (2.386), and (2.388) imply that

$$\sum_{t=S}^{S+Q-1} v(x_t, x_{t+1}) < Qv(\bar{x}, \bar{x}) - M. \tag{2.389}$$

It follows from (2.381), (2.386), assumption (A4), the choice of $\tilde{L}_1$, $\tilde{L}_2$, $\tilde{L}_3$, and (2.387) that there exists a program $\{y_t\}_{t=S}^{S+Q}$ such that

$$y_S = x_S, \ y_{S+L_0} = \bar{x}, \ y_t = \bar{x}. \ t = S + L_0, \ldots, S + Q - \tilde{L}_3, \ y_{S+Q} = x_{S+Q}. \tag{2.390}$$

By (2.389), (2.390), (2.382), and (2.384), we have

$$Qv(\bar{x}, \bar{x}) - M \geq \sum_{t=S}^{S+Q-1} v(x_t, x_{t+1}) \geq \sum_{t=S}^{S+Q-1} v(y_t, y_{t+1}) - 1$$

$$\geq (Q - L_0 - \tilde{L}_3)v(\bar{x}, \bar{x}) - (L_0 + \tilde{L}_3)\|v\| - 1$$

$$\geq Qv(\bar{x}, \bar{x}) - \|v\|(2L_0 + 2\tilde{L}_3) - 1$$

and

$$\|v\|(2L_0 + 2\tilde{L}_3) + 1 > M.$$

This contradicts (2.380). The contradiction we have reached proves that there is an integer $j \in \{S + 1, \ldots, S + Q\}$ such that $x_j \in \bar{Y}_{L_0}$.

Thus we have shown that the following property holds:

(P30)   for each integer $S \in [0, T - Q]$ satisfying $x_S \in \bar{Y}_{L_0}$ there exists an integer $j \in \{S + 1, \ldots, S + Q\}$ such that $x_j \in \bar{Y}_{L_0}$.

In view of property (P30), (2.382), and (2.383), there exists an integer

$$S \in [T - L_1 - Q, T - L_1], \ x_S \in \bar{Y}_{L_0}. \tag{2.391}$$

It follows from (2.391) that there exists a program $\{z_t\}_{t=S}^{T}$ such that

$$z_S = x_S, \ z_t = \bar{x}, \ t = L_0 + S, \ldots, T. \tag{2.392}$$

Relations (2.392), (2.391), (2.382), (2.385), and (2.380) imply that

$$\sum_{t=S}^{T-1} v(x_t, x_{t+1}) \geq -1 + \sum_{t=S}^{T-1} v(y_t, y_{t+1}) \geq -1 + (T - L_0 - S)v(\bar{x}, \bar{x}) - L_0\|v\|$$

$$\geq (T - S)v(\bar{x}, \bar{x}) - 1 - 2L_0\|v\| \geq (T - S)v(\bar{x}, \bar{x}) - M.$$

In view of this relation, (2.391), and property (P29) there exists an integer j such that

$$j \in \{S + 1, \ldots, S + L_1\}, \quad \rho(x_j, \bar{x}) \leq \delta. \tag{2.393}$$

It is clear that $x_j \in Y_{L_0}$. In view of (2.393), (2.391), and (2.382), we have

$$j \geq T - L_1 - Q > 2\tilde{L} - L_1 - Q > 2L. \tag{2.394}$$

By relations (2.382), (2.384), (2.383), (2.393), (2.394) and the choice of L and δ we apply Theorem 2.41 to the program $\{x_t\}_{t=0}^{j}$ and obtain that there are integers $\tau_1 \in [0, L]$, $\tau_2 \in [j - L, j] \subset [T - \tilde{L}, T]$ such that $\rho(x_t, \bar{x}) \leq \epsilon$, $t = \tau_1, \ldots, \tau_2$ and if $\rho(x_0, \bar{x}) \leq \delta$, then $\tau_1 = 0$. This completes the proof of Theorem 2.42 with $L = \tilde{L}$.

2.24 An Example

In this section we present optimal control systems which satisfy assumptions (A1)–(A4).

Let the Euclidean space R^n be equipped with norm $|\cdot|$ induced by the scalar product $\langle \cdot, \cdot \rangle$. We say that $x \leq y$, where $x = (x_1, \ldots, x_n)$, $y = (y_1, \ldots, y_n) \in R^n$ if $x_i \leq y_i$, $i = 1, \ldots, n$, and $x << y$ if $x_i < y_i$ for all integers $i = 1, \ldots, n$. Let $e = (1, 1, \ldots, 1)$, $M > 0$, $X = \{y \in R^n : 0 \leq y \leq Me\}$. Ω be a nonempty closed subset of $X \times X$, $\bar{x} \in R^n$ satisfy $0 << \bar{x} \leq Me$, and there is $\bar{r} \in (0, 1)$ such that

$$\{(x, y) \in X \times X : |\bar{x} - x|, |y - \bar{x}| \leq \bar{r}\} \subset \Omega. \tag{2.395}$$

Assume that

if $(x_1, y_1) \in \Omega$, $x_2, y_2 \in X$, $x_2 \geq x_1$, $y_2 \leq y_1$, then

$$(x_2, y_2) \in \Omega, \tag{2.396}$$

$$a(x) = \{0\} \text{ for each } x = (x_1, \ldots, x_n) \text{ satisfying } \min\{x_i : i = 1, \ldots, n\} = 0. \tag{2.397}$$

Equation (2.397) implies the following result.

Lemma 2.45. *Let ϵ be a positive number. Then there exists a positive number δ such that for each $(x, y) \in \Omega$ satisfying $\min\{x_i : i = 1, \ldots, n\} \leq \delta$ the inequality $y \leq \epsilon e$ holds.*

Proposition 2.46. *Assume that $\pi : X \to R^1$ is a continuous function, α is a real number, and $L : X \times X \to [0, \infty)$ is a continuous function such that for each $(x, y) \in X \times X$, the equality $L(x, y) = 0$ holds if and only if $(x, y) = (\bar{x}, \bar{x})$. Set $v(x, y) = \alpha - L(x, y) + \pi(x) - \pi(y)$ for all $x, y \in X$. Assume that*

$$\text{for each } \gamma \in (0, M) \text{ there is a program } \{x_t\}_{t=0}^m$$

$$\text{such that } x_0 = \gamma e, \ x_m = \bar{x}. \tag{2.398}$$

Then (A1)–(A4) hold.

Proof. Assumptions (A1)–(A3) hold in view of Example 2.10. Assumption (A4) follows from Lemma 2.45 and relations (2.396) and (2.398). $\qquad\qquad\square$

Proposition 2.47. *Assume that $v : X \times X \to R^1$ is a strictly concave continuous function such that $v(\bar{x}, \bar{x}) = \sup\{v(z, z) : z \in X \text{ and } (z, z) \in \Omega\}$. Assume that (2.398) holds. Then (A1)–(A4) hold.*

Proof. Assumptions (A1)–(A3) hold in view of Example 2.11. Assumption (A4) follows from Lemma 2.45 and relations (2.396) and (2.398). $\qquad\qquad\square$

2.25   Optimal Control Systems with Discounting

Let (X, ρ) be a compact metric space and Ω be a nonempty closed subset of $X \times X$. Recall that $\mathcal{M}$ is the set of all bounded functions $u : \Omega \to R^1$ and that for each function $w \in \mathcal{M}$,

$$\|w\| = \sup\{|w(x, y)| : (x, y) \in \Omega\}. \tag{2.399}$$

In this section which is based on [71] we use the notation and definitions of Sects. 2.1, 2.13, and 2.18 [see (2.209)–(2.211), (2.289)–(2.291)].

Assume that $v \in \mathcal{M}$ is an upper semicontinuous function and set $v(x, y) = -\|v\| - 1$ for all $(x, y) \in (X \times X) \setminus \Omega$.

We suppose that there exist a point $\bar{x} \in X$ and a positive constant $\bar{c}$ such that assumptions (A1)–(A3) introduced in Sect. 2.1 hold.

In this section we consider the maximization problems with discounting

$$\sum_{t=0}^{T-1} \alpha_t u_t(x_t, x_{t+1}) \to \max\{x_t\}_{t=0}^T \text{ is program such that } x_0 = z,$$

where T is a natural number, $z \in X$, $\{\alpha_t\}_{t=0}^{T-1} \subset (0, 1]$ is a sequence of discount coefficients, and u_t, $t = 0, \ldots, T-1$ is a sequence of objective functions belonging to a neighborhood of v.

Assumption (A1) implies that there exists a real number $\bar{r} \in (0, 1)$ such that

$$\{(x, y) \in X \times X : \rho(x, \bar{x}), \rho(y, \bar{x}) \leq \bar{r}\} \subset \Omega \tag{2.400}$$

and

$$|v(x, y) - v(\bar{x}, \bar{x})| \leq 1/8 \text{ for all } x, y \in X \text{ satisfying } \rho(x, \bar{x}), \rho(y, \bar{x}) \leq \bar{r}. \tag{2.401}$$

It is clear that for each pair of integers T_1, T_2 satisfying $0 \leq T_1 < T_2$, each sequence of functions $\{w_t\}_{t=T_1}^{T_2-1} \subset \mathcal{M}$, and each pair of points $x, y \in X$ satisfying $\rho(x, \bar{x}), \rho(y, \bar{x}) \leq \bar{r}$, the value $\sigma(\{w_t\}_{t=T_1}^{T_2-1}, T_1, T_2, x, y)$ is finite.

Let $T \geq 1$ be an integer. Recall that we denote by Y_T the set of all points $x \in X$ for which there exists a program $\{x_t\}_{t=0}^{T}$ such that $x_0 = \bar{x}$ and $x_T = x$ and denote by $\bar{Y}_T$ the set of all points $x \in X$ for which there exists a program $\{x_t\}_{t=0}^{T}$ such that $x_0 = x$ and $x_T = \bar{x}$.

Denote by $\mathcal{M}_0$ the set of all upper semicontinuous functions $u \in \mathcal{M}$.

Since all the functions belonging to $\mathcal{M}_0$ are upper semicontinuous and bounded it is easy to see that the following proposition holds.

Proposition 2.48. *Let integers T_1, T_2 satisfy $0 \leq T_1 < T_2$, $\{u_t\}_{t=T_1}^{T_2-1} \subset \mathcal{M}_0$, and $x \in \cup\{X_M : M \in (0, \infty)\}$. Then there exists a program $\{x_t\}_{t=T_1}^{T_2}$ such that $x_0 = x$ and*

$$\sum_{t=T_1}^{T_2-1} u_t(x_t, x_{t+1}) = \sigma(\{u_t\}_{t=T_1}^{T_2-1}, T_1, T_2, x).$$

The results of this section were obtained in [71].

Theorem 2.49. *Let $\epsilon \in (0, \bar{r})$ and $L_0 \geq 1$ be an integer. Then there exist an integer $L > L_0$ and real numbers $\delta \in (0, \epsilon)$ and $\lambda \in (0, 1)$ such that for each integer $T > 2L$, each sequence of functions $\{u_t\}_{t=0}^{T-1} \subset \mathcal{M}_0$ satisfying*

$$\|u_t - v\| \leq \delta, \ t = 0, \ldots, T - 1, \tag{2.402}$$

each sequence $\{\alpha_t\}_{t=0}^{T-1} \subset (0, 1]$ such that for each integer $\tau \in [0, T - L]$,

$$\alpha_i \alpha_j^{-1} \geq \lambda \text{ for each } i, j \in \{\tau, \ldots, \tau + L\}, \tag{2.403}$$

and each program $\{x_t\}_{t=0}^{T}$ which satisfies the inclusion

$$x_0 \in \bar{Y}_{L_0}$$

and the inequality

$$\sum_{t=0}^{T-1} \alpha_t u_t(x_t, x_{t+1}) = \sigma(\{\alpha_t u_t\}_{t=0}^{T-1}, 0, T, x_0)$$

the following inequality holds:

$$\rho(x_t, \bar{x}) \leq \epsilon, \ t = L, \ldots, T - L.$$

Moreover, if $\rho(x_0, \bar{x}) \leq \delta$, then

$$\rho(x_t, \bar{x}) \leq \epsilon, \ t = 0, \ldots, T - L.$$

Theorem 2.49 establishes the turnpike property in the case of discounting. Roughly speaking, the turnpike property holds if discount coefficients $\{\alpha_t\}_{t=0}^{T-1} \subset (0, 1]$ are changed rather slowly.

Let $\{w_t\}_{t=0}^{\infty} \subset \mathcal{M}_0$ be given. A program $\{x_t\}_{t=0}^{\infty}$ is called $(\{w_t\}_{t=0}^{\infty})$-overtaking optimal if for each program $\{y_t\}_{t=0}^{\infty}$ satisfying $x_0 = y_0$, we have

$$\limsup_{T \to \infty} \left[\sum_{t=0}^{T-1} w_t(y_t, y_{t+1}) - \sum_{t=0}^{T-1} w_t(x_t, x_{t+1}) \right] \leq 0.$$

The following result establishes the turnpike property for overtaking optimal programs.

Theorem 2.50. *Let $\epsilon \in (0, \bar{r})$ and $L_0 \geq 1$ be an integer. Then there exist an integer $L > L_0$ and real numbers $\delta \in (0, \epsilon)$ and $\lambda \in (0, 1)$ such that for each sequence of functions $\{u_t\}_{t=0}^{\infty} \subset \mathcal{M}_0$ satisfying*

$$\|u_t - v\| \leq \delta, \ t = 0, 1, \ldots,$$

each $\{\alpha_t\}_{t=0}^{\infty} \subset (0, 1]$ such that

$$\alpha_i \alpha_j^{-1} \geq \lambda \text{ for each } i, j \in \{0, 1, \ldots,\} \text{ satisfying } |i - j| \leq L,$$

and each $(\{\alpha_t u_t\}_{t=0}^{\infty})$-overtaking optimal program $\{x_t\}_{t=0}^{\infty}$ which satisfies the inclusion

$$x_0 \in \bar{Y}_{L_0},$$

the following inequality holds:

$$\rho(x_t, \bar{x}) \leq \epsilon \text{ for all integers } t \geq L.$$

Moreover, if $\rho(x_0, \bar{x}) \le \delta$, then

$$\rho(x_t, \bar{x}) \le \epsilon \text{ for all integers } t \ge 0.$$

Let $L_0 \ge 1$ be an integer, $\epsilon = \bar{r}/4$, and let $\delta \in (0, \bar{r}/4)$, an integer $L > L_0$, and $\lambda \in (0, 1)$ be as guaranteed by Theorem 2.49.

Let

$$u_t \in \mathcal{M}_0 \text{ and } \|u_t - v\| \le \delta, \ t = 0, 1, \ldots, \qquad (2.404)$$

and let $\{\alpha_t\}_{t=0}^{\infty} \subset (0, 1]$ satisfy the relations

$$\lim_{t \to \infty} \alpha_t = 0,$$

$$\alpha_i \alpha_j^{-1} \ge \lambda \text{ for all nonnegative integers } i, j \text{ satisfying } |i - j| \le L. \qquad (2.405)$$

Theorem 2.51. *For each $z \in \bar{Y}_{L_0}$ there is a program $\{x_t^{(z)}\}_{t=0}^{\infty} \subset X$ such that $x_0^{(z)} = z$ and that the following property holds:*

For each real number $\gamma > 0$ there exists an integer $n_0 \ge 1$ such that for each integer $T \ge n_0$ and each point $z \in \bar{Y}_{L_0}$, the inequality

$$\left| \sigma\left(\{\alpha_t u_t\}_{t=0}^{T-1}, 0, T, z\right) - \sum_{t=0}^{T-1} \alpha_t u_t \left(x_t^{(z)}, x_{t+1}^{(z)}\right) \right| \le \gamma$$

holds.

It is clear that Theorem 2.51 establishes the existence of $(\{\alpha_t u_t\}_{t=0}^{\infty})$-overtaking optimal program when (2.404) and (2.405) hold. Roughly speaking, an $(\{\alpha_t u_t\}_{t=0}^{\infty})$-overtaking optimal program exists if the objective functions $u_t, t = 0, 1, \ldots$ belong to the δ-neighborhood of v in the topology of the uniform convergence on the set Ω and the sequence of the discount coefficients $\{\alpha_t\}_{t=0}^{\infty}$ tends to zero slowly.

Note that the existence of an $(\{\alpha_t u_t\}_{t=0}^{\infty})$-overtaking optimal program when the discount coefficients $\{\alpha_t\}_{t=0}^{\infty}$ tend to zero rapidly is a well-known fact. Here we present a version of this result.

Theorem 2.52. *Let $\{w_t\}_{t=0}^{\infty} \subset \mathcal{M}_0$ satisfy*

$$\sum_{t=0}^{\infty} \|w_t\| < \infty. \qquad (2.406)$$

Then for each point $x_0 \in \cup_{L=1}^{\infty} \bar{Y}_L$ there exists a $(\{w_t\}_{t=0}^{\infty})$-overtaking optimal program $\{x_t\}_{t=0}^{\infty}$.

Proof. Let $x_0 \in \cup_{L=1}^{\infty} \bar{Y}_L$ be given. For each program $\{y_t\}_{t=0}^{\infty}$,

$$\sum_{t=0}^{\infty} |w_t(y_t, y_{t+1})| < \infty.$$

Set

$$\Delta = \sup \left\{ \sum_{t=0}^{\infty} w_t(y_t, y_{t+1}) : \{y_t\}_{t=0}^{\infty} \text{ is a program and } y_0 = x_0 \right\}. \qquad (2.407)$$

Evidently, the value Δ is finite. In order to prove Theorem 2.52 it is sufficient to show that there exists a program $\{x_i\}_{i=0}^{\infty}$ such that

$$\sum_{i=0}^{\infty} w_t(x_t, x_{t+1}) = \Delta \text{ and } x_0 = x.$$

For each natural number k there exists a program $\{x_t^k\}_{t=0}^{\infty}$ such that

$$x_0^k = x_0,$$

$$\sum_{t=0}^{\infty} w_t(x_t^k, x_{t+1}^k) \geq \Delta - 1/k. \qquad (2.408)$$

Extracting a subsequence and re-indexing if necessary we may assume without loss of generality that for any nonnegative integer t there exists a limit

$$x_t = \lim_{k\to\infty} x_t^k. \qquad (2.409)$$

Let $\epsilon > 0$ be given. In view of (2.406) there exists an integer $T_0 \geq 1$ such that

$$\sum_{t=T_0}^{\infty} \|w_t\| < \epsilon. \qquad (2.410)$$

Relations (2.408), (2.409), and (2.410) imply that for all integers $T \geq T_0$, we have

$$\sum_{t=0}^{T-1} w_t(x_t, x_{t+1}) \geq \limsup_{k\to\infty} \sum_{t=0}^{T-1} w_t(x_t^k, x_{t+1}^k)$$

$$\geq \limsup_{k\to\infty} \left(\sum_{t=0}^{\infty} w_t(x_t^k, x_{t+1}^k) - \epsilon \right) = \Delta - \epsilon$$

and

$$\sum_{t=0}^{\infty} w_t(x_t, x_{t+1}) \geq \Delta - \epsilon.$$

Since ϵ is any positive number we conclude that $\sum_{t=0}^{\infty} w_t(x_t, x_{t+1}) = \Delta$. Theorem 2.52 is proved. $\qquad\square$

2.26   Proof of Theorems 2.49 and 2.50

We prove Theorems 2.49 and 2.50 simultaneously.

Fix a real number $M > 4$. In view of Lemma 2.15 there exists a real number $\delta_0 \in (0, \epsilon)$ such that the following property holds:

(P31)   for each natural number T and each program $\{x_t\}_{t=0}^T$ satisfying

$$\rho(x_0, \bar{x}), \ \rho(x_T, \bar{x}) \le \delta_0, \ \sum_{t=0}^{T-1} v(x_t, x_{t+1}) \ge \sigma(v, T, x_0, x_T) - 2\delta_0$$

the inequality $\rho(x_t, \bar{x}) \le \epsilon$ holds for all $t = 0, \ldots, T$.

Lemma 2.16 implies that there exists a natural number $L_1 > 2L_0 + 1$ such that the following property holds:

(P32)   for each natural number $T \ge L_1$, each program $\{x_t\}_{t=0}^T$ which satisfies the inequality

$$\sum_{t=0}^{T-1} v(x_t, x_{t+1}) \ge Tv(\bar{x}, \bar{x}) - M - 2$$

and each integer $S \in [0, T - L_1]$ we have

$$\min\{\rho(x_t, \bar{x}) : \ t = S + 1, \ldots, S + L_1\} \le \delta_0.$$

Choose a natural number k such that

$$8^{-1}k > 4L_1(\|v\| + 1). \tag{2.411}$$

Put

$$L_2 = kL_1 \tag{2.412}$$

and choose an integer

$$L > L_2. \tag{2.413}$$

Choose

$$\delta \in (0, 2^{-1}) \tag{2.414}$$

for which

$$8\delta L_1 \le 1, \ 8L_2\delta < \delta_0 \tag{2.415}$$

and choose

$$\lambda \in (2^{-1}, 1) \tag{2.416}$$

such that

$$8L_1(1 - \lambda)(\|v\| + 1) < 1, \tag{2.417}$$

$$8L_2(\|v\| + 1)(1 - \lambda) < \delta_0.$$

Lemma 2.53. *Let $T > L$ be an integer, a sequence of functions $\{u_t\}_{t=0}^{T-1} \subset \mathcal{M}_0$ satisfy*

$$\|u_t - v\| \leq \delta, \ t = 0, 1, \ldots, T - 1, \tag{2.418}$$

$\{\alpha_t\}_{t=0}^{T-1} \subset (0, 1]$ *be such that for each integer $t \in [0, T - L]$,*

$$\alpha_i \alpha_j^{-1} \geq \lambda \text{ for each } i, j \in [t, t + L], \tag{2.419}$$

and let a program $\{x_t\}_{t=0}^{T}$ and an integer $\tau \in [0, T - L_1]$ satisfy

$$\min\{\rho(x_t, \bar{x}) : \ t = \tau + 1, \ldots, \tau + L_1\} > \delta_0. \tag{2.420}$$

Then

$$\sum_{t=\tau}^{\tau+L_1-1} \alpha_t u_t(x_t, x_{t+1}) - \sum_{t=\tau}^{\tau+L_1-1} \alpha_t u_t(\bar{x}, \bar{x}) \leq \alpha_\tau(-M - 1).$$

Proof. In view of (2.420) and property (P32),

$$\sum_{t=\tau}^{\tau+L_1-1} v(x_t, x_{t+1}) < L_1 v(\bar{x}, \bar{x}) - M - 2. \tag{2.421}$$

By (2.421), (2.418), (2.419), (2.416), (2.417), and (2.415), we have

$$\sum_{t=\tau}^{\tau+L_1-1} \alpha_t u_t(x_t, x_{t+1}) - \sum_{t=\tau}^{\tau+L_1-1} \alpha_t u_t(\bar{x}, \bar{x})$$

$$= \alpha_\tau \left[\sum_{t=\tau}^{\tau+L_1-1} \alpha_t \alpha_\tau^{-1} u_t(x_t, x_{t+1}) - \sum_{t=\tau}^{\tau+L_1-1} \alpha_t \alpha_\tau^{-1} u_t(\bar{x}, \bar{x}) \right]$$

$$\leq \alpha_\tau \left[\sum_{t=\tau}^{\tau+L_1-1} v(x_t, x_{t+1}) - L_1 v(\bar{x}, \bar{x}) + 2 \sum_{t=\tau}^{\tau+L_1-1} \|\alpha_t \alpha_\tau^{-1} u_t - v\| \right]$$

$$\leq \alpha_\tau(-M-2) + 2\alpha_\tau \sum_{t=\tau}^{\tau+L_1-1} [\|\alpha_t \alpha_\tau^{-1} u_t - \alpha_t \alpha_\tau^{-1} v\| + \|\alpha_t \alpha_\tau^{-1} v - v\|]$$

$$\leq -\alpha_\tau(M+2) + 2\alpha_\tau L_1(\delta\lambda^{-1} + (\lambda^{-1} - 1)\|v\|)$$

$$\leq \alpha_\tau(-M-2+4L_1\delta + 4L_1(1-\lambda)\|v\|) \leq \alpha_\tau(-M-1).$$

This completes the proof of Lemma 2.53.   □

In the case of Theorem 2.49 we assume that an integer $T > 2L$, $\{u_t\}_{t=0}^{T-1} \subset \mathcal{M}_0$ satisfies

$$\|u_t - v\| \leq \delta, \tag{2.422}$$

for all $t = 0, \dots, T-1$, $\{\alpha_t\}_{t=0}^{T-1} \subset (0, 1]$ satisfies for each integer $\tau \in [0, T - L]$,

$$\alpha_i \alpha_j^{-1} \geq \lambda \text{ for each } i, j \in \{\tau, \dots, \tau + L\}, \tag{2.423}$$

and a program $\{x_t\}_{t=0}^{T}$ satisfies the inclusion

$$x_0 \in \bar{Y}_{L_0} \tag{2.424}$$

and the inequality

$$\sum_{t=0}^{T-1} \alpha_t u_t(x_t, x_{t+1}) = \sigma(\{\alpha_t u_t\}_{t=0}^{T-1}, 0, T, x_0). \tag{2.425}$$

In the case of Theorem 2.50 we assume that $\{u_t\}_{t=0}^{\infty} \subset \mathcal{M}_0$, (2.422) holds for all nonnegative integers t, $\{\alpha_t\}_{t=0}^{\infty} \subset (0, 1]$, (2.423) holds for all nonnegative integers τ, and $\{x_t\}_{t=0}^{\infty}$ is a $(\{\alpha_t u_t\}_{t=0}^{\infty})$-overtaking optimal program which satisfies (2.424).
    Set

$$J = [0, T] \text{ in the case of Theorem 2.49 and}$$

$$J = [0, \infty) \text{ in the case of Theorem 2.50.} \tag{2.426}$$

Assume that an integer S satisfies

$$S \geq 0, \ S + L_2 \in J \text{ and } x_S \in \bar{Y}_{L_0}. \tag{2.427}$$

We claim that there is an integer $j \in \{S+1, \dots, S+L_2\}$ such that

$$\rho(x_j, \bar{x}) \leq \delta_0.$$

Assume the contrary. Then

$$\rho(x_t, \bar{x}) > \delta_0, \ t = S+1, \dots, S+L_2. \tag{2.428}$$

There are two cases:

$$\rho(x_t, \bar{x}) > \delta_0 \text{ for all integers } t \in J \text{ such that } t > S; \tag{2.429}$$

$$\rho(x_t, \bar{x}) \leq \delta_0 \text{ for some integer } t \in J \text{ such that } t > S + L_2. \tag{2.430}$$

Assume that relation (2.429) is true. In view of (2.427) there exists a program $\{y_t\}_{t \in J}$ such that

$$y_t = x_t, \ t = 0, \ldots, S, \ y_t = \bar{x} \text{ for all integers } t \in J \text{ such that } t \geq S + L_0. \tag{2.431}$$

Relation (2.431) implies that for each $Q \in J$ satisfying $Q > S$, we have

$$\sum_{t=S}^{Q-1} \alpha_t u_t(x_t, x_{t+1}) - \sum_{t=S}^{Q-1} \alpha_t u_t(y_t, y_{t+1})$$

$$= \sum_{t=0}^{Q-1} \alpha_t u_t(x_t, x_{t+1}) - \sum_{t=0}^{Q-1} \alpha_t u_t(y_t, y_{t+1}). \tag{2.432}$$

Now we choose an integer $p \geq 1$. In the case of Theorem 2.50, we choose an integer $p \geq k$. Consider the case of Theorem 2.49. It follows from relations (2.427) and (2.412) that there exists a natural number p such that

$$T - S \in [pL_1, (p+1)L_1). \tag{2.433}$$

By (2.433), (2.432), and (2.437),

$$p \geq k.$$

Thus in both cases $p \geq k$.

In view of Lemma 2.53, the choice of p [see (2.433)], (2.439), (2.422), and (2.423) for each $i = 0, \ldots, p - 1$, we have

$$\sum_{t=S+iL_1}^{S+(i+1)L_1-1} \alpha_t u_t(x_t, x_{t+1}) - \sum_{t=S+iL_1}^{S+(i+1)L_1-1} \alpha_t u_t(\bar{x}, \bar{x}) \leq \alpha_{S+iL_1}(-M - 1).$$

This inequality implies that

$$\sum_{t=S}^{S+pL_1-1} \alpha_t u_t(x_t, x_{t+1}) - \sum_{t=S}^{S+pL_1-1} \alpha_t u_t(\bar{x}, \bar{x}) \leq (-M - 1) \sum_{i=0}^{p-1} \alpha_{S+iL_1}. \tag{2.434}$$

Set

$$\tilde{T} = T \text{ in the case of Theorem 2.49 and}$$

$$\tilde{T} = S + pL_1 \text{ in the case of Theorem 2.50.} \tag{2.435}$$

It follows from (2.411)–(2.413), (2.416), (2.435), (2.433), (2.434), (2.431), (2.432), (2.433) and the inequality $p \geq k$ that

$$\sum_{t=S}^{\tilde{T}-1} \alpha_t u_t(x_t, x_{t+1}) - \sum_{t=S}^{\tilde{T}-1} \alpha_t u_t(y_t, y_{t+1}) \leq \sum_{t=S}^{S+pL_1-1} \alpha_t u_t(\bar{x}, \bar{x})$$

$$-(M+1)\sum_{i=0}^{p-1} \alpha_{S+iL_1} + \sum \{\alpha_t \|u_t\| : t \text{ is an integer and } S + pL_1 \leq t < \tilde{T}\}$$

$$+ \sum_{t=S}^{S+L_0-1} \alpha_t \|u_t\| - \sum_{t=S+L_0}^{\tilde{T}-1} \alpha_t u_t(\bar{x}, \bar{x})$$

$$\leq -(M+1)\sum_{i=0}^{p-1} \alpha_{S+iL_1} + \sum_{t=S}^{S+L_0-1} \alpha_t(\|v\| + 1)$$

$$+2(\|v\| + 1)\sum \{\alpha_t : t \text{ is an integer and } S + pL_1 \leq t < T\}$$

$$+ \sum_{t=S}^{S+L_0-1} \alpha_t(\|v\| + 1) \leq (-M - 1)\sum_{i=0}^{p-1} \alpha_{S+iL_1}$$

$$+2(\|v\| + 1)L_0\lambda^{-1}\alpha_S + 2(\|v\| + 1)L_1\lambda^{-1}\alpha_{S+pL_1}$$

$$\leq 2^{-1}(-M - 1)\sum_{i=0}^{p-1} \alpha_{S+iL_1} + 2(\|v\| + 1)L_1\lambda^{-1}\alpha_S$$

$$+2^{-1}(-M - 1)\sum_{i=0}^{p-1} \alpha_{S+iL_1} + 2(\|v\| + 1)L_1\lambda^{-1}\alpha_{S+pL_1}$$

$$\leq 2^{-1}(-M - 1)\sum_{i=0}^{k-1} \alpha_{S+iL_1} + 2(\|v\| + 1)L_1\lambda^{-1}\alpha_S$$

$$+2^{-1}(-M - 1)\sum_{i=p-k}^{p-1} \alpha_{S+iL_1} + 2(\|v\| + 1)L_1\lambda^{-1}\alpha_{S+pL_1}$$

$$\leq 2^{-1}(-M - 1)k\lambda\alpha_S + 2(\|v\| + 1)L_1\lambda^{-1}\alpha_S$$

$$+2^{-1}(-M - 1)k\lambda\alpha_{S+pL_1} + 2(\|v\| + 1)L_1\lambda^{-1}\alpha_{S+pL_1}$$

$$= \alpha_S\lambda^{-1}(2^{-1}(-M - 1)k\lambda^2 + 2(\|v\| + 1)L_1)$$

$$+\alpha_{S+pL_1}\lambda^{-1}(2^{-1}(-M - 1)k\lambda^2 + 2(\|v\| + 1)L_1)$$

$$= (\alpha_S + \alpha_{S+pL_1})\lambda^{-1}(8^{-1}(-M - 1)k + 2(\|v\| + 1)L_1)$$

$$< -\lambda^{-1}\alpha_S(16^{-1}k). \tag{2.436}$$

In the case of Theorem 2.49, (2.436), (2.435), (2.432), and (2.431) contradict (2.425). In this case of Theorem 2.50, (2.436), (2.435), (2.432), and (2.431) contradict ($\{\alpha_t u_t\}_{t=0}^\infty$)-overtaking optimality of the program $\{x_t\}_{t=0}^\infty$ because p is any integer satisfying $p \geq k$.

The contradiction we have reached proves that (2.429) does not hold. Thus (2.430) holds. We may assume without loss of generality that there is an integer $\tilde{S}$ such that

$$\tilde{S} \in J, \ \tilde{S} > S + L_2, \ \rho(x_{\tilde{S}}, \bar{x}) \leq \delta_0,$$

$$\rho(x_t, \bar{x}) > \delta_0 \text{ for all integers } t \text{ satisfying } S < t < \tilde{S}. \tag{2.437}$$

It follows from relations (2.427), (2.437), and (2.400) that there exists a program $\{y_t\}_{t \in J}$ such that

$$y_t = x_t, \ t = 0, \ldots, S, \ y_t = \bar{x}, \ t = S + L_0, \ldots, \tilde{S} - 1,$$

$$y_t = x_t \text{ for all integers } t \in J \text{ satisfying } \tilde{S} \leq t. \tag{2.438}$$

In view of (2.438) for any integer $Q \in J$ satisfying $Q \geq \tilde{S}$, we have

$$\sum_{t=S}^{\tilde{S}-1} \alpha_t u_t(x_t, x_{t+1}) - \sum_{t=S}^{\tilde{S}-1} \alpha_t u_t(y_t, y_{t+1}) = \sum_{t=0}^{T-1} \alpha_t u_t(x_t, x_{t+1}) - \sum_{t=0}^{T-1} \alpha_t u_t(y_t, y_{t+1}). \tag{2.439}$$

Relations (2.412) and (2.437) imply that there exists a natural number p such that

$$\tilde{S} - S - 1 \in [pL_1, (p+1)L_1). \tag{2.440}$$

In view of (2.440), (2.412), and (2.437),

$$p \geq k. \tag{2.441}$$

In view of Lemma 2.53, the choice of the natural number p [see (2.440)], (2.437), (2.422), and (2.423), for each $i = 0, \ldots, p - 1$, we have

$$\sum_{t=S+iL_1}^{S+(i+1)L_1-1} \alpha_t u_t(x_t, x_{t+1}) - \sum_{t=S+iL_1}^{S+(i+1)L_1-1} \alpha_t u_t(\bar{x}, \bar{x}) \leq \alpha_{S+iL_1}(-M - 1).$$

This inequality implies (2.434). It follows from relations (2.440), (2.434), (2.437), (2.438), (2.422), (2.423), (2.441), and (2.416) that

$$\sum_{t=S}^{\tilde{S}-1} \alpha_t u_t(x_t, x_{t+1}) - \sum_{t=S}^{\tilde{S}-1} \alpha_t u_t(y_t, y_{t+1}) \leq \sum_{t=S}^{S+pL_1-1} \alpha_t u_t(\bar{x}, \bar{x})$$

$$-(M+1) \sum_{i=0}^{p-1} \alpha_{S+iL_1} + \sum \{\alpha_t \|u_t\| : t \text{ is an integer and } S + pL_1 \leq t < \tilde{S}\}$$

$$+ \sum_{t=S}^{S+L_0-1} \alpha_t \|u_t\| - \sum_{t=S+L_0}^{S+pL_1-1} \alpha_t u_t(\bar{x}, \bar{x})$$

$$+ \sum \{\alpha_t \|u_t\| : t \text{ is an integer and } S + pL_1 \leq t < \tilde{S}\}$$

$$\leq -(M+1) \sum_{i=0}^{p-1} \alpha_{S+iL_1} + +2 \sum_{t=S}^{S+L_0-1} \alpha_t (\|v\| + 1)$$

$$+ \sum \{2\alpha_t (\|v\| + 1) : t \text{ is an integer and } S + pL_1 \leq t < \tilde{S}\}$$

$$\leq -(M+1) \sum_{i=0}^{p-1} \alpha_{S+iL_1} + +2(\|v\| + 1)L_0 \lambda^{-1} \alpha_S + +2(\|v\| + 1)L_1 \lambda^{-1} \alpha_{S+pL_1}$$

$$\leq 2^{-1}(-M-1)k\lambda\alpha_S + 2(\|v\| + 1)L_1\lambda^{-1}\alpha_S$$

$$+2^{-1}(-M-1)k\lambda\alpha_{S+pL_1} + 2(\|v\| + 1)L_1\lambda^{-1}\alpha_{S+pL_1}$$

$$\leq \alpha_S \lambda^{-1}(2^{-1}(-M-1)k\lambda^2 + 2(\|v\| + 1)L_1)$$

$$+\alpha_{S+pL_1} \lambda^{-1}(2^{-1}(-M-1)k\lambda^2 + 2(\|v\| + 1)L_1)$$

$$\leq (\alpha_S + \alpha_{S+pL_1})\lambda^{-1}(8^{-1}(-M-1)k + 2(\|v\| + 1)L_1) < 0.$$

Combined with relation (2.438) this contradicts (2.425) in the case of Theorem 2.49 and contradicts the $(\{\alpha_t u_t\}_{t=0}^{\infty})$-overtaking optimality of $\{x_t\}_{t=0}^{\infty}$ in the case of Theorem 2.50.

The contradiction we have reached proves that there exists an integer $j \in [S+1, S+L_2]$ such that

$$\rho(x_j, \bar{x}) \leq \delta_0.$$

Thus we have shown that the following property holds:

(P33) For each integer S satisfying (2.427) there is an integer $j \in [S+1, S+L_2]$ such that

$$\rho(x_j, \bar{x}) \leq \delta_0.$$

It follows from property (P33), (2.427), and (2.424) that in the case of Theorem 2.49 there exists a sequence of integers $\{S_i\}_{i=0}^{q}$ such that

$$S_0 \in [1, L_2],$$

for each integer $i \in \{0, \ldots, q-1\}$,

$$S_{i+1} - S_i \in [0, L_2], \tag{2.442}$$

$S_q \in (T - L_2, T]$, and

$$\rho(x_{S_i}, \bar{x}) \le \delta_0, \ i = 1, \ldots, q. \tag{2.443}$$

It follows from property (P33), (2.427), and (2.424) that in the case of Theorem 2.50 there exists a sequence $\{S_i\}_{i=0}^{\infty}$ of integers such that $S_0 \in [1, L_2]$ and that for all integers $i \ge 0$ equations (2.442) and (2.443) hold. Evidently, we may assume that $S_0 = 0$ if $\rho(x_0, \bar{x}) \le \delta_0$. Let $i \ge 0$ be an integer and let $i \le q - 1$ in the case of Theorem 2.49. Then

$$\sum_{t=S_i}^{S_{i+1}-1} \alpha_{S_i}^{-1} \alpha_t u_t(x_t, x_{t+1}) = \sigma\left(\{\alpha_{S_i}^{-1} \alpha_t u_t\}_{t=S_i}^{S_{i+1}-1}, S_i, S_{i+1}, x_{S_i}, x_{S_{i+1}}\right). \tag{2.444}$$

In view of (2.442), (2.423), (2.422), and (2.416), for all $t \in \{S_i, \ldots, S_{i+1} - 1\}$, we have

$$\|\alpha_{S_i}^{-1} \alpha_t u_t - v\| \le \alpha_{S_i}^{-1} \alpha_t \|u_t - v\| + \|v\| |\alpha_{S_i}^{-1} \alpha_t - 1|$$

$$\le \lambda^{-1} \delta + \|v\|(\lambda^{-1} - 1) \le \lambda^{-1}(\delta + \|v\|(1 - \lambda)) \le 2(\delta + \|v\|(1 - \lambda)). \tag{2.445}$$

Relations (2.445), (2.442), (2.444), (2.415), and (2.417) imply that

$$\sum_{t=S_i}^{S_{i+1}-1} v(x_t, x_{t+1}) \ge \sum_{t=S_i}^{S_{i+1}-1} \alpha_{S_i}^{-1} \alpha_t u_t(x_t, x_{t+1}) - 2L_2(\delta + \|v\|(1 - \lambda))$$

$$= \sigma\left(\{\alpha_{S_i}^{-1} \alpha_t u_t\}_{t=S_i}^{S_{i+1}-1}, S_i, S_{i+1}, x_{S_i}, x_{S_{i+1}}\right) - 2L_2(\delta + \|v\|(1 - \lambda))$$

$$\ge \sigma(v, S_{i+1} - S_i, x_{S_i}, x_{S_{i+1}}) - 4L_2(\delta + \|v\|(1 - \lambda))$$

$$\ge \sigma(v, S_{i+1} - S_i, x_{S_i}, x_{S_{i+1}}) - \delta_0.$$

Combined with (2.443) and property (P31) this implies that

$$\rho(x_t, \bar{x}) \le \epsilon, \ t = S_i, \ldots, S_{i+1}.$$

Theorems 2.49 and 2.50 are proved.

2.27 Proof of Theorem 2.51

Note that in the proof of Theorem 2.51 we use the natural number $L > L_0$ and $\lambda \in (0, 1)$ which is given by Theorem 2.49.

In the proof we use the following auxiliary result.

Lemma 2.54. *Let γ be a positive number. Then there exists an integer $n_0 \geq 1$ such that for each pair of integers $T > S \geq n_0$ and each program $\{x_t\}_{t=0}^T$ satisfying the inclusion*

$$x_0 \in \bar{Y}_{L_0} \tag{2.446}$$

and the inequality

$$\sum_{t=0}^{T-1} \alpha_t u_t(x_t, x_{t+1}) = \sigma\left(\{\alpha_t u_t\}_{t=0}^{T-1}, 0, T, x_0\right), \tag{2.447}$$

the following inequality holds:

$$\sum_{t=0}^{S-1} \alpha_t u_t(x_t, x_{t+1}) \geq \sigma\left(\{\alpha_t u_t\}_{t=0}^{S-1}, 0, S, x_0\right) - \gamma. \tag{2.448}$$

Proof. Since $\lim_{t \to \infty} \alpha_t = 0$ [see (2.405)] there exists a natural number

$$n_0 > 4L + 4 \tag{2.449}$$

such that for all integers $t > n_0 - L - 4$, we have

$$\alpha_t \leq \gamma(8L + 8)^{-1}(\|v\| + 1)^{-1}. \tag{2.450}$$

Assume that integers $T > S \geq n_0$ and that a program $\{x_t\}_{t=0}^T$ satisfies relations (2.446) and (2.447). There exists a program $\{\tilde{x}_t\}_{t=0}^S$ such that

$$\tilde{x}_0 = x_0, \quad \sum_{t=0}^{S-1} \alpha_t u_t(\tilde{x}_t, \tilde{x}_{t+1}) = \sigma\left(\{\alpha_t u_t\}_{t=0}^{S-1}, 0, S, x_0\right). \tag{2.451}$$

In view of the choice of the numbers δ and L, Theorem 2.49, (2.404), (2.405), (2.449), (2.451), (2.447), and (2.446), we have

$$\rho(x_t, \bar{x}) \leq \bar{r}/4, \ t = L, \ldots, T - L, \tag{2.452}$$

$$\rho(\tilde{x}_t, \bar{x}) \leq \bar{r}/4, \ t = L, \ldots, S - L. \tag{2.453}$$

It follows from (2.400), (2.449), (2.452), and (2.453) that there exists a program $\{y_t\}_{t=0}^T$ such that

$$y_t = \tilde{x}_t, \ t = 0, \ldots, S - L, \ y_t = x_t, \ t = S - L + 1, \ldots, T. \tag{2.454}$$

By (2.454), (2.451), (2.447), (2.404), and (2.450), we have

$$0 \le \sum_{t=0}^{T-1} \alpha_t u_t (x_t, x_{t+1}) - \sum_{t=0}^{T-1} \alpha_t u_t (y_t, y_{t+1})$$

$$= \sum_{t=0}^{S-L} \alpha_t u_t (x_t, x_{t+1}) - \sum_{t=0}^{S-L} \alpha_t u_t (y_t, y_{t+1})$$

$$\le \sum_{t=0}^{S-L-1} \alpha_t u_t (x_t, x_{t+1}) - \sum_{t=0}^{S-L-1} \alpha_t u_t (\tilde{x}_t, \tilde{x}_{t+1}) + 2\alpha_{S-L}(\|v\| + 1)$$

$$\le \sum_{t=0}^{S-1} \alpha_t u_t (x_t, x_{t+1}) + (\|v\| + 1) \sum_{t=S-L}^{S-1} \alpha_t$$

$$-\sigma\left(\{\alpha_t u_t\}_{t=0}^{S-1}, 0, S, x_0\right) + (\|v\| + 1) \sum_{t=S-L}^{S-1} \alpha_t + 2\alpha_{S-L}(\|v\| + 1)$$

$$\le \sum_{t=0}^{S-1} \alpha_t u_t (x_t, x_{t+1}) - \sigma\left(\{\alpha_t u_t\}_{t=0}^{S-1}, 0, S, x_0\right) + \gamma.$$

Lemma 2.54 is proved. $\qquad \square$

Completion of the proof of Theorem 2.51. Let a point $z \in \bar{Y}_{L_0}$ be given. For each natural number T there exists a program $\{x_t^{(z,T)}\}_{t=0}^{T}$ such that

$$x_0^{(z,T)} = z, \tag{2.455}$$

$$\sum_{t=0}^{T-1} \alpha_t u_t \left(x_t^{(z,T)}, x_{t+1}^{(z,T)}\right) = \sigma\left(\{\alpha_t u_t\}_{t=0}^{T-1}, 0, T, x_0\right).$$

Evidently, there exists a strictly increasing sequence of natural numbers $\{T_j\}_{j=1}^{\infty}$ such that for any nonnegative integer t there exists a limit

$$x_t^{(z)} = \lim_{j \to \infty} x_t^{(z,T_j)}. \tag{2.456}$$

It is clear that $\{x_t^{(z)}\}_{t=0}^{\infty}$ is a program and

$$x_0^{(z)} = z. \tag{2.457}$$

Let $\gamma > 0$ be given. It follows from Lemma 2.54 that there exists an integer $n_0 \ge 1$ such that the following property holds:

(P34) For each pair of integers $T > S \ge n_0$ and each program $\{x_t\}_{t=0}^{T}$ satisfying (2.446) and (2.447), equation (2.448) holds.

Let $S \geq n_0$ be an integer. In view of property (P34) and (2.455) for each natural number j satisfying $T_j > S$, we have

$$\sum_{t=0}^{S-1} \alpha_t u_t \left(x_t^{(z,T)}, x_{t+1}^{(z,T)} \right) \geq \sigma(\{\alpha_t u_t\}_{t=0}^{S-1}, 0, S, z) - \gamma.$$

Combined with (2.456) this inequality implies that

$$\sum_{t=0}^{S-1} \alpha_t u_t \left(x_t^{(z)}, x_{t+1}^{(z)} \right) \geq \sigma(\{\alpha_t u_t\}_{t=0}^{S-1}, 0, S, z) - \gamma$$

for all integers $S \geq n_0$ and $z \in \bar{Y}_{L_0}$. Theorem 2.51 is proved. $\square$

2.28   Nonautonomous Discrete-Time Control System

In this section which is based on [55] we analyze the structure of solutions of the optimization problems

$$\sum_{i=m_1}^{m_2-1} v_i(z_i, z_{i+1}) \to \min, \ \{z_i\}_{i=m_1}^{m_2} \subset X \text{ and } z_{m_1} = x, \ z_{m_2} = y, \qquad (P)$$

where $v_i : X \times X \to R^1, i = 0, \pm 1, \pm 2, \ldots$ is a continuous function defined on a metric space X and $x, y \in X$.

Let $\mathbf{Z} = \{0, \pm 1, \pm 2, \ldots\}$ be the set of all integers, (X, ρ) be a compact metric space, and $v_i : X \times X \to R^1, i = 0, \pm 1, \pm 2, \ldots$ be a sequence of continuous functions such that

$$\sup\{|v_i(x, y)| : \ x, y \in X, \ i \in \mathbf{Z}\} < \infty \qquad (2.458)$$

and which satisfy the following assumption:

(A) For each positive number ϵ there exists a positive number δ such that if $i \in \mathbf{Z}$ and if points $x_1, x_2, y_1, y_2 \in X$ satisfy $\rho(x_j, y_j) \leq \delta, \ j = 1, 2$, then $|v_i(x_1, x_2) - v_i(y_1, y_2)| \leq \epsilon$.

For each pair of points $y, z \in X$ and each pair of integers $n_1, n_2 > n_1$, put

$$\sigma(n_1, n_2, y, z) = \inf \left\{ \sum_{i=n_1}^{n_2-1} v_i(x_i, x_{i+1}) : \ \{x_i\}_{i=n_1}^{n_2} \subset X, \ x_{n_1} = y, \ x_{n_2} = z \right\},$$
$$\qquad (2.459)$$

$$\sigma(n_1, n_2) = \inf \left\{ \sum_{i=n_1}^{n_2-1} v_i(x_i, x_{i+1}) : \ \{x_i\}_{i=n_1}^{n_2} \subset X \right\}. \qquad (2.460)$$

Choose a real number $d_0 > 0$ such that

$$|v_i(x, y)| \leq d_0, \; x, y \in X, \; i \in \mathbf{Z}. \tag{2.461}$$

A sequence $\{y_i\}_{i=-\infty}^{\infty} \subset X$ is called good if there exists a positive number c such that for each pair of integers $m_1, m_2 > m_1$,

$$\sum_{i=m_1}^{m_2-1} v_i(y_i, y_{i+1}) \leq \sigma(m_1, m_2, y_{m_1}, y_{m_2}) + c.$$

We say that the sequence $\{v_i\}_{i=-\infty}^{\infty}$ has the turnpike property (TP) if there exists a sequence $\{\hat{x}_i\}_{i=-\infty}^{\infty} \subset X$ which satisfies the following condition:

For each $\epsilon > 0$ there are $\delta > 0$ and a natural number N such that for each pair of integers $T_1, T_2 \geq T_1 + 2N$ and each sequence $\{y_i\}_{i=T_1}^{T_2} \subset X$ which satisfies

$$\sum_{i=T_1}^{T_2-1} v_i(y_i, y_{i+1}) \leq \sigma(T_1, T_2, y_{T_1}, y_{T_2}) + \delta$$

there are integers $\tau_1 \in \{T_1, \dots, T_1 + N\}, \tau_2 \in \{T_2 - N, \dots, T_2\}$ such that:

(i) $\rho(y_i, \hat{x}_i) \leq \epsilon, \; i = \tau_1, \dots, \tau_2;$
(ii) if $\rho(y_{T_1}, \hat{x}_{T_1}) \leq \delta$, then $\tau_1 = T_1$, and if $\rho(y_{T_2}, \hat{x}_{T_2}) \leq \delta$, then $\tau_2 = T_2$.

The sequence $\{\hat{x}_i\}_{i=-\infty}^{\infty} \subset X$ is called the turnpike of $\{v_i\}_{i=-\infty}^{\infty}$.

Assume that $\{\hat{x}_i\}_{i=-\infty}^{\infty} \subset X$. How to verify if the sequence of cost functions $\{v_i\}_{i=-\infty}^{\infty}$ has (TP) and $\{\hat{x}_i\}_{i=-\infty}^{\infty}$ is its turnpike? In this section we introduce three properties (P35)–(P37) and show that $\{v_i\}_{i=-\infty}^{\infty}$ has (TP) if and only if $\{v_i\}_{i=-\infty}^{\infty}$ possesses the properties (P35)–(P37). Property (P35) means that all good sequences have the same asymptotic behavior. Property (P36) means that for each pair of integers $m_1, m_2 > m_1$ the sequence $\{\hat{x}_i\}_{i=m_1}^{m_2}$ is a unique solution of problem (P) with $x = \hat{x}_{m_1}, y = \hat{x}_{m_2}$ and that if a sequence $\{y_i\}_{i=-\infty}^{\infty} \subset X$ is a solution of problem (P) for each pair of integers $m_1, m_2 > m_1$ with $x = y_{m_1}, y = y_{m_2}$, then $y_i = \hat{x}_i$ for all integers i. Property (P37) means that if a sequence $\{y_i\}_{i=m_1}^{m_2} \subset X$ is an approximate solution of problem (P) and $m_2 - m_1$ is large enough, then there is $j \in [m_1, m_2]$ such that y_j is close to $\hat{x}_j$.

The next theorem was obtained in [55].

Theorem 2.55. *Let $\{\hat{x}_i\}_{i=-\infty}^{\infty} \subset X$. Then the sequence $\{v_i\}_{i=-\infty}^{\infty}$ possesses the turnpike property and $\{\hat{x}_i\}_{i=-\infty}^{\infty}$ is its turnpike if and only if the following properties hold:*

(P35)   If $\{y_i\}_{i=-\infty}^{\infty} \subset X$ is good, then

$$\lim_{i \to \infty} \rho(y_i, \hat{x}_i) = 0 \, , \; \lim_{i \to -\infty} \rho(y_i, \hat{x}_i) = 0.$$

(P36)   For each pair of integers $m_1, m_2 > m_1$,

$$\sum_{i=m_1}^{m_2-1} v_i(\hat{x}_i, \hat{x}_{i+1}) = \sigma(m_1, m_2, \hat{x}_{m_1}, \hat{x}_{m_2})$$

and if a sequence $\{y_i\}_{i=-\infty}^{\infty} \subset X$ satisfies

$$\sum_{i=m_1}^{m_2-1} v_i(y_i, y_{i+1}) = \sigma(m_1, m_2, y_{m_1}, y_{m_2}) \tag{2.462}$$

for each pair of integers $m_1, m_2 > m_1$, then $y_i = \hat{x}_i$, $i \in \mathbf{Z}$.

(P37)   For each positive number ϵ there exist a positive number δ and an integer $L \geq 1$ such that for each integer m and each sequence $\{y_i\}_{i=m}^{m+L} \subset X$ which satisfies

$$\sum_{i=m}^{m+L-1} v_i(y_i, y_{i+1}) \leq \sigma(m, m+L, y_m, y_{m+L}) + \delta$$

there exists an integer $j \in \{m, \ldots, m+L\}$ for which $\rho(y_j, \hat{x}_j) \leq \epsilon$.

It should be mentioned that properties (P35)–(P37) easily follow from the turnpike property. However it is very nontrivial to show that properties (P35)–(P37) are sufficient for this property.

Assume that the sequence $\{v_i\}_{i=-\infty}^{\infty}$ has the turnpike property, a sequence $\{\hat{x}_i\}_{i=-\infty}^{\infty}$ is its turnpike, and $v_i = v_0$ for all integers i. Let k be an integer and $y_i = \hat{x}_{i+k}$ for all integers i. Then (2.462) is valid for each pair of integers $m_1, m_2 > m_1$. In view of Theorem 2.55 $\hat{x}_i = y_i = \hat{x}_{i+k}$ for all integers i. Since k is an arbitrary integer we conclude that $\hat{x}_i = \hat{x}_0$ for all integers i.

2.29   Auxiliary Results for Theorem 2.55

Lemma 2.56. *Let ϵ be a positive number. Then there exists a positive number δ such that for each pair of integers $m_1, m_2 > m_1$ and each $x_1, x_2, y_1, y_2 \in X$ which satisfy $\rho(x_i, y_i) \leq \delta$, $i = 1, 2$ the inequality*

$$|\sigma(m_1, m_2, x_1, x_2) - \sigma(m_1, m_2, y_1, y_2)| \leq \epsilon$$

holds.

Proof. In view of assumption (A) there exists a positive number δ such that for each $x_1, x_2, y_1, y_2 \in X$ satisfying

$$\rho(x_i, y_i) \leq \delta, \ i = 1, 2 \tag{2.463}$$

and for all integers j, we have

$$|v_j(x_1, x_2) - v_j(y_1, y_2)| \leq \epsilon/4. \tag{2.464}$$

Assume that $m_1, m_2 \in \mathbf{Z}, m_2 > m_1$, points $x_1, x_2, y_1, y_2 \in X$, and (2.463) holds. In order to prove the lemma it is sufficient to show that

$$\sigma(m_1, m_2, y_1, y_2) \leq \sigma(m_1, m_2, x_1, x_2) + \epsilon. \tag{2.465}$$

Clearly, if $m_2 = m_1 + 1$, then (2.465) follows from (2.463) and the definition of δ [see (2.464)]. Consider the case with $m_2 > m_1 + 1$. There exists a sequence $\{z_i\}_{i=m_1}^{m_2} \subset X$ such that

$$z_{m_1} = x_1, \; z_{m_2} = x_2, \; \sum_{i=m_1}^{m_2-1} v_i(z_i, z_{i+1}) = \sigma(m_1, m_2, x_1, x_2). \tag{2.466}$$

Define a sequence $\{\tilde{z}_i\}_{i=m_1}^{m_2} \subset X$ by

$$\tilde{z}_{m_1} = y_1, \; \tilde{z}_{m_2} = y_2, \; \tilde{z}_i = z_i, \; i \in \{m_1,, \ldots, m_2\} \setminus \{m_1, m_2\}. \tag{2.467}$$

By the definition of δ, (2.463), and (2.467), we have

$$|v_{m_1}(x_1, z_{m_1+1}) - v_{m_1}(y_1, z_{m_1+1})| \leq \epsilon/4,$$

$$|v_{m_2-1}(z_{m_2-1}, x_2) - v_{m_2}(z_{m_2-1}, y_2)| \leq \epsilon/4. \tag{2.468}$$

In view of (2.466)–(2.468)

$$\left| \sum_{i=m_1}^{m_2-1} v_i(z_i, z_{i+1}) - \sum_{i=m_1}^{m_2-1} v_i(\tilde{z}_i, \tilde{z}_{i+1}) \right|$$

$$= |v_{m_1}(z_{m_1}, z_{m_1+1}) + v_{m_2-1}(z_{m_2-1}, z_{m_2}) - v_{m_1}(\tilde{z}_{m_1}, \tilde{z}_{m_1+1})$$

$$- v_{m_2-1}(\tilde{z}_{m_2-1}, \tilde{z}_{m_2})|$$

$$\leq |v_{m_1}(z_{m_1}, z_{m+1}) - v_{m_1}(\tilde{z}_{m_1}, \tilde{z}_{m_1+1})|$$

$$+ |-v_{m_2-1}(\tilde{z}_{m_2-1}, \tilde{z}_{m_2}) + v_{m_2-1}(z_{m_2-1}, z_{m_2})| \leq \epsilon/2.$$

It follows from these inequalities, (2.467), and (2.466) that

$$\sigma(m_1, m_2, y_1, y_2) \leq \sum_{i=m_1}^{m_2-1} v_i(\tilde{z}_i, \tilde{z}_{i+1})$$

$$\leq \sum_{i=m_1}^{m_2-1} v_i(z_i, z_{i+1}) + \epsilon = \sigma(m_1, m_2, x_1, x_2) + \epsilon.$$

Lemma 2.56 is proved.   □

Lemma 2.57. *For each pair of integers $m_1, m_2 > m_1$ and each pair of points $x, y \in X$ the inequality*

$$\sigma(m_1, m_2, x, y) \le \sigma(m_1, m_2) + 4d_0$$

holds.

Proof. Let $m_1, m_2 > m_1$ be integers and let points $x, y \in X$ be given. We may consider only the case with $m_2 > m_1 + 1$. There exists a sequence $\{z_i\}_{i=m_1}^{m_2} \subset X$ such that

$$\sum_{i=m_1}^{m_2-1} v_i(z_i, z_{i+1}) = \sigma(m_1, m_2). \tag{2.469}$$

Define a sequence $\{\tilde{z}_i\}_{i=m_1}^{m_2} \subset X$ by

$$\tilde{z}_{m_1} = x, \ \tilde{z}_{m_2} = y, \ \tilde{z}_i = z_i, \ i \in \{m_1, \ldots, m_2\} \setminus \{m_1, m_2\}. \tag{2.470}$$

In view of (2.459), (2.470), (2.461), and (2.469), we have

$$\sigma(m_1, m_2, x, y) \le \sum_{i=m_1}^{m_2-1} v_i(\tilde{z}_i, \tilde{z}_{i+1})$$

$$\le \sum_{i=m_1}^{m_2-1} v_i(z_i, z_{i+1}) + |v_{m_1}(z_{m_1}, z_{m_1+1}) - v_{m_1}(\tilde{z}_{m+1}, \tilde{z}_{m_1+1})|$$

$$+ |v_{m_2-1}(z_{m_2-1}, z_{m_2}) - v_{m_2-1}(\tilde{z}_{m_2-1}, \tilde{z}_{m_2})|$$

$$\le \sum_{i=m_1}^{m_2-1} v_i(z_i, z_{i+1}) + 4d_0 = \sigma(m_1, m_2) + 4d_0.$$

Lemma 2.57 is proved.   □

Lemma 2.58. *Assume that a sequence $\{y_i\}_{i=-\infty}^{\infty} \subset X$ is good and that ϵ is a positive number. Then there exists a pair of natural numbers L_1, L_2 such that the following properties hold:*

(a) For each pair of integers $m_1 \ge L_1, m_2 > m_1$,

$$\sum_{i=m_1}^{m_2-1} v_i(y_i, y_{i+1}) \le \sigma(m_1, m_2, y_{m_1}, y_{m_2}) + \epsilon. \tag{2.471}$$

(b) For each pair of integers $m_2 \le -L_2, m_1 < m_2$, the inequality (2.471) is true.

Proof. We claim that property (a) holds. Assume the contrary. Then there exist sequences of natural numbers $\{i_k\}_{k=1}^\infty$ and $\{j_k\}_{k=1}^\infty$ such that for each natural number k, we have

$$i_k < j_k < i_{k+1} - 8,$$

$$\sum_{i=i_k}^{j_k-1} v_i(y_i, y_{i+1}) > \sigma(i_k, j_k, y_{i_k}, y_{j_k}) + \epsilon. \tag{2.472}$$

For each natural number k there exists a sequence $\{z_i^{(k)}\}_{i=i_k}^{j_k} \subset X$ such that

$$z_{i_k}^{(k)} = y_{i_k}, \; z_{j_k}^{(k)} = y_{j_k},$$

$$\sum_{i=i_k}^{j_k-1} v_i\left(z_i^{(k)}, z_{i+1}^{(k)}\right) \le \sum_{i=i_k}^{j_k-1} v_i(y_i, y_{i+1}) - \epsilon. \tag{2.473}$$

Define a sequence $\{x_i\}_{i=-\infty}^\infty \subset X$ by

$$x_i = z_i^{(k)}, \; i \in \{i_k, \dots, j_k\}, \; k = 1, 2, \dots$$

$$x_i = y_i, \; i \in \mathbf{Z} \setminus \cup_{k=1}^\infty \{i_k, \dots, j_k\}. \tag{2.474}$$

By (2.474) and (2.473), for each integer $n \ge 1$, we have

$$y_0 = x_0, \; x_{j_n} = y_{j_n},$$

$$\sum_{i=0}^{j_n-1} [v_i(y_i, y_{i+1}) - v_i(x_i, x_{i+1})]$$

$$= \sum_{k=1}^{n} \left[\sum_{i=i_k}^{j_k-1} \left(v_i(y_i, y_{i+1}) - v_i\left(z_i^{(k)}, z_{i+1}^{(k)}\right) \right) \right] \ge n\epsilon \to \infty \text{ as } n \to \infty.$$

Since the sequence $\{y_i\}_{i=-\infty}^\infty$ is good we have reached a contradiction which proves property (a). Analogously we can show that property (b) holds. Lemma 2.58 is proved. $\qquad\square$

2.30   (TP) Implies Properties (P35), (P36), and (P37)

Proposition 2.59. *There exists a sequence $\{y_i\}_{i=-\infty}^\infty \subset X$ such that for each pair of integers $m_1, m_2 > m_1$ the equality*

$$\sum_{i=m_1}^{m_2-1} v_i(y_i, y_{i+1}) = \sigma(m_1, m_2, y_{m_1}, y_{m_2}) \tag{2.475}$$

holds.

Proof. Let $\{n_k\}_{k=1}^{\infty}$ be a strictly increasing sequence of natural numbers. For each natural number k there exists a sequence $\{y_i^{(k)}\}_{i=-n_k}^{n_k} \subset X$ such that

$$\sum_{i=-n_k}^{n_k-1} v_i \left(y_i^{(k)}, y_{i+1}^{(k)} \right) = \sigma(-n_k, n_k, y_{-n_k}, y_{n_k}). \tag{2.476}$$

We may assume without loss of generality that for each integer i there exists the limit

$$y_i := \lim_{k \to \infty} y_i^{(k)}. \tag{2.477}$$

By (2.476), (2.477), the continuity of v_i, $i \in \mathbf{Z}$, and Lemma 2.56, the equality (2.475) is true for each pair of integers $m_1, m_2 > m_1$. Proposition 2.59 is proved. $\square$

Proposition 2.60. *Assume that* $\{v_i\}_{i=-\infty}^{\infty}$ *possesses (TP) and that*

$$\{\hat{x}_i\}_{i=-\infty}^{\infty} \subset X$$

is the turnpike. Then properties (P35)–(P37) hold.

Proof. It is not difficult to see that property (P37) holds. Property (P35) follows from (TP) and Lemma 2.58. We claim that property (P36) holds.

By (TP), if a sequence $\{y_i\}_{i=-\infty}^{\infty} \subset X$ satisfies the equality

$$\sum_{i=m_1}^{m_2-1} v_i(y_i, y_{i+1}) = \sigma(m_1, m_2, y_{m_1}, y_{m_2})$$

for each pair of integers $m_1, m_2 > m_1$, then $y_i = \hat{x}_i, i \in \mathbf{Z}$. Property (P36) follows from this implication and Proposition 2.59. Proposition 2.60 is proved. $\square$

2.31   A Basic Lemma for Theorem 2.55

In this section we assume that $\{\hat{x}_i\}_{i=-\infty}^{\infty} \subset X$ and that properties (P35)–(P37) (see Theorem 2.55) hold.

Lemma 2.61. *Let ϵ be a positive number. Then the following properties hold:*

(a) *There exist a positive number δ and an integer $L \geq 1$ such that for each pair of integers $m_1 \geq L_1$, $m_2 > m_1$ and each sequence $\{y_i\}_{i=m_1}^{m_2} \subset X$ which satisfy*

$$\rho(\hat{x}_j, y_j) \leq \delta, \; j = m_1, m_2, \; \sum_{i=m_1}^{m_2-1} v_i(y_i, y_{i+1}) \leq \sigma(m_1, m_2, y_{m_1}, y_{m_2}) + \delta,$$

$$\tag{2.478}$$

the inequality

$$\rho(y_i, \hat{x}_i) \leq \epsilon, \ i = m_1, \ldots, m_2 \tag{2.479}$$

holds.

(b) *There exist a positive number $\tilde{\delta}$ and an integer $\tilde{L} \geq 1$ such that for each pair of integers $m_2 \leq -\tilde{L}$, $m_1 < m_2$ and each sequence $\{y_i\}_{i=m_1}^{m_2} \subset X$ which satisfies*

$$\rho(\hat{x}_j, y_j) \leq \tilde{\delta}, \ j = m_1, m_2, \ \sum_{i=m_1}^{m_2-1} v_i(y_i, y_{i+1}) \leq \sigma(m_1, m_2, y_{m_1}, y_{m_2}) + \tilde{\delta}$$

the inequality (2.479) is true.

Proof. In view of assumption (A) there exists a strictly increasing sequence $\{\delta_k\}_{k=0}^{\infty} \subset (0, 1)$ such that

$$\delta_0 < \epsilon, \ \delta_k < 4^{-k}, \ k = 1, 2, \ldots \tag{2.480}$$

and that for each natural number k and each $x_1, x_2, y_1, y_2 \in X$ which satisfy

$$\rho(x_j, y_j) \leq \delta_k, \ j = 1, 2 \tag{2.481}$$

the inequality

$$|v_i(x_1, x_2) - v_i(y_1, y_2)| \leq 4^{-k} \tag{2.482}$$

holds for all $i \in \mathbf{Z}$.

We claim that property (a) holds. Assume the contrary. Then there exist sequences of natural numbers $\{i_k\}_{k=1}^{\infty}$, $\{j_k\}_{k=1}^{\infty}$ such that for each integer $k \geq 1$,

$$i_k < j_k < i_{k+1} - 8 \tag{2.483}$$

and there exists $\{y_i^{(k)}\}_{i=i_k}^{j_k} \subset X$ which satisfies

$$\rho\left(y_{i_k}^{(k)}, \hat{x}_{i_k}\right), \rho\left(y_{j_k}^{(k)}, \hat{x}_{j_k}\right) \leq \delta_k, \tag{2.484}$$

$$\sum_{i=i_k}^{j_k-1} v_i\left(y_i^{(k)}, y_{i+1}^{(k)}\right) \leq \sigma\left(i_k, j_k, y_{i_k}^{(k)}, y_{j_k}^{(k)}\right) + \delta_k, \tag{2.485}$$

$$\max\left\{\rho\left(y_i^{(k)}, \hat{x}_i\right) : \ i = i_k, \ldots, j_k\right\} > \epsilon. \tag{2.486}$$

It is easy to see that

$$j_k > i_k + 1, \ldots, k = 1, 2, \ldots. \tag{2.487}$$

Define a sequence $\{y_i\}_{i=-\infty}^{\infty} \subset X$ by

$$y_i = y_i^{(k)}, \; i \in \{i_k, \ldots, j_k\}, \; k = 1, 2, \ldots$$
$$y_i = \hat{x}_i, \; i \in \mathbf{Z} \setminus \cup_{k=1}^{\infty} \{i_k, \ldots, j_k\}. \tag{2.488}$$

Let $k \geq 1$ be an integer. We estimate

$$\sum_{i=i_k-1}^{j_k} v_i(y_i, y_{i+1}) - \sum_{i=i_k-1}^{j_k} v_i(\hat{x}_i, \hat{x}_{i+1}).$$

Put

$$h_{i_k} = y_{i_k}^{(k)}, \; h_{j_k} = y_{j_k}^{(k)}, \; h_i = \hat{x}_i, \; i \in \{i_k, \ldots, j_k\} \setminus \{i_k, j_k\}. \tag{2.489}$$

In view of (2.485) and (2.489), we have

$$\sum_{i=i_k}^{j_k-1} v_i \left(y_i^{(k)}, y_{i+1}^{(k)}\right) \leq \sigma \left(i_k, j_k, y_{i_k}^{(k)}, y_{j_k}^{(k)}\right) + \delta_k$$

$$\leq \delta_k + \sum_{i=i_k}^{j_k-1} v_i(h_i, h_{i+1}) \leq \delta_k + \sum_{i=i_k}^{j_k-1} v_i(\hat{x}_i, \hat{x}_{i+1})$$

$$+ \left| v_{i_k}(\hat{x}_{i_k}, \hat{x}_{i_k+1}) - v_{i_k}\left(y_{i_k}^{(k)}, \hat{x}_{i_k+1}\right) \right|$$

$$+ \left| v_{j_k-1}(\hat{x}_{j_k-1}, \hat{x}_{j_k}) - v_{j_k-1}\left(\hat{x}_{j_k-1}, y_{j_k}^{(k)}\right) \right|. \tag{2.490}$$

By the choice of δ_k [see (2.481) and (2.482)] and (2.484), we have

$$\left| v_{i_k}(\hat{x}_{i_k}, \hat{x}_{i_k+1}) - v_{i_k}\left(y_{i_k}^{(k)}, \hat{x}_{i_k+1}\right) \right| \leq 4^{-k}, \tag{2.491}$$

$$\left| v_{i_k-1}(\hat{x}_{i_k-1}, \hat{x}_{i_k}) - v_{i_k-1}\left(\hat{x}_{i_k-1}, y_{i_k}^{(k)}\right) \right| \leq 4^{-k}, \tag{2.492}$$

$$\left| v_{j_k-1}(\hat{x}_{j_k-1}, \hat{x}_{j_k}) - v_{j_k-1}\left(\hat{x}_{j_k-1}, y_{j_k}^{(k)}\right) \right| \leq 4^{-k}, \tag{2.493}$$

$$\left| v_{j_k}(\hat{x}_{j_k}, \hat{x}_{j_k+1}) - v_{j_k}\left(y_{j_k}^{(k)}, \hat{x}_{j_k+1}\right) \right| \leq 4^{-k}. \tag{2.494}$$

It follows from (2.490), (2.491), (2.493), and (2.480) that

$$\sum_{i=i_k}^{j_k-1} v_i \left(y_i^{(k)}, y_{i+1}^{(k)}\right) \leq \delta_k + \sum_{i=i_k}^{j_k-1} v_i(\hat{x}_i, \hat{x}_{i+1}) + 4^{-k} + 4^{-k}$$

$$\leq \sum_{i=i_k}^{j_k-1} v_i(\hat{x}_i, \hat{x}_{i+1}) + 3 \cdot 4^{-k}. \tag{2.495}$$

Relations (2.488), (2.483), (2.487), (2.492), (2.494), and (2.495) imply that

$$
\sum_{i=i_k-1}^{j_k} v_i(y_i, y_{i+1}) = v_{i_k-1}\left(\hat{x}_{i_k-1}, y_{i_k}^{(k)}\right) + \sum_{i=i_k}^{j_k-1} v_i\left(y_i^{(k)}, y_{i+1}^{(k)}\right)
$$

$$
+ v_{j_k}\left(y_{j_k}^{(k)}, \hat{x}_{j_k+1}\right) \le [v_{i_k-1}(\hat{x}_{i_k-1}, \hat{x}_{i_k}) + 4^{-k}]
$$

$$
+ \left[\sum_{i=i_k}^{j_k-1} v_i(\hat{x}_i, \hat{x}_{i+1}) + 3 \cdot 4^{-k}\right] + [v_{j_k}(\hat{x}_{j_k}, \hat{x}_{j_k+1}) + 4^{-k}]
$$

$$
= 5 \cdot 4^{-k} + \sum_{i=i_k-1}^{j_k} v_i(\hat{x}_i, \hat{x}_{i+1}).
$$

Hence

$$
\sum_{i=i_k-1}^{j_k} v_i(y_i, y_{i+1}) \le \sum_{i=i_k-1}^{j_k} v_i(\hat{x}_i, \hat{x}_{i+1}) + 5 \cdot 4^{-k}
$$

for any integer $k \ge 1$. Together with (2.488) and property (P36) this inequality implies that the sequence $\{y_i\}_{i=-\infty}^{\infty}$ is good. In view of property (P35), we have

$$
\lim_{i \to \infty} \rho(y_i, \hat{x}_i) = 0.
$$

On the other hand it follows from (2.488) and (2.486) that

$$
\limsup_{i \to \infty} \rho(y_i, \hat{x}_i) \ge \epsilon.
$$

The contradiction we have reached proves property (a). Analogously we can prove property (b). Lemma 2.61 is proved.   □

Lemma 2.62.

1. Let $s \in \mathbf{Z}$, $\{y_i\}_{i=-\infty}^{s} \subset X$, $y_s = \hat{x}_s$ and let

$$
\sum_{i=i_1}^{i_2-1} v_i(y_i, y_{i+1}) = \sigma(i_1, i_2, y_{i_1}, y_{i_2}) \tag{2.496}
$$

for each pair of integers $i_2 \le s$, $i_1 < i_2$. Then $y_i = \hat{x}_i$ for all integers $i \le s$.
2. Let $s \in \mathbf{Z}$, $\{y_i\}_{i=s}^{\infty} \subset X$, $y_s = \hat{x}_s$ and let

$$
\sum_{i=i_1}^{i_2-1} v_i(y_i, y_{i+1}) = \sigma(i_1, i_2, y_{i_1}, y_{i_2})
$$

for each pair of integers $i_1 \ge s$, $i_2 > i_1$. Then $y_i = \hat{x}_i$ for all integers $i \ge s$.

Proof. We prove assertion 1. Put

$$\tilde{y}_i = y_i \text{ for all integers } i \le s, \ \tilde{y}_i = \hat{x}_i \text{ for all integers } i > s. \tag{2.497}$$

Let i_1, i_2 be integers and $i_1 < s < i_2$. By (2.497), (2.496), and Lemma 2.57, we have

$$\sum_{i=i_1}^{i_2-1} v_i(\tilde{y}_i, \tilde{y}_{i+1}) = \sum_{i=i_1}^{s-1} v_i(y_i, y_{i+1}) + \sum_{i=s}^{i_2-1} v_i(\hat{x}_i, \hat{x}_{i+1})$$

$$= \sigma(i_1, s, y_{i_1}, \hat{x}_s) + \sigma(s, i_2, \hat{x}_s, \hat{x}_{i_2})$$

$$\le \sigma(i_1, s) + 4d_0 + \sigma(s, i_2) + 4d_0 \le \sigma(i_1, i_2) + 8d_0.$$

Therefore $\{\tilde{y}_i\}_{i=-\infty}^{\infty}$ is a good sequence. It follows from property (P35) and (2.497) that

$$\lim_{i \to -\infty} \rho(y_i, \hat{x}_i) = 0. \tag{2.498}$$

Let $\epsilon > 0$ be given. We claim that there exists an integer $i_0 < s$ such that for each pair of integers $k_1 < i_0$ and $k_2 > s$, we have

$$\sum_{i=k_1}^{k_2-1} v_i(\tilde{y}_i, \tilde{y}_{i+1}) \le \sigma(k_1, k_2, \tilde{y}_{k_1}, \tilde{y}_{k_2}) + \epsilon. \tag{2.499}$$

In view of assumption (A) and Lemma 2.56 there exists a positive number $\delta < \epsilon$ such that:

(i) for each $i \in \mathbf{Z}$ and each $x_1, x_2, y_1, y_2 \in X$ which satisfy $\rho(x_j, y_j) \le \delta$, $j = 1, 2$, we have

$$|v_i(x_1, x_2) - v_i(y_1, y_2)| \le \epsilon/32;$$

(ii) for each pair of integers $m_1, m_2 > m_1$ and each $x_1, x_2, y_1, y_2 \in X$ which satisfy $\rho(x_1, y_1), \rho(x_2, y_2) \le \delta$, we have

$$|\sigma(m_1, m_2, x_1, x_2) - \sigma(m_1, m_2, y_1, y_2)| \le \epsilon/32.$$

Relation (2.498) implies that there exists an integer $i_0 < s$ such that

$$\rho(\tilde{y}_i, \hat{x}_i) \le \delta \text{ for all integers } i \le i_0. \tag{2.500}$$

Let $k_1 < i_0$ and $k_2 > s$ be integers. We estimate

$$\sum_{i=k_1}^{k_2-1} v_i(\tilde{y}_i, \tilde{y}_{i+1}) - \sigma(k_1, k_2, \tilde{y}_{k_1}, \tilde{y}_{k_2}).$$

By property (ii), (2.497), and (2.500),

$$|\sigma(k_1, k_2, \tilde{y}_{k_1}, \tilde{y}_{k_2}) - \sigma(k_1, k_2, \hat{x}_{k_1}, \hat{x}_{k_2})| \leq \epsilon/32. \tag{2.501}$$

Put

$$h_{k_1} = \tilde{y}_{k_1}, \; h_{k_2} = \tilde{y}_{k_2}, \; h_i = \hat{x}_i, \; i \in \{k_1, \ldots, k_2\} \setminus \{k_1, k_2\}. \tag{2.502}$$

By (2.502), (2.497), and (2.496),

$$\sum_{i=k_1}^{k_2-1} v_i(\tilde{y}_i, \tilde{y}_{i+1}) - \sum_{i=k_1}^{k_2-1} v_i(h_i, h_{i+1})$$

$$= \sum_{i=k_1}^{s-1} v_i(\tilde{y}_i, \tilde{y}_{i+1}) - \sum_{i=k_1}^{s-1} v_i(h_i, h_{i+1}) = \sum_{i=k_1}^{s-1} v_i(y_i, y_{i+1}) - \sum_{i=k_1}^{s-1} v_i(h_i, h_{i+1})$$

$$= \sigma(k_1, s, y_{k_1}, \hat{x}_s) - \sum_{i=k_1}^{s-1} v_i(h_i, h_{i+1}) \leq 0. \tag{2.503}$$

It follows from (2.502), (2.500), (2.497) and property (i) that

$$\sum_{i=k_1}^{k_2-1} v_i(h_i, h_{i+1}) - \sum_{i=k_1}^{k_2-1} v_i(\hat{x}_i, \hat{x}_{i+1})$$

$$= v_{k_1}(\tilde{y}_{k_1}, \hat{x}_{k_1+1}) - v_{k_1}(\hat{x}_{k_1}, \hat{x}_{k_1+1}) \leq \epsilon/32.$$

Together with (2.503), property (P36), and (2.501) this inequality implies that

$$\sum_{i=k_1}^{k_2-1} v_i(\tilde{y}_i, \tilde{y}_{i+1}) \leq \sum_{i=k_1}^{k_2-1} v_i(h_i, h_{i+1}) \leq \sum_{i=k_1}^{k_2-1} v_i(\hat{x}_i, \hat{x}_{i+1}) + \epsilon/32$$

$$= \sigma(k_1, k_2, \hat{x}_{k_1}, \hat{x}_{k_2}) + \epsilon/32 \leq \sigma(k_1, k_2, \tilde{y}_{k_1}, \tilde{y}_{k_2}) + \epsilon/16.$$

Hence

$$\sum_{i=k_1}^{k_2-1} v_i(\tilde{y}_i, \tilde{y}_{i+1}) \leq \sigma(k_1, k_2, \tilde{y}_k, \tilde{y}_{k_2}) + \epsilon/16$$

for each pair of integers $k_1 < i_0$ and $k_2 > s$. Since ϵ is an arbitrary positive number we conclude that

$$\sum_{i=k_1}^{k_2-1} v_i(\tilde{y}_i, \tilde{y}_{i+1}) = \sigma(k_1, k_2, \tilde{y}_{k_1}, \tilde{y}_{k_2})$$

for each pair of integers $k_1, k_2 > k_1$. In view of property (P36) and (2.497), we have

$$\tilde{y}_i = \hat{x}_i, \ i \in \mathbf{Z}, \ y_i = \hat{x}_i \text{ for all integers } i \leq s.$$

Assertion 1 is proved. Analogously we can prove assertion 2 and the lemma itself.

$\square$

Lemma 2.63 (Basic lemma). *Let ϵ be a positive number. Then there exists a positive number δ such that for each pair of integers $m_1, m_2 > m_1$ and for each sequence $\{y_i\}_{i=m_1}^{m_2} \subset X$ which satisfies*

$$\rho(\hat{x}_{m_1}, y_{m_1}), \ \rho(\hat{x}_{m_2}, y_{m_2}) \leq \delta, \tag{2.504}$$

$$\sum_{i=m_1}^{m_2-1} v_i(y_i, y_{i+1}) \leq \sigma(m_1, m_2, y_{m_1}, y_{m_2}) + \delta, \tag{2.505}$$

the inequality

$$\rho(y_i, \hat{x}_i) \leq \epsilon, \ i = m_1, \ldots, m_2 \tag{2.506}$$

holds.

Proof. In view of Lemma 2.61 there exist a real number

$$\gamma_0 \in (0, \min\{1, \epsilon\}) \tag{2.507}$$

and an integer $L_0 \geq 1$ such that the following properties hold:

(P38)   If $m_1, m_2 \in \mathbf{Z}, m_2 > m_1 \geq L_0$ and if $\{y_i\}_{i=m_1}^{m_2} \subset X$ satisfies

$$\rho(\hat{x}_j, y_j) \leq \gamma_0, \ j = m_1, m_2, \tag{2.508}$$

$$\sum_{i=m_1}^{m_2-1} v_i(y_i, y_{i+1}) \leq \sigma(m_1, m_2, y_{m_1}, y_{m_2}) + \gamma_0, \tag{2.509}$$

then

$$\rho(\hat{x}_i, y_i) \leq \epsilon, \ i = m_1, \ldots, m_2. \tag{2.510}$$

(P39)   If $m_1, m_2 \in \mathbf{Z}, m_1 < m_2 \leq -L_0$ and if $\{y_i\}_{i=m_1}^{m_2} \subset X$ satisfies (2.508) and (2.509), then (2.510) is true.

In view of property (P37) there exist a real number

$$\gamma_1 \in (0, \gamma_0) \tag{2.511}$$

and an integer $L_1 \geq 1$ such that the following property holds:

(P40)   For each integer m and each sequence $\{y_i\}_{i=m}^{m+L_1} \subset X$ which satisfy

$$\sum_{i=m}^{m+L_1-1} v_i(y_i, y_{i+1}) \le \sigma(m, m + L_1, y_m, y_{m+L_1}) + \gamma_1 \qquad (2.512)$$

there exists an integer $j \in \{m, \ldots, m + L_1\}$ such that $\rho(y_j, \hat{x}_j) \le \gamma_0$.

Choose a strictly decreasing sequence $\{\delta_i\}_{i=0}^{\infty} \subset (0, 1)$ such that

$$\delta_1 < \gamma_1, \ \lim_{i \to \infty} \delta_i = 0. \qquad (2.513)$$

Assume that the lemma is not true. Then for each integer $k \ge 1$ there exist integers

$$m_k, p_k > m_k, \ j_k \in \{m_k, \ldots, p_k\} \qquad (2.514)$$

and a sequence $\{y_i^{(k)}\}_{i=m_k}^{p_k} \subset X$ such that

$$\rho\left(y_{m_k}^{(k)}, \hat{x}_{m_k}\right) \le \delta_k, \ \rho\left(y_{p_k}^{(k)}, \hat{x}_{p_k}\right) \le \delta_k, \qquad (2.515)$$

$$\sum_{i=m_k}^{p_k-1} v_i\left(y_i^{(k)}, y_{i+1}^{(k)}\right) \le \sigma\left(m_k, p_k, y_{m_k}^{(k)}, y_{p_k}^{(k)}\right) + \delta_k, \qquad (2.516)$$

$$\rho\left(\hat{x}_{j_k}, y_{j_k}^{(k)}\right) > \epsilon. \qquad (2.517)$$

Let $k \ge 1$ be an integer. If $m_k \ge L_0$, then property (P38) and relations (2.515), (2.513), (2.511), and (2.516) imply that

$$\rho\left(\hat{x}_i, y_i^{(k)}\right) \le \epsilon, \ i = m_k, \ldots, p_k.$$

Since this inequality contradicts (2.517) we conclude that

$$m_k < L_0. \qquad (2.518)$$

If $p_k \le -L_0$, then property (P39) and relations (2.515), (2.513), (2.516), and (2.511) imply that

$$\rho\left(\hat{x}_i, y_i^{(k)}\right) \le \epsilon, \ i = m_k, \ldots, p_k.$$

Since this inequality contradicts (2.517) we conclude that

$$p_k > -L_0. \qquad (2.519)$$

We claim that

$$-2L_1 - L_0 - 2 \le j_k \le 2L_1 + 2 + L_0. \qquad (2.520)$$

Assume that

$$j_k > 2L_1 + L_0 + 2 \qquad (2.521)$$

and consider the sequence $\{y_i^{(k)}\}_{i=j_k-2L_1}^{j_k-L_1}$. In view of (2.521) and (2.518),

$$j_k - 2L_1 > L_0 + 2 > m_k.$$

By property (P40), (2.516), and (2.513), there exists an integer $s \in \{j_k - 2L_1, \ldots, j_k - L_1\}$ such that

$$\rho\left(y_s^{(k)}, \hat{x}_s\right) \le \gamma_0. \qquad (2.522)$$

By (2.521), we have

$$s \ge j_k - 2L_1 > L_0 + 2.$$

This inequality, (2.522), (2.515), (2.516), (2.513), (2.514), (2.511), and property (P38) imply that

$$\rho\left(\hat{x}_i, y_i^{(k)}\right) \le \epsilon, \ i = s, \ldots, p_k.$$

Since $s < j_k \le p_k$ we have $\rho(\hat{x}_{j_k}, y_{j_k}^{(k)}) \le \epsilon$, a contradiction [see (2.517)]. The contradiction we have reached proves that

$$j_k \le 2L_1 + L_0 + 2. \qquad (2.523)$$

Assume that

$$j_k < -2L_1 - L_0 - 2 \qquad (2.524)$$

and consider a sequence $\{y_i^{(k)}\}_{i=j_k+L_1}^{j_k+2L_1}$. Relations (2.524) and (2.519) imply that

$$j_k + 2L_1 < -L_0 - 2 < p_k.$$

It follows from property (P40), (2.516), and (2.513) that there exists an integer $s \in \{j_k + L_1, \ldots, j_k + 2L_1\}$ such that

$$\rho\left(y_s^{(k)}, \hat{x}_s\right) \le \gamma_0. \qquad (2.525)$$

In view of (2.524),

$$s \le j_k + 2L_1 < -L_0 - 2.$$

By this inequality, (2.525), (2.515), (2.516), (2.513), (2.511), and property (P39), we have

$$\rho\left(\hat{x}_i, y_i^{(k)}\right) \le \epsilon, \ i = m_k, \dots, s.$$

Since $s \ge j_k + L_1 \ge j_k \ge m_k$ we conclude that

$$\rho\left(\hat{x}_{j_k}, y_{j_k}^{(k)}\right) \le \epsilon,$$

a contradiction [see (2.517)]. The contradiction we have reached proves that

$$j_k \ge -2L_1 - L_0 - 2.$$

Therefore relation (2.520) is valid.

We can assume by extracting a subsequence and re-indexing that

$$j_k = j_1 \text{ for all natural numbers } k \tag{2.526}$$

and that one of the following conditions holds [see (2.518) and (2.519)]:

(a) the sequences $\{m_k\}_{k=1}^{\infty}$, $\{p_k\}_{k=1}^{\infty}$ are bounded;
(b) the sequence $\{m_k\}_{k=1}^{\infty}$ is bounded and $\lim_{k\to\infty} p_k = \infty$;
(c) the sequence $\{p_k\}_{k=1}^{\infty}$ is bounded and $\lim_{k\to\infty} m_k = -\infty$;
(d) $\lim_{k\to\infty} m_k = -\infty$ and $\lim_{k\to\infty} p_k = \infty$.

Assume that condition (a) holds. Then we can assume by extracting a subsequence and re-indexing that

$$m_k = m_1, \ p_k = p_1 \text{ for all integers } k \ge 1 \tag{2.527}$$

and that there exists a sequence $\{y_i\}_{i=m_1}^{p_1} \subset X$ such that

$$y_i = \lim_{k\to\infty} y_i^{(k)}, \ i = m_1, \dots, p_1. \tag{2.528}$$

By (2.528), the continuity of v_i, $i \in \mathbf{Z}$, (2.516), (2.513), (2.527), and Lemma 2.56, we have

$$\sum_{i=m_1}^{p_1-1} v_i(y_i, y_{i+1}) = \lim_{k\to\infty} \sum_{i=m_1}^{p_1-1} v_i\left(y_i^{(k)}, y_{i+1}^{(k)}\right)$$

$$= \lim_{k\to\infty} \sigma(m_1, p_1, y_{m_1}^{(k)}, y_{p_1}^{(k)})$$

$$= \sigma\left(m_1, p_1, y_{m_1}, y_{p_1}\right). \tag{2.529}$$

It follows from (2.527), (2.528), and (2.513) that

$$y_{m_1} = \hat{x}_{m_1}, \ y_{p_1} = \hat{x}_{p_1}. \tag{2.530}$$

By (2.526), (2.528), (2.517), we have

$$\rho(y_{j_1}, \hat{x}_{j_1}) \geq \epsilon. \tag{2.531}$$

Put

$$z_i = y_i, \ i = m_1 \ldots, p_1, z_i = \hat{x}_i, \ i \in I \setminus \{m_1, \ldots, p_1\}. \tag{2.532}$$

It follows from (2.532), (2.530), (2.529) and property (P36) that for each pair of integers $i_1, i_2 > i_1$,

$$\sum_{i=i_1}^{i_2-1} v_i(z_i, z_{i+1}) = \sigma(i_1, i_2, z_{i_1}, z_{i_2}).$$

By property (P36) and (2.532), $z_i = \hat{x}_i$, $i \in \mathbf{Z}$, $y_i = \hat{x}_i$, $i \in \{m_1, \ldots, p_1\}$, and in particular $y_{j_1} = \hat{x}_{j_1}$. This contradicts (2.531). The contradiction we have reached proves that condition (a) does not hold.

Thus one of the conditions (b), (c), (d) holds. We can assume by extracting a subsequence and re-indexing that

if the case (b) holds, then $m_k = m_1$ for all natural numbers k;   (2.533)

if the case (c) holds, then $p_k = p_1$ for all natural numbers k.   (2.534)

Define

$$I = \{i \in \mathbf{Z}, \ i \geq m_1\} \text{ in the case (b)}, \ I = \{i \in \mathbf{Z}, \ i \leq p_1\}$$

$$\text{in the case (c)}, \ I = \mathbf{Z} \text{ in the case (d).} \tag{2.535}$$

We can assume by extracting a subsequence and re-indexing that there exists a sequence y_i, $i \in I$ such that

$$y_i = \lim_{k \to \infty} y_i^{(k)} \text{ for all } k \in I. \tag{2.536}$$

By (2.536), (2.526), and (2.517),

$$\rho(y_{j_1}, \hat{x}_{j_1}) \geq \epsilon. \tag{2.537}$$

In view of (2.536), the continuity of v_i, $i \in \mathbf{Z}$, (2.516), (2.513), and Lemma 2.56, for each pair of integers $i_1, i_2 \in I$ satisfying $i_2 > i_1$, we have

$$\sum_{i=i_1}^{i_2-1} v_i(y_i, y_{i+1}) = \lim_{k\to\infty} \sum_{i=i_1}^{i_2-1} v_i\left(y_i^{(k)}, y_{i+1}^{(k)}\right)$$

$$= \lim_{k\to\infty} \sigma\left(i_1, i_2, y_{i_1}^{(k)}, y_{i_2}^{(k)}\right)$$

$$= \sigma(i_1, i_2, y_{i_1}, y_{i_2}). \tag{2.538}$$

Assume that condition (b) holds. Then it follows from (2.513), (2.536), and (2.515) that

$$y_{m_1} = \hat{x}_{m_1}.$$

By the equality above, (2.538), and assertion 2 of Lemma 2.62, $y_i = \hat{x}_i$ for all $i \in I$ and in particular $y_{j_1} = \hat{x}_{j_1}$. This contradicts (2.537). The contradiction we have reached proves that condition (b) does not hold.

Assume that condition (c) holds. Then it follows from (2.513), (2.536), (2.515), and (2.534) that

$$y_{p_1} = \hat{x}_{p_1}.$$

By this equality, (2.538), and assertion 1 of Lemma 2.62, $y_i = \hat{x}_i$ for all $i \in I$ and in particular $y_{j_1} = \hat{x}_{j_1}$. This contradicts (2.537). Hence condition (c) does not hold. Assume that condition (d) holds. Then in view of (2.538) and property (P36), we have

$$y_i = \hat{x}_i, \ i \in \mathbf{Z}$$

and in particular $y_{j_1} = \hat{x}_{j_1}$. This contradicts (2.537). The obtained contradiction shows that condition (d) does not hold. Thus conditions (a), (b), and (d) do not hold, a contradiction. The contradiction we have reached proves Lemma 2.63. $\qquad\square$

2.32 Proof of Theorem 2.55

Let $\{\hat{x}_i\}_{i=-\infty}^{\infty} \subset X$. Assume that $\{v_i\}_{i=-\infty}^{\infty}$ possesses (TP) and that $\{\hat{x}_i\}_{i=-\infty}^{\infty}$ is the turnpike. In view of Proposition 2.60 properties (P35)–(P37) hold.

Assume now that (P35)–(P37) hold. We claim that $\{\hat{x}_i\}_{i=-\infty}^{\infty}$ is the turnpike.

Let a positive number ϵ be given. In view of Lemma 2.63 there exists a real number $\delta_0 \in (0, \epsilon)$ such that the following property holds:

(P41)   for each pair of integers $m_1, m_2 > m_1$ and each sequence $\{y_i\}_{i=m_1}^{m_2} \subset X$
which satisfies

$$\rho(\hat{x}_{m_1}, y_{m_1}), \ \rho(\hat{x}_{m_2}, y_{m_2}) \le \delta_0,$$

$$\sum_{i=m_1}^{m_2-1} v_i(y_i, y_{i+1}) \le \sigma(m_1, m_2, y_{m_1}, y_{m_2}) + \delta_0,$$

we have

$$\rho(y_i, \hat{x}_i) \le \epsilon, \ i = m_1, \ldots, m_2.$$

Property (P37) implies that there exist a real number $\delta \in (0, \delta_0)$ and an integer
$N \ge 1$ such that for each integer m and each sequence $\{y_i\}_{i=m}^{m+N} \subset X$ which
satisfies

$$\sum_{i=m}^{m+N-1} v_i(y_i, y_{i+1}) \le \sigma(m, m+N, y_m, y_{m+N}) + \delta \qquad (2.539)$$

there exists an integer $j \in \{m, \ldots, m+N\}$ such that $\rho(y_j, \hat{x}_j) \le \delta_0$.
   Assume that $T_1, T_2 \ge T_1 + 2N$ are integers and that a sequence $\{y_i\}_{i=T_1}^{T_2} \subset X$
satisfies

$$\sum_{i=T_1}^{T_2-1} v_i(y_i, y_{i+1}) \le \sigma(T_1, T_2, y_{T_1}, y_{T_2}) + \delta. \qquad (2.540)$$

It follows from (2.540) and the choice of δ, N that there exist

$$\tau_1 \in \{T_1, \ldots, T_1 + N\}, \ \tau_2 \in \{T_2 - N, \ldots, T_2\}$$

such that

$$\rho(y_j, \hat{x}_j) \le \delta_0, \ j = \tau_1, \ \tau_2. \qquad (2.541)$$

Evidently, if $\rho(y_{T_1}, \hat{x}_{T_1}) \le \delta$, then $\tau_1 = T_1$, and if $\rho(y_{T_2}, \hat{x}_{T_2}) \le \delta$, then $\tau_2 = T_2$.
We claim that

$$\rho(y_i, \hat{x}_i) \le \epsilon, \ i = \tau_1, \ldots, \tau_2.$$

Assume that an integer j satisfies $\tau_1 < j < \tau_2$. We show that $\rho(y_j, \hat{x}_j) \le \epsilon$.
By (2.540) and the choice of δ, N [see (2.539)], there exist integers m_1, m_2 for
which

$$\tau_1 \le m_1 \le j \le m_2 \le \tau_2, \ j \le m_1 + N, \ m_2 \le j + N,$$

$$\rho(y_{m_i}, \hat{x}_{m_i}) \le \delta_0, \ i = 1, 2.$$

In view of these relations, (2.540), and property (P41), we have

$$\rho(y_i, \hat{x}_i) \le \epsilon, \ i = m_1, \ldots, m_2.$$

Therefore $\rho(y_j, \hat{x}_j) \le \epsilon$. We have shown that the sequence $\{v_i\}_{i=-\infty}^{\infty}$ has (TP) and $\{\hat{x}_i\}_{i=-\infty}^{\infty}$ is the turnpike. This completes the proof of the theorem.

2.33 An Example

Let $u_i : X \times X \to R^1$, $i = 0, \pm 1, \pm 2, \ldots$ be a sequence of continuous functions such that (2.458) is true and assumption (A) holds with $v_i = u_i$, $i \in \mathbf{Z}$. Proposition 2.59 implies that there exists a sequence $\{y_i\}_{i=-\infty}^{\infty} \subset X$ such that for each pair of integers $m_1, m_2 > m_1$, we have

$$\sum_{i=m_1}^{m_2-1} u_i(y_i, y_{i+1}) = \inf \left\{ \sum_{i=m_1}^{m_2-1} u_i(z_i, z_{i+1}) : \ \{z_i\}_{i=m_1}^{m_2} \subset X, \right.$$

$$\left. z_{m_1} = y_{m_1}, \ z_{m_2} = y_{m_2} \right\}. \tag{2.542}$$

Fix a positive number r and define

$$v_i(x, y) = u_i(x, y) + r\rho(x, y_i), \ x, y \in X, \ i \in \mathbf{Z}. \tag{2.543}$$

It is clear that (2.458) and assumption (A) hold. We denote by Card(A) the cardinality of a set A. In this section we use the notation and definitions introduced in Sect. 2.28. We show that $\{v_i\}_{i=-\infty}^{\infty}$ possesses the turnpike property with the turnpike $\{y_i\}_{i=-\infty}^{\infty}$. In view of Theorem 2.55 it is sufficient to show that properties (P35)–(P37) hold.

It follows from (2.542) and (2.543) that for each pair of integers $m_1, m_2 > m_1$,

$$\sum_{i=m_1}^{m_2-1} v_i(y_i, y_{i+1}) = \sigma(m_1, m_2, y_{m_1}, y_{m_2}). \tag{2.544}$$

We claim that property (P35) holds. Assume that a sequence $\{z_i\}_{i=-\infty}^{\infty} \subset X$ is good and ϵ is a positive number. There exists a positive constant c such that for each pair of integers $m_1, m_2 > m_1$ we have

$$\sum_{i=m_1}^{m_2-1} v_i(z_i, z_{i+1}) \le \sigma(m_1, m_2, z_{m_1}, z_{m_2}) + c. \tag{2.545}$$

By (2.543), Lemma 2.57, (2.545), and (2.461), for each pair of integers m_1, $m_2 > m_1$,

$$\sum_{i=m_1}^{m_2-1} u_i(z_i, z_{i+1}) + r \epsilon \operatorname{Card}\{i \in \{m_1, \ldots, m_2 - 1\} : \rho(z_i, y_i) \geq \epsilon\}$$

$$\leq \sum_{i=m_1}^{m_2-1} u_i(z_i, z_{i+1}) + r \sum_{i=m_1}^{m_2-1} \rho(z_i, y_i)$$

$$= \sum_{i=m_1}^{m_2-1} v_i(z_i, z_{i+1}) \leq \sigma(m_1, m_2, z_{m_1}, z_{m_2}) + c \leq c + 4d_0 + \sigma(m_1, m_2)$$

$$\leq c + 4d_0 + \sum_{i=m_1}^{m_2-1} v_i(y_i, y_{i+1}) = c + 4d_0 + \sum_{i=m_1}^{m_2-1} u_i(y_i, y_{i+1}).$$

When combined with (2.542), Lemma 2.57, (2.543), and (2.461) this relation implies that for each pair of integers $m_1, m_2 > m_1$ we have

$$\sum_{i=m_1}^{m_2-1} u_i(z_i, z_{i+1}) + r \epsilon \operatorname{Card}\{i \in \{m_1, \ldots, m_2 - 1\} :$$

$$\rho(z_i, y_i) \geq \epsilon\} \leq c + 4d_0 + \sum_{i=m_1}^{m_2-1} u_i(z_i, z_{i+1})$$

$$+ 4(d_0 + r \sup\{\rho(x_1, x_2) : x_1, x_2 \in X\}),$$

$$\operatorname{Card}\{i \in \{m_1, \ldots, m_2 - 1\} : \rho(z_i, y_i) \geq \epsilon\}$$

$$\leq (r\epsilon)^{-1}[c + 4d_0 + 8d_0 + 8r \sup\{\rho(x_1, x_2) : x_1, x_2 \in X\}].$$

Since this relation is valid for each pair of integers $m_1, m_2 > m_1$ we conclude that (P35) holds.

We claim that property (P36) holds. Assume that a sequence $\{z_i\}_{i=-\infty}^{\infty} \subset X$ satisfies

$$\sum_{i=m_1}^{m_2-1} v_i(z_i, z_{i+1}) = \sigma(m_1, m_2, z_{m_1}, z_{m_2}) \qquad (2.546)$$

for each pair of integers $m_1, m_2 > m_1$. Then the sequence $\{z_i\}_{i=-\infty}^{\infty}$ is good and by property (P35) we have

$$\lim_{i \to \infty} \rho(y_i, z_i) = 0, \quad \lim_{i \to -\infty} \rho(y_i, z_i) = 0. \qquad (2.547)$$

Let $\epsilon > 0$ be given. By Lemma 2.56, (2.547), (2.544), and (2.542), there exists an integer $L_0 \geq 1$ such that for each pair of integers $m_1 \leq -L_0$, $m_2 \geq L_0$

$$\left| \sum_{i=m_1}^{m_2-1} v_i(z_i, z_{i+1}) - \sum_{i=m_1}^{m_2-1} v_i(y_i, y_{i+1}) \right| \leq \epsilon, \tag{2.548}$$

$$\sum_{i=m_1}^{m_2-1} u_i(y_i, y_{i+1}) \leq \sum_{i=m_1}^{m_2-1} u_i(z_i, z_{i+1}) + \epsilon. \tag{2.549}$$

Let $m_1 \leq -L_0$ and $m_2 \geq L_0$ be integers. It follows from (2.548), (2.543), and (2.549) that

$$\sum_{i=m_1}^{m_2-1} u_i(z_i, z_{i+1}) + \sum_{i=m_1}^{m_2-1} r\rho(y_i, z_i) = \sum_{i=m_1}^{m_2-1} v_i(z_i, z_{i+1})$$

$$\leq \sum_{i=m_1}^{m_2-1} v_i(y_i, y_{i+1}) + \epsilon = \sum_{i=m_1}^{m_2-1} u_i(y_i, y_{i+1}) + \epsilon$$

$$\leq \sum_{i=m_1}^{m_2-1} u_i(z_i, z_{i+1}) + 2\epsilon, \quad \sum_{i=m_1}^{m_2-1} \rho(y_i, z_i) \leq \epsilon/r.$$

Thus $\rho(y_i, z_i) \leq \epsilon/r$ for all $i \in \mathbf{Z}$. Since ϵ is an arbitrary positive number we conclude that $y_i = z_i$, $i \in \mathbf{Z}$. Therefore (P36) holds.

We claim that (P37) holds. It follows from (2.542) and Lemma 2.57 that there exists a positive number d_1 such that for each pair of integers $m_1, m_2 > m_1$ and each $\{x_i\}_{i=m_1}^{m_2} \subset X$, we have

$$\sum_{i=m_1}^{m_2-1} u_i(y_i, y_{i+1}) \leq \sum_{i=m_1}^{m_2-1} u_i(x_i, x_{i+1}) + d_1. \tag{2.550}$$

Let $\epsilon > 0$ be given. Choose an integer $L_0 \geq 1$ such that

$$L_0 > \epsilon^{-1} r^{-1}(1 + 4d_0 + d_1) + 1. \tag{2.551}$$

Assume that $m \in \mathbf{Z}$ and that a sequence $\{z_i\}_{i=m}^{m+L_0}$ satisfies

$$\sum_{i=m}^{m+L_0-1} v_i(z_i, z_{i+1}) \leq \sigma(m, m + L_0, z_m, z_{m+L_0}) + 1. \tag{2.552}$$

By (2.543), (2.552), Lemma 2.57, and (2.550),

$$\sum_{i=m}^{m+L_0-1} u_i(z_i, z_{i+1}) + \sum_{i=m}^{m+L_0-1} r\rho(z_i, y_i) = \sum_{i=m}^{m+L_0-1} v_i(z_i, z_{i+1})$$

$$\le \sigma(m, m+L_0, z_m, z_{m+L_0}) + 1 \le 1 + 4d_0 + \sum_{i=m}^{m+L_0-1} v_i(y_i, y_{i+1})$$

$$= 1 + 4d_0 + \sum_{i=m}^{m+l_0-1} u_i(y_i, y_{i+1}) \le 1 + 4d_0 + d_1 + \sum_{i=m}^{m+L_0-1} u_i(z_i, z_{i+1}),$$

$$\sum_{i=m}^{m+L_0-1} \rho(z_i, y_i) \le r^{-1}(1 + 4d_0 + d_1).$$

Together with (2.551) this relation implies that there exists an integer $j \in \{m, \ldots, m + L_0 - 1\}$ for which $\rho(z_j, y_j) \le \epsilon$. Therefore property (P37) holds.

Chapter 3
Variational Problems with Extended-Valued Integrands

In this chapter we study turnpike properties of approximate solutions of an autonomous variational problem with a lower semicontinuous integrand $f : R^n \times R^n \to R^1 \cup \{\infty\}$, where R^n is the n-dimensional Euclidean space. More precisely, we consider the following variational problems:

$$\int_0^T f(v(t), v'(t))dt \to \min, \qquad (P_1)$$

$v : [0, T] \to R^n$ is an absolutely continuous (a.c.) function such that

$$v(0) = x, \ v(T) = y$$

and

$$\int_0^T f(v(t), v'(t))dt \to \min, \qquad (P_2)$$

$v : [0, T] \to R^n$ is an a. c. function such that $v(0) = x$,

where $x, y \in R^n$. Here R^n is the n-dimensional Euclidean space with the Euclidean norm $|\cdot|$ and $f : R^n \times R^n \to R^1 \cup \{\infty\}$ is an extended-valued integrand.

3.1 Turnpike Results for Variational Problems

We denote by $\mathrm{mes}(E)$ the Lebesgue measure of a Lebesgue measurable set $E \subset R^1$, by $|\cdot|$ the Euclidean norm of the space R^n, and by $\langle \cdot, \cdot \rangle$ the inner product of R^n. For each function $f : X \to R^1 \cup \{\infty\}$, where X is a nonempty, set

$$\mathrm{dom}(f) = \{x \in X : \ f(x) < \infty\}.$$

© Springer International Publishing Switzerland 2014
A.J. Zaslavski, *Turnpike Phenomenon and Infinite Horizon Optimal Control*,
Springer Optimization and Its Applications 99, DOI 10.1007/978-3-319-08828-0_3

Let a be a real positive number, $\psi : [0, \infty) \to [0, \infty)$ be an increasing function such that

$$\lim_{t \to \infty} \psi(t) = \infty \qquad (3.1)$$

and let $f : R^n \times R^n \to R^1 \cup \{\infty\}$ be a lower semicontinuous function such that the set

$$\mathrm{dom}(f) = \{(x, y) \in R^n \times R^n : f(x, y) < \infty\} \qquad (3.2)$$

is nonempty, convex, and closed and that

$$f(x, y) \geq \max\{\psi(|x|), \ \psi(|y|)|y|\} - a \text{ for each } x, y \in R^n. \qquad (3.3)$$

For each pair of points $x, y \in R^n$ and each positive number T define

$$\sigma(f, T, x) = \inf\{\int_0^T f(v(t), v'(t))dt \ : \ v : [0, T] \to R^n$$

is an absolutely continuous (a.c.) function satisfying $v(0) = x\}$, $\qquad (3.4)$

$$\sigma(f, T, x, y) = \inf\{\int_0^T f(v(t), v'(t))dt \ : \ v : [0, T] \to R^n$$

is an a. c. function satisfying $v(0) = x, \ v(T) = y\}$, $\qquad (3.5)$

$$\sigma(f, T) = \inf\{\int_0^T f(v(t), v'(t))dt \ : \ v : [0, T] \to R^n \text{ is an a.c. function}\}.$$
$$(3.6)$$

(Here we assume that infimum over an empty set is infinity.)

We suppose that there exists a point $\bar{x} \in R^n$ such that

$$f(\bar{x}, 0) \leq f(x, 0) \text{ for each } x \in R^n \qquad (3.7)$$

and that the following assumptions hold:

(A1) $(\bar{x}, 0)$ is an interior point of the set $\mathrm{dom}(f)$ and the function f is continuous at the point $(\bar{x}, 0)$;

(A2)   for each positive number M there exists a positive number c_M such that

$$\sigma(f, T, x) \geq Tf(\bar{x}, 0) - c_M$$

for each point $x \in R^n$ satisfying $|x| \leq M$ and each real number $T > 0$;

(A3)   for each point $x \in R^n$ the function $f(x, \cdot) : R^n \to R^1 \cup \{\infty\}$ is convex.

Assumption (A2) implies that for each a.c. function $v : [0, \infty) \to R^n$ the function

$$T \to \int_0^T f(v(t), v'(t))dt - Tf(\bar{x}, 0), \ T \in (0, \infty)$$

is bounded from below.

It should be mentioned that inequality (3.7) and assumptions (A1)–(A3) are common in the literature and hold for many infinite horizon optimal control problems [16, 56]. In particular, we need inequality (3.7) and assumption (A2) in the cases when the problem (P) possesses the turnpike property and the point $\bar{x}$ is its turnpike. Assumption (A2) means that the constant function $\bar{v}(t) = \bar{x}$, $t \in [0, \infty)$ is an approximate solution of the infinite horizon variational problem with the integrand f related to the problems (P_1) and (P_2).

We say that an a.c. function $v : [0, \infty) \to R^n$ is (f)-good [16, 56] if

$$\sup\{| \int_0^T f(v(t), v'(t))dt - Tf(\bar{x}, 0)| : \ T \in (0, \infty)\} < \infty.$$

The following result obtained in [59] will be proved in Sect. 3.3.

Proposition 3.1. *Let $v : [0, \infty) \to R^n$ be an a.c. function. Then either the function v is (f)-good or*

$$\int_0^T f(v(t), v'(t))dt - Tf(\bar{x}, 0) \to \infty \text{ as } T \to \infty.$$

Moreover, if the function v is (f)-good, then $\sup\{|v(t)| : \ t \in [0, \infty)\} < \infty$.

For each pair of number $T_1 \in R^1$, $T_2 > T_1$ and each a.c. function $v : [T_1, T_2] \to R^n$ put

$$I^f(T_1, T_2, v) = \int_{T_1}^{T_2} f(v(t), v'(t))dt. \tag{3.8}$$

For each positive number M denote by X_M the set of all points $x \in R^n$ such that $|x| \leq M$ and there exists an a.c. function $v : [0, \infty) \to R^n$ which satisfies

$$v(0) = x, \ I^f(0, T, v) - Tf(\bar{x}, 0) \leq M \text{ for each } T \in (0, \infty). \tag{3.9}$$

It is clear that $\cup\{X_M : \ M \in (0, \infty)\}$ is the set of all points $x \in X$ for which there exists an (f)-good function $v : [0, \infty) \to R^n$ such that $v(0) = x$.

We suppose that the following assumption holds:

(A4) (the asymptotic turnpike property) for each (f)-good function $v : [0, \infty) \to R^n$, $\lim_{t \to \infty} |v(t) - \bar{x}| = 0$.

The following turnpike result for the problem (P_2) was established in [59].

Theorem 3.2. *Let ϵ, M be positive numbers. Then there exist an integer $L \geq 1$ and a real number $\delta > 0$ such that for each real number $T > 2L$ and each a.c. function $v : [0, T] \to R^n$ which satisfies*

$$v(0) \in X_M \text{ and } I^f(0, T, v) \leq \sigma(f, T, v(0)) + \delta$$

there exist a pair of numbers $\tau_1 \in [0, L]$ and $\tau_2 \in [T - L, T]$ such that

$$|v(t) - \bar{x}| \leq \epsilon \text{ for all } t \in [\tau_1, \tau_2]$$

and if $|v(0) - \bar{x}| \leq \delta$, then $\tau_1 = 0$.

Theorem 3.2 will be proved in Sect. 3.5.

In the sequel we use a notion of an overtaking optimal function [16, 56].

An a.c. function $v : [0, \infty) \to R^n$ is called (f)-overtaking optimal if for each a.c. function $u : [0, \infty) \to R^n$ satisfying $u(0) = v(0)$ the inequality

$$\limsup_{T \to \infty} [I^f(0, T, v) - I^f(0, T, u)] \leq 0$$

holds.

The following result which establishes the existence of an overtaking optimal function was obtained in [59].

Theorem 3.3. *Assume that $x \in R^n$ and that there exists an (f)-good function $v : [0, \infty) \to R^n$ satisfying $v(0) = x$. Then there exists an (f)-overtaking optimal function $u_* : [0, \infty) \to R^n$ such that $u_*(0) = x$.*

Theorem 3.3 will be proved in Sect. 3.6.

Denote by $\text{Card}(A)$ the cardinality of the set A.

Let M be a positive number. Denote by Y_M the set of all points $x \in R^n$ for which there exist a number $T \in (0, M]$ and an a. c. function $v : [0, T] \to R^n$ such that $v(0) = \bar{x}$, $v(T) = x$ and $I^f(0, T, v) \leq M$.

The following turnpike results for the problems (P_1) were established in [67]. They are proved in Sects. 3.7 and 3.8, respectively.

Theorem 3.4. *Let $\epsilon, M_0, M_1, M_2 > 0$. Then there exist an integer $Q \geq 1$ and a positive number L such that for each number $T > L$, each point $z_0 \in X_{M_0}$, and each point $z_1 \in Y_{M_1}$, the value $\sigma(f, T, z_0, z_1)$ is finite and for each a.c. function $v : [0, T] \to R^n$ which satisfies*

$$v(0) = z_0, \ v(T) = z_1, \ I^f(0, T, v) \leq \sigma(f, T, z_0, z_1) + M_2,$$

there exists a finite sequence of closed intervals $[a_i, b_i] \subset [0, T]$, $i = 1, \ldots, q$ such that

$$q \leq Q, \ b_i - a_i \leq L, \ i = 1, \ldots, q,$$
$$|v(t) - \bar{x}| \leq \epsilon, \ t \in [0, T] \setminus \cup_{i=1}^q [a_i, b_i].$$

Theorem 3.5. *Let $\epsilon, M_0, M_1 > 0$. Then there exist numbers $L, \delta > 0$ such that for each number $T > 2L$, each point $z_0 \in X_{M_0}$, and each point $z_1 \in Y_{M_1}$, the value $\sigma(v, T, z_0, z_1)$ is finite and for each a.c. function $v : [0, T] \to R^n$ which satisfies*

$$v(0) = z_0, \ v(T) = z_1, \ I^f(0, T, v) \leq \sigma(f, T, z_0, z_1) + \delta$$

there exists a pair of numbers $\tau_1 \in [0, L], \tau_2 \in [T - L, T]$ such that

$$|v(t) - \bar{x}| \leq \epsilon, \ t \in [\tau_1, \tau_2].$$

Moreover if $|v(0) - \bar{x}| \leq \delta$, then $\tau_1 = 0$ and if $|v(T) - \bar{x}| \leq \delta$, then $\tau_2 = T$.

Examples of integrands f which satisfy assumptions (A1)–(A4) are considered in Sect. 3.9.

3.2 Three Propositions

Proposition 3.6. *Let $M_0, M_1 > 0$. Then there exists a positive number M_2 such that for each positive number T and each a.c. function $v : [0, T] \to R^n$ which satisfies*

$$|v(0)| \leq M_0, \ I^f(0, T, v) \leq Tf(\bar{x}, 0) + M_1, \tag{3.10}$$

the following inequality holds:

$$|v(t)| \leq M_2 \text{ for all } t \in [0, T]. \tag{3.11}$$

Proof. In view of relation (3.1) there exists a number $\Gamma > M_0 + 1$ such that

$$\psi(\Gamma) > 2|f(\bar{x}, 0)| + 4 + a. \tag{3.12}$$

Assumption (A2) implies that there exists a positive number $c(\Gamma)$ such that

$$\sigma(f, T, x) \geq Tf(\bar{x}, 0) - c(\Gamma) \text{ for each } T > 0$$

and each $x \in R^n$ satisfying $|x| \leq \Gamma$. $\tag{3.13}$

Choose number $M_2 > 0$ such that

$$M_2 > 4\Gamma + 4 + (M_1 + 2c(\Gamma))(4\Gamma + 2a + 1 + |f(\bar{x}, 0)|). \tag{3.14}$$

Assume that T is positive number and that an a.c. function $v : [0, T] \to R^n$ satisfies (3.10). We claim that inequality (3.11) holds. Assume the contrary. Then there exists a number $t_0 \in [0, T]$ such that

$$|v(t_0)| > M_2. \tag{3.15}$$

By (3.15), (3.10), (3.14), and the inequality $\Gamma > M_0 + 1$, we have

$$t_0 \in (0, T]. \tag{3.16}$$

It follows from (3.15), (3.16), (3.10), (3.14), and the inequality $\Gamma > M_0 + 1$ that there exists a number $t_1 \in (0, t_0)$ such that

$$|v(t_1)| = \Gamma \text{ and } |v(t)| > \Gamma \text{ for each } t \in (t_1, t_0). \tag{3.17}$$

There are two cases:

$$|v(t)| \geq \Gamma, \ t \in [t_0, T]; \tag{3.18}$$

$$\inf\{|v(t)| : t \in [t_0, T]\} < \Gamma. \tag{3.19}$$

If relation (3.18) holds, then we put $t_2 = T$. If relation (3.19) is valid, then there exists a number

$$t_2 \in (t_0, T) \tag{3.20}$$

for which

$$|v(t_2)| = \Gamma \text{ and } |v(t)| > \Gamma \text{ for each } t \in (t_0, t_2). \tag{3.21}$$

In view of (3.4), the choice of t_2, (3.21), and (3.13), we have

$$I^f(t_2, T, v) \geq \sigma(f, T - t_2, v(t_2)) \geq (T - t_2) f(\bar{x}, 0) - c(\Gamma). \tag{3.22}$$

Relations (3.4), (3.17), and (3.13) imply that

$$I^f(0, t_1, v) \geq \sigma(f, t_1, v(0)) \geq t_1 f(\bar{x}, 0) - c(\Gamma). \tag{3.23}$$

By (3.10) and (3.22),

$$I^f(0, t_2, v) - t_2 f(\bar{x}, 0) = I^f(0, T, v) - T f(\bar{x}, 0) - I^f(t_2, T, v) + (T - t_2) f(\bar{x}, 0)$$

$$\leq M_1 - [I^f(t_2, T, v) - (T - t_2) f(\bar{x}, 0)] \leq M_1 + c(\Gamma). \tag{3.24}$$

In view of (3.24) and (3.23),

$$I^f(t_1, t_2, v) - (t_2 - t_1) f(\bar{x}, 0) = I^f(0, t_2, v) - t_2 f(\bar{x}, 0)$$

$$- [I^f(0, t_1, v) - t_1 f(\bar{x}, 0)] \leq M_1 + 2c(\Gamma). \tag{3.25}$$

It follows from (3.17) and the choice of t_2 (see (3.18), (3.21)) that

$$|v(t)| \geq \Gamma \text{ for all } t \in [t_1, t_2]. \tag{3.26}$$

When combined with (3.3) and (3.12) the inequality above implies that for all numbers $t \in [t_1, t_2]$ (a.e.), we have

$$f(v(t), v'(t)) \geq \psi(|v(t)|) - a \geq \psi(\Gamma) - a \geq 2|f(\bar{x}, 0)| + 4$$

and

$$I^f(t_1, t_2, v) - (t_2 - t_1)f(\bar{x}, 0) \geq 4(t_2 - t_1). \tag{3.27}$$

Together with (3.27) this inequality implies that

$$t_2 - t_1 \leq M_1 + c(\Gamma). \tag{3.28}$$

Put

$$E_1 = \{t \in [t_1, t_0] : |v'(t)| \geq \Gamma\}, \ E_2 = [t_1, t_0] \setminus E_1. \tag{3.29}$$

It follows from (3.17) and (3.29) that

$$M_2 - \Gamma \leq |v(t_0)| - |v(t_1)| \leq |v(t_0) - v(t_1)| \leq \int_{t_1}^{t_0} |v'(t)|t = \int_{E_1} |v'(t)| dt$$

$$+ \int_{E_2} |v'(t)| dt \leq \int_{E_1} |v'(t)| dt + (t_0 - t_1)\Gamma.$$

When combined with (3.28) and the choice of t_2 this relation implies that

$$\int_{E_1} |v'(t)| dt \geq M_2 - \Gamma - \Gamma(M_1 + c(\Gamma)). \tag{3.30}$$

In view of (3.3), (3.29), (3.12), (3.28), the choice of t_2, and (3.30), we have

$$\int_{E_1} f(v(t), v'(t)) dt \geq \int_{E_1} [\psi(|v'(t)|)|v'(t)| - a] dt$$

$$\geq \int_{E_1} \psi(|v'(t)|)|v'(t)| dt - a(t_0 - t_1)$$

$$\geq 4 \int_{E_1} |v'(t)| dt - a(M_1 + c(\Gamma))$$

$$\geq 4(M_2 - \Gamma) - 4\Gamma(M_1 + c(\Gamma)) - a(M_1 + c(\Gamma)). \tag{3.31}$$

By the choice of t_2 (see (3.18), (3.20)), (3.29), (3.3), (3.14), (3.31), and (3.28),

$$\int_{t_1}^{t_2} f(v(t), v'(t))dt = \int_{E_1} f(v(t), v'(t))dt$$

$$+ \int_{E_2} f(v(t), v'(t))dt + \int_{t_0}^{t_2} f(v(t), v'(t))dt$$

$$\geq \int_{E_1} f(v(t), v'(t))dt - a(\text{mes}(E_2)) - a(t_2 - t_0)$$

$$\geq \int_{E_1} f(v(t), v'(t))dt - a(t_2 - t_1)$$

$$\geq 2M_2 - (M_1 + c(\Gamma))(4\Gamma + a) - a(M_1 + c(\Gamma)).$$

By this inequality, (3.25) and (3.28),

$$2M_2 \leq (M_1 + c(\Gamma))(4\Gamma + 2a) + I^f(t_1, t_2, v)$$

$$\leq (M_1 + c(\Gamma))(4\Gamma + 2a) + M_1 + 2c(\Gamma) + (t_2 - t_1)f(\bar{x}, 0)$$

$$\leq (M_1 + 2c(\Gamma))(4\Gamma + 2a + 1) + |f(\bar{x}, 0)|(M_1 + c(\Gamma))$$

$$\leq (M_1 + 2c(\Gamma))(4\Gamma + 2a + 1 + |f(\bar{x}, 0)|).$$

This inequality contradicts (3.14). The contradiction we have reached proves that (3.11) holds. Proposition 3.6 is proved.   □

Proposition 3.7 (Chap. 10 of [9,10,19]). . Let T be a positive number and let v_k : $[0, T] \to R^n$, $k = 1, 2, \ldots$ be a sequence of a.c. functions such that the sequence $\{I^f(0, T, v_k)\}_{k=1}^{\infty}$ is bounded and that the sequence $\{v_k(0)\}_{k=1}^{\infty}$ is bounded. Then there exist a strictly increasing sequence of natural numbers $\{k_i\}_{i=1}^{\infty}$ and an a.c. function $v : [0, T] \to R^n$ such that

$$v_{k_i}(t) \to v(t) \text{ as } i \to \infty \text{ uniformly on } [0, T],$$

$$I^f(0, T, v) \leq \liminf_{i \to \infty} I^f(0, T, v_{k_i}).$$

Proposition 3.8. Let ϵ be a positive number. Then there exists a positive number δ such that if an a.c. function $v : [0, 1] \to R^n$ satisfies $|v(0) - \bar{x}|$, $|v(1) - \bar{x}| \leq \delta$, then

$$I^f(0, 1, v) \geq f(\bar{x}, 0) - \epsilon.$$

Proof. Assumption (A2) implies that the following property holds:

(P1) $I^f(0, 1, u) \geq f(\bar{x}, 0)$ for each a.c. function $u : [0, 1] \to R^n$ satisfying $u(0) = u(1) = \bar{x}$.

Assume that the proposition is wrong. Then for each natural number i there exists an a.c. function $v_i : [0, 1] \to R^n$ such that

$$|v_i(0) - \bar{x}|, \ |v_i(1) - \bar{x}| \leq 1/i, \ I^f(0, 1, v_i) < f(\bar{x}, 0) - \epsilon. \tag{3.32}$$

In view of Proposition 3.7 extracting a subsequence and re-indexing if necessary we may assume that there exists an a.c. function $v : [0, 1] \to R^n$ such that

$$v_i(t) \to v(t) \text{ as } i \to \infty \text{ uniformly on } [0, 1],$$

$$I^f(0, 1, v) \leq \liminf_{i \to \infty} I^f(0, 1, v_i) \leq f(\bar{x}, 0) - \epsilon.$$

When combined with (3.32) this implies that

$$v(0) = \bar{x}, \ v(1) = \bar{x}, \ I^f(0, 1, v) \leq f(\bar{x}, 0) - \epsilon.$$

These relations contradict property (P1). The contradiction we have reached proves Proposition 3.8. $\qquad\square$

3.3   Proof of Proposition 3.1

In view of assumption (A2) there exists a positive number c_0 such that

$$\int_0^T f(v(t), v'(t))dt - Tf(\bar{x}, 0) \geq -c_0 \text{ for each } T > 0. \tag{3.33}$$

Assume that there exists a strictly increasing sequence of numbers $\{T_k\}_{k=1}^\infty$ such that

$$T_k \geq k \text{ for each natural number } k, \tag{3.34}$$

$$\sup\{I^f(0, T_k, v) - T_k f(\bar{x}, 0) : \ k \text{ is a natural number }\} < \infty. \tag{3.35}$$

In order to prove the proposition it is sufficient to show that the function v is (f)-good and that $\sup\{|v(t)| : \ t \in [0, \infty)\} < \infty$. In view of (3.1) and (3.3) there exists a positive number M_0 such that

$$M_0 > |v(0)| + 1, \tag{3.36}$$

$$f(y, z) \geq 2(|f(\bar{x}, 0)| + 1) \text{ for each } y, z \in R^n \text{ satisfying } |y| \geq 4^{-1}M_0.$$

We claim that

$$\liminf_{t \to \infty} |v(t)| < M_0.$$

Assume the contrary. Then there exists a positive number S_0 such that

$$|v(t)| \geq 2^{-1} M_0 \text{ for each } t \geq S_0. \tag{3.37}$$

It follows from (3.3), (3.37), and (3.36) that for each integer $k \geq 1$ such that $T_k > S_0$, we have

$$\int_0^{T_k} f(v_k(t), v_k'(t)) dt - T_k f(\bar{x}, 0) = \int_0^{S_0} f(v(t), v'(t)) dt - S_0 f(\bar{x}, 0)$$
$$+ \int_{S_0}^{T_k} f(v(t), v'(t)) dt - (T_k - S_0) f(\bar{x}, 0)$$
$$\geq S_0(-a - f(\bar{x}, 0)) + (T_k - S_0)[2(|f(\bar{x}, 0)| + 1)$$
$$- f(\bar{x}, 0)] \rightarrow \infty \text{ as } k \rightarrow \infty.$$

This relation contradicts inequality (3.35). The contradiction we have reached proves that

$$\liminf_{t \rightarrow \infty} |v(t)| < M_0, \tag{3.38}$$

as claimed.

Assumption (A2) implies that there exists a positive number c_1 such that

$$\sigma(f, T, x) \geq T f(\bar{x}, 0) - c_1 \text{ for each } T > 0$$

$$\text{and each } x \in R^n \text{ satisfying } |x| \leq M_0. \tag{3.39}$$

By (3.35) there exists a positive number c_2 such that

$$I^f(0, T_k, v) - T_k f(\bar{x}, 0) \leq c_2 \text{ for each integer } k \geq 1. \tag{3.40}$$

Let $T > 0$ be given. In view of (3.38) there exists a number $\tau \geq T$ such that

$$|v(\tau)| \leq M_0;$$

$$\text{if a number } t \text{ satisfies } T \leq t < \tau, \text{ then } |v(t)| > M_0. \tag{3.41}$$

It follows from (3.41) that

$$I^f(0, T, v) - T f(\bar{x}, 0) = I^f(0, \tau, v) - \tau f(\bar{x}, 0) - I^f(T, \tau, v)$$
$$+ (\tau - T) f(\bar{x}, 0)$$
$$\leq I^f(0, \tau, v) - \tau f(\bar{x}, 0)$$
$$- (\tau - T)[2(|f(\bar{x}, 0)| + 1) - f(\bar{x}, 0)]$$
$$\leq I^f(0, \tau, v) - \tau f(\bar{x}, 0). \tag{3.42}$$

Choose an integer $k \geq 1$ such that

$$T_k > \tau + 1. \tag{3.43}$$

In view of (3.41) and (3.39),

$$I^f(\tau, T_k, v) \geq \sigma(f, T_k - \tau, v(\tau)) \geq (T_k - \tau) f(\bar{x}, 0) - c_1. \tag{3.44}$$

By (3.42), (3.40), and (3.44), we have

$$I^f(0, T, v) - Tf(\bar{x}, 0) \leq I^f(0, \tau, v) - \tau f(\bar{x}, 0)$$
$$\leq I^f(0, T_k, v) - T_k f(\bar{x}, 0) - I^f(\tau, T_k, v) + (T_k - \tau) f(\bar{x}, 0)$$
$$\leq c_2 - I^f(\tau, T_k, v) + (T_k - \tau) f(\bar{x}, 0)$$
$$\leq c_2 - (T_k - \tau) f(\bar{x}, 0) + c_1 + (T_k - \tau) f(\bar{x}, 0) = c_2 + c_1.$$

Thus we have shown that for each positive number T,

$$I^f(0, T, v) - Tf(\bar{x}, 0) \leq c_2 + c_1 \tag{3.45}$$

and the function v is (f)-good. By (3.45) and Proposition 3.6,

$$\sup\{|v(t)| : t \in [0, \infty)\} < \infty.$$

Proposition 3.1 is proved.

3.4 Auxiliary Results

In view of assumption (A4), for each (f)-good function $v : [0, \infty) \to R^n$, we have

$$\lim_{t \to \infty} |v(t) - \bar{x}| = 0. \tag{3.46}$$

Lemma 3.9. *Let M, ϵ be positive numbers. Then there exists a positive number T such that for each a.c. function $v : [0, T] \to R^n$ which satisfies*

$$|v(0)| \leq M, \ I^f(0, T, v) \leq Tf(\bar{x}, 0) + M$$

the following inequality holds:

$$\min\{|v(t) - \bar{x}| : t \in [0, T]\} \leq \epsilon.$$

Proof. Assume the contrary. Then for each natural number k there exists an a.c. function $v_k : [0, k] \to R^n$ such that

$$|v_k(0)| \leq M, \ I^f(0, k, v_k) \leq k f(\bar{x}, 0) + M, \tag{3.47}$$

$$\min\{|v_k(t) - \bar{x}| : \ t \in [0, k]\} > \epsilon. \tag{3.48}$$

It follows from Proposition 3.6 and (3.47) that there exists a positive number M_1 such that for each natural number k, we have

$$|v_k(t)| \leq M_1, \ t \in [0, k]. \tag{3.49}$$

Assumption (A2) implies that there exists a positive number c_1 such that

$$\sigma(f, T, x) \geq T f(\bar{x}, 0) - c_1 \text{ for each } T > 0$$

and each point $x \in R^n$ satisfying $|x| \leq M_1$. \hfill (3.50)

Let q be a natural number. By (3.47), (3.49), and (3.50), for each integer $k > q$, we have

$$I^f(0, q, v_k) - q f(\bar{x}, 0) = I^f(0, k, v_k) - k f(\bar{x}, 0) - [I^f(q, k, v_k) - (k - q) f(\bar{x}, 0)]$$

$$\leq M - [I^f(q, k, v_k) - (k - q) f(\bar{x}, 0)]$$

$$\leq M - [\sigma(f, k - q, v_k(q)) - (k - q) f(\bar{x}, 0)] \leq M + c_1$$

and

$$I^f(0, q, v_k) \leq q f(\bar{x}, 0) + M + c_1 \text{ for each integer } k > q. \tag{3.51}$$

Relations (3.51), (3.49) and Proposition 3.7 imply that there exist a subsequence $\{v_{k_i}\}_{i=1}^{\infty}$ and an a.c. function $v : [0, \infty) \to R^n$ such that for each integer $q \geq 1$,

$$v_{k_i}(t) \to v(t) \text{ as } i \to \infty \text{ uniformly on } [0, q], \tag{3.52}$$

$$I^f(0, q, v) \leq q f(\bar{x}, 0) + M + c_1. \tag{3.53}$$

In view of Proposition 3.1 and (3.53), v is an (f)-good function. By assumption (A4),

$$\lim_{t \to \infty} v(t) = \bar{x}.$$

Thus there exists a positive number τ such that

$$|v(\tau) - \bar{x}| < \epsilon/4.$$

Together with relation (3.52) this inequality implies that there exists a natural number i such that $k_i > \tau$ and

$$|v_{k_i}(\tau) - v(\tau)| < \epsilon/4.$$

Now we have

$$|v_{k_i}(\tau) - \bar{x}| \leq |v_{k_i}(\tau) - v(\tau)| + |v(\tau) - \bar{x}| < \epsilon/2.$$

This inequality contradicts (3.48). The contradiction we have reached proves Lemma 3.9.   □

Lemma 3.10. *Let M, ϵ be positive numbers. Then there exists a positive number L_0 such that for each number $T \geq L_0$, each a.c. function $v : [0, T] \rightarrow R^n$ satisfying*

$$|v(0)| \leq M, \ I^f(0, T, v) \leq Tf(\bar{x}, 0) + M, \tag{3.54}$$

and each number $s \in [0, T - L_0]$ the inequality

$$\min\{|v(t) - \bar{x}| : \ t \in [s, s + L_0]\} \leq \epsilon$$

holds.

Proof. In view of Proposition 3.6 there exists a number $M_0 > M$ such that for each positive number T and each a.c. function $v : [0, T] \rightarrow R^n$ which satisfies (3.54), we have

$$|v(t)| \leq M_0, \ t \in [0, T]. \tag{3.55}$$

Assumption (A2) implies that there exists a positive number c_0 such that

$$\sigma(f, T, x) \geq Tf(\bar{x}, 0) - c_0 \text{ for each positive number } T$$

$$\text{and each } x \in R^n \text{ satisfying } |x| \leq M_0. \tag{3.56}$$

It follows from Lemma 3.9 that there exists a positive number L_0 such that for each a.c. function $v : [0, L_0] \rightarrow R^n$ which satisfies

$$|v(0)| \leq M_0, \ I^f(0, L_0, v) \leq L_0 f(\bar{x}, 0) + M + 2c_0,$$

we have

$$\min\{|v(t) - \bar{x}| : \ t \in [0, L_0]\} \leq \epsilon. \tag{3.57}$$

Assume that $T \geq L_0$, an a.c. function $v : [0, T] \rightarrow R^n$ satisfies (3.54) and that a number $S \in [0, T - L_0]$. In view of the choice of M_0,

$$|v(S)| \leq M_0, \ |v(S + L_0)| \leq M_0. \tag{3.58}$$

It follows from the choice of c_0, (3.56), (3.54), and (3.58) that

$$I^f(0, S, v) \geq \sigma(f, S, v(0)) \geq Sf(\bar{x}, 0) - c_0, \tag{3.59}$$

$$I^f(S + L_0, T, v) \geq \sigma(f, T - (S + L_0), v(S + L_0)) \geq (T - (S + L_0))f(\bar{x}, 0) - c_0. \tag{3.60}$$

It follows from (3.54), (3.59), and (3.60) that

$$\begin{aligned}
I^f(S, S + L_0, v) &= I^f(0, T, v) - I^f(0, S, v) - I^f(S + L_0, T, v) \\
&\leq Tf(\bar{x}, 0) + M - Sf(\bar{x}, 0) + c_0 - (T - S - L_0)f(\bar{x}, 0) + c_0 \\
&= L_0 f(\bar{x}, 0) + M + 2c_0. \tag{3.61}
\end{aligned}$$

By (3.61), (3.58), and the choice of L_0 (see (3.57)),

$$\min\{|v(t) - \bar{x}| : t \in [S, L_0 + S]\} \leq \epsilon.$$

Lemma 3.10 is proved.   □

Assumption (A1) implies that there exists a number $\bar{r} \in (0, 1)$ such that

$$\Omega_0 := \{(x, y) \in R^n \times R^n : |x - \bar{x}| \leq \bar{r} \text{ and } |y| \leq \bar{r}\} \subset \text{dom}(f); \tag{3.62}$$

$$\Delta_0 := \sup\{|f(z_1, z_2)| : (z_1, z_2) \in \Omega_0\} < \infty. \tag{3.63}$$

It is easy to see that the value $\sigma(f, T, x, y)$ is finite for each number $T \geq 1$ and each pair of points $x, y \in R^n$ such that $|x - \bar{x}|, |y - \bar{x}| \leq \bar{r}/2$.

Lemma 3.11. *Let ϵ be a positive number. Then there exists a number $\delta \in (0, \bar{r}/2)$ such that for each number $T \geq 2$ and each a.c. function $v : [0, T] \to R^n$ which satisfies*

$$|v(0) - \bar{x}|, \ |v(T) - \bar{x}| \leq \delta,$$

$$I^f(0, T, v) \leq \sigma(f, T, v(0), v(T)) + \delta,$$

the inequality $|v(t) - \bar{x}| \leq \epsilon$ is true for all numbers $t \in [0, T]$.

Proof. In view of assumption (A1), for each integer $k \geq 1$, there exists a number

$$\delta_k \in (0, 4^{-k}\bar{r}) \tag{3.64}$$

such that

$$|f(x, y) - f(\bar{x}, 0)| \leq 4^{-k} \tag{3.65}$$

for each pair of points $x, y \in R^n$ satisfying

$$|x - \bar{x}|, \ |y| \le 2\delta_k. \tag{3.66}$$

We may assume without loss of generality that $\delta_{k+1} < \delta_k$ for all natural numbers k.

Assume that the lemma is wrong. Then for each integer $k \ge 1$ there exist a number $T_k \ge 2$ and an a.c. function $v_k : [0, T_k] \to R^n$ such that

$$|v_k(0) - \bar{x}|, \ |v_k(T_k) - \bar{x}| \le \delta_k, \tag{3.67}$$

$$I^f(0, T_k, v_k) \le \sigma(f, T_k, v_k(0), v_k(T_k)) + \delta_k, \tag{3.68}$$

$$\max\{|v_k(t) - \bar{x}| : t \in [0, T_k]\} > \epsilon. \tag{3.69}$$

Let k be a natural number. Define an a.c. function $u_k : [0, T_k] \to R^n$ by

$$u_k(t) = v_k(0) + t(\bar{x} - v_k(0)), \ t \in [0, 1], \ u_k(t) = \bar{x}, \ t \in (1, T_k - 1],$$

$$u_k(t) = \bar{x} + (T_k - t - 1)(v_k(T_k) - \bar{x}), \ t \in (T_k - 1, T_k]. \tag{3.70}$$

It follows from (3.70) and (3.69) that for each number $t \in [0, 1] \cup [T_k - 1, T_k]$, we have

$$|u_k(t) - \bar{x}|, \ |u'_k(t)| \le \delta_k. \tag{3.71}$$

By (3.71) and (3.65), for $t \in [0, 1] \cup [T_k - 1, T_k]$ a.e.,

$$|f(u_k(t), u'_k(t)) - f(\bar{x}, 0)| \le 4^{-k}. \tag{3.72}$$

In view of (3.68), (3.70), (3.72), and (3.54),

$$I^f(0, T_k, v_k) \le \sigma(f, T_k, v_k(0), v_k(T_k)) + \delta_k \le I^f(0, T_k, u_k) + \delta_k$$

$$= I^f(0, 1, u_k) + I^f(T_k - 1, T_k, u_k) + (T_k - 2)f(\bar{x}, 0) + \delta_k$$

$$\le T_k f(\bar{x}, 0) + 2 \cdot 4^{-k} + \delta_k \le T_k f(\bar{x}, 0) + 3 \cdot 4^{-k}. \tag{3.73}$$

Put

$$\bar{v}_k(t) = v_k(t), \ t \in [0, T_k], \ \bar{v}_k(T_k + t) = v_k(T_k) + t(v_{k+1}(0) - v_k(T_k)), \ t \in (0, 1]. \tag{3.74}$$

It is clear that $\bar{v}_k : [0, T_k + 1] \to R^n$ is an a.c. function,

$$\bar{v}_k(0) = v_k(0), \ \bar{v}_k(T_k + 1) = v_{k+1}(0). \tag{3.75}$$

It follows from (3.74), (3.67) and the inequality $\delta_{k+1} < \delta_k$, for $t \in [T_k, T_k + 1]$, we have

$$
\begin{aligned}
|\bar{v}_k(t) - \bar{x}| &= |(1 - t + T_k)v_k(T_k) + (t - T_k)v_{k+1}(0) - \bar{x}| \\
&\le (1 - t + T_k)|v_k(T_k) - \bar{x}| + (t - T_k)|v_{k+1}(0) - \bar{x}| \\
&\le (1 - t + T_k)\delta_k + (t - T_k)\delta_{k+1} \le \delta_k,
\end{aligned}
\tag{3.76}
$$

$$
|\bar{v}_k'(t)| = |v_{k+1}(0) - v_k(T_k)| \le |v_{k+1}(0) - \bar{x}| + |\bar{x} - v_k(T_k)| \le \delta_{k+1} + \delta_k \le 2\delta_k.
\tag{3.77}
$$

By relations (3.74), (3.77), (3.65), and (3.66), for $t \in [T_k, T_k + 1]$ a.e.,

$$
|f(\bar{v}_k(t), \bar{v}_k'(t)) - f(\bar{x}, 0)| \le 4^{-k}.
\tag{3.78}
$$

Relations (3.74), (3.73), and (3.78) imply that

$$
\begin{aligned}
I^f(0, T_k + 1, \bar{v}_k) &= I^f(0, T_k, v_k) + I^f(T_k, T_k + 1, \bar{v}_k) \\
&\le T_k f(\bar{x}, 0) + 3 \cdot 4^{-k} + f(\bar{x}, 0) + 4^{-k} \\
&= (T_k + 1) f(\bar{x}, 0) + 4^{-k+1}.
\end{aligned}
\tag{3.79}
$$

In view of relation (3.75), there exists an a.c. function $u : [0, \infty) \to R^n$ such that

$$
u(t) = \bar{v}_1(t), \ t \in [0, T_1 + 1]
\tag{3.80}
$$

and that for each natural number k

$$
u\left(\sum_{i=1}^{k}(T_i + 1) + t\right) = \bar{v}_{k+1}(t), \ t \in [0, T_{k+1} + 1].
\tag{3.81}
$$

By (3.80), (3.81), and (3.79) for each natural number k

$$
\begin{aligned}
I^f\left(0, \sum_{i=1}^{k+1}(T_i + 1), u\right) &= \sum_{i=1}^{k} I^f\left(\sum_{j=1}^{i}(T_j + 1), \sum_{j=1}^{i+1}(T_j + 1), u\right) + I^f(0, T_1 + 1, u) \\
&= \sum_{i=1}^{k} I^f(0, 1 + T_{i+1}, \bar{v}_{i+1}) + I^f(0, T_1 + 1, \bar{v}_1) \\
&= \sum_{i=1}^{k+1} I^f(0, T_i + 1, \bar{v}_i) \le \sum_{i=1}^{k+1}[(T_i + 1) f(\bar{x}, 0) + 4^{-i+1}] \\
&\le \sum_{i=1}^{k+1}(T_i + 1) f(\bar{x}, 0) + 4.
\end{aligned}
$$

Since this relation holds for any natural number k Proposition 3.1 implies that the function u is (f)-good. When combined with assumption (A4) this implies that

$$\lim_{t \to \infty} |u(t) - \bar{x}| = 0.$$

On the other hand it follows from (3.80), (3.81), (3.74), and (3.69) that $\limsup_{t \to \infty} |u(t) - \bar{x}| \geq \epsilon$. The contradiction we have reached proves Lemma 3.11. $\qquad \square$

Lemma 3.12. *Let M_0, M_1 be positive numbers. Then there exist positive numbers T_0, M_2 such that for each number $T \geq T_0$, each point $z_0 \in X_{M_0}$, and each point $z_1 \in Y_{M_1}$,*

$$\sigma(f, T, z_0, z_1) \leq Tf(\bar{x}, 0) + M_2.$$

Proof. Lemma 3.10 implies that there exists a positive number L_0 such that the following property holds:

(P2) For each number $T \geq L_0$, each a.c. function $u : [0, T] \to R^n$ satisfying

$$|u(0)| \leq M_0, \quad I^f(0, T, u) \leq Tf(\bar{x}, 0) + M_0,$$

and each number $s \in [0, T - L_0]$, we have

$$\min\{|u(t) - \bar{x}| : t \in [s, s + L_0]\} \leq \bar{r}/8.$$

Choose real numbers

$$T_0 > 2M_1 + 2L_0 + 2, \tag{3.82}$$

$$M_2 > M_0 + \Delta_0 + |f(\bar{x}, 0)| + M_1(1 + |f(\bar{x}, 0)|). \tag{3.83}$$

Let

$$z_0 \in X_{M_0}, \; z_1 \in Y_{M_1}, \; T \geq T_0 \tag{3.84}$$

In view of (3.84),

$$|z_0| \leq M_0 \tag{3.85}$$

and there exists an a.c. function $v_0 : [0, \infty) \to R^n$ such that

$$v_0(0) = z_0, \; I^f(0, S, v_0) - Sf(\bar{x}, 0) \leq M_0 \text{ for all numbers } S \in (0, \infty). \tag{3.86}$$

By relation (3.84) there exist a number

$$T_1 \in (0, M_1] \tag{3.87}$$

and an a.c. function $v_1 : [0, T_1] \to R^n$ such that

$$v_1(0) = \bar{x}, \ v_1(T_1) = z_1, \ I^f(0, T_1, v_1) \le M_1. \tag{3.88}$$

It follows from property (P2), (3.86), and (3.85) that there exists a number

$$t_0 \in [L_0, 2L_0] \tag{3.89}$$

such that

$$|v_0(t_0) - \bar{x}| \le \bar{r}/8. \tag{3.90}$$

Relations (3.84), (3.87), and (3.89) imply that

$$T - T_1 > t_0 + 1 \ge L_0 + 1. \tag{3.91}$$

Define an a.c. function $v : [0, T] \to R^n$ by

$$v(t) = v_0(t), \ t \in [0, t_0], \tag{3.92}$$

$$v(t) = (t - t_0)\bar{x} + (1 + t_0 - t)v_0(t_0), \ t \in (t_0, t_0 + 1],$$

$$v(t) = \bar{x}, \ t \in (t_0 + 1, T - T_1],$$

$$v(t) = v_1(t - T + T_1), \ t \in (T - T_1, T].$$

By (3.91) and (3.88), the function v is well defined. Relations (3.92), (3.86), and (3.88) imply that

$$v(0) = v_0(0) = z_0,$$

$$v(T) = v_1(T_1) = z_1. \tag{3.93}$$

It follows from (3.93), (3.91), and (3.92) that

$$\sigma(f, T, z_0, z_1) - Tf(\bar{x}, 0) \le I^f(0, T, v) - Tf(\bar{x}, 0)$$

$$= I^f(0, t_0, v) + I^f(t_0, t_0 + 1, v) + I^f(t_0 + 1, T - T_1, v)$$

$$+ I^f(T - T_1, T, v) - Tf(\bar{x}, 0)$$

$$= I^f(0, t_0, v_0) + I^f(t_0, t_0 + 1, v)$$

$$+ (T - T_1 - t_0 - 1)f(\bar{x}, 0) + I^f(0, T_1, v_1) - Tf(\bar{x}, 0)$$

$$= I^f(0, t_0, v_0) - t_0 f(\bar{x}, 0) + I^f(t_0, t_0 + 1, v)$$

$$- f(\bar{x}, 0) + I^f(0, T_0, v_1) - T_1 f(\bar{x}, 0). \tag{3.94}$$

In view of (3.86),

$$I^f(0, t_0, v_0) - t_0 f(\bar{x}, 0) \le M_0. \tag{3.95}$$

Relations (3.88) and (3.87) imply that

$$I^f(0, T_1, v_1) - T_1 f(\bar{x}, 0) \le M_1 + M_1|f(\bar{x}, 0)|. \tag{3.96}$$

By (3.92) and (3.90), for all numbers $t \in [t_0, t_0 + 1]$,

$$|v(t) - \bar{x}| = |(1 + t_0 - t)v_0(t_0) - (1 - t + t_0)\bar{x}| \le |v_0(t_0) - \bar{x}| \le \bar{r}/8. \tag{3.97}$$

It follows from (3.89) and (3.92) that

$$|v'(t)| \le |\bar{x} - v_0(t_0)| \le \bar{r}/8 \tag{3.98}$$

for all numbers $t \in (t_0, t_0 + 1)$. By (3.87), (3.98), (3.62), and (3.63), we have

$$f(v(t), v'(t)) \le \Delta_0, \ t \in (t_0, t_0 + 1)$$

and that

$$I^f(t_0, t_0 + 1, v) - f(\bar{x}, 0) \le \Delta_0 + |f(\bar{x}, 0)|. \tag{3.99}$$

It follows from (3.94), (3.95), (3.99), (3.96), and (3.83) that

$$\sigma(f, T, z_0, z_1) - T f(\bar{x}, 0) \le M_0 + \Delta_0 + |f(\bar{x}, 0)| + M_1 + M_1|f(\bar{x}, 0)| \le M_2.$$

Lemma 3.12 is proved.   □

3.5 Completion of the Proof of Theorem 3.2

Let a number $\bar{r} \in (0, 1)$ satisfy (3.62) and (3.63). We may assume without loss of generality that $\epsilon < \bar{r}/2$. In view of Lemma 3.11 there exists a number $\delta \in (0, \epsilon/2)$ such that the following property holds:

(P3)   For each number $T \ge 2$ and each a.c. function $v : [0, T] \to R^n$ which satisfies $|v(0) - \bar{x}|, |v(T) - \bar{x}| \le \delta$ and

$$I^f(0, T, v) \le \sigma(f, T, v(0), v(T)) + \delta$$

the inequality $|v(t) - \bar{x}| \le \epsilon$ is valid for all numbers $t \in [0, T]$.

Lemma 3.10 implies that there exists a positive number L_0 such that the following property holds:

(P4)   For each number $T \geq L_0$, each a.c. function $v : [0, T] \to R^n$ satisfying

$$|v(0)| \leq M, \ I^f(0, T, v) \leq Tf(\bar{x}, 0) + M + 1,$$

and each number $S \in [0, T - L_0]$, we have

$$\min\{|v(t) - \bar{x}| : t \in [S, S + L_0]\} \leq \delta.$$

Choose an integer

$$L > 4L_0 + 4. \tag{3.100}$$

Assume that a number $T > 2L$ and an a.c. function $v : [0, T] \to R^n$ satisfy

$$v(0) \in X_M, \ I^f(0, T, v) \leq \sigma(f, T, v(0)) + \delta. \tag{3.101}$$

In view of (3.101)

$$|v(0)| \leq M \tag{3.102}$$

and that there exists an ac. function $u : [0, \infty) \to R^n$ such that

$$u(0) = v(0), \ I^f(0, \tau, u) - \tau f(\bar{x}, 0) \leq M \text{ for each } \tau \in (0, \infty). \tag{3.103}$$

Relations (3.101) and (3.103) imply that

$$I^f(0, T, v) \leq \delta + \sigma(f, T, v(0)) \leq 1 + I^f(0, T, u) \leq Tf(\bar{x}, 0) + M + 1. \tag{3.104}$$

By (3.102), (3.104), (3.100) and the choice of L_0, there exist numbers

$$\tau_1 \in [0, L_0], \ \tau_2 \in [T - L_0, T] \tag{3.105}$$

such that

$$|v(\tau_i) - \bar{x}| \leq \delta, \ i = 1, 2. \tag{3.106}$$

If $|v(0) - \bar{x}| \leq \delta$, then set $\tau_1 = 0$. It is clear that $\tau_2 - \tau_1 \geq T - 2L_0 > 4$. By (3.101)

$$I^f(\tau_1, \tau_2, v) \leq \sigma(f, \tau_2 - \tau_1, v(\tau_1), v(\tau_2)) + \delta. \tag{3.107}$$

By (3.107), (3.106), and the inequality $\tau_2 - \tau_1 > 4$, we have $|v(t) - \bar{x}| \leq \epsilon$ for all $t \in [\tau_1, \tau_2]$. Theorem 3.2 is proved.

3.6 Proof of Theorem 3.3

Let a point $x \in R^n$ and let $v : [0, \infty) \to R^n$ be an (f)-good function satisfying $v(0) = x$. Let $\{T_k\}_{k=1}^{\infty}$ be a strictly increasing sequence of natural numbers. By definition, there exists a positive number c_0 such that

$$|I^f(0, S, v) - Sf(\bar{x}, 0)| \leq c_0 \text{ for each } T \in (0, \infty). \tag{3.108}$$

In view of Proposition 3.7, for each natural number T_k, there exists an a.c. function $v_k : [0, T_k] \to R^n$ such that

$$v_k(0) = x, \quad I^f(0, T_k, v_k) = \sigma(f, T_k, x). \tag{3.109}$$

Relations (3.108) and (3.109) imply that for each natural number k

$$I^f(0, T_k, v_k) \leq I^f(0, T_k, v) \leq T_k f(\bar{x}, 0) + c_0. \tag{3.110}$$

By (3.110), (3.109), and Proposition 3.6, there exists a positive number M_0 such that for each natural number k

$$|v_k(t)| \leq M_0, \ t \in [0, T_k]. \tag{3.111}$$

Assumption (A2) implies that there exists a positive number c_1 such that

$$\sigma(f, S, z) \geq Sf(\bar{x}, 0) - c_1$$

for each $S > 0$ and each point $z \in R^n$ satisfying $|z| \leq M_0$. $\tag{3.112}$

By (3.111) and (3.112), for each natural number k and each number $S \in [0, T_k)$,

$$I^f(S, T_k, v_k) \geq (T_k - S)f(\bar{x}, 0) - c_1. \tag{3.113}$$

It follows from (3.113) and (3.110) that for each natural number k and each number $S \in (0, T_k)$,

$$I^f(0, S, v_k) = I^f(0, T_k, v_k) - I^f(S, T_k, v_k) \leq T_k f(\bar{x}, 0) + c_0$$

$$- (T_k - S)f(\bar{x}, 0) + c_1 = Sf(\bar{x}, 0) + c_0 + c_1. \tag{3.114}$$

In view of (3.114), for each natural number m, the sequence $\{I^f(0, m, v_k)\}_{k=m}^{\infty}$ is bounded. When combined with Proposition 3.7 this implies that there exist a strictly increasing sequence of natural numbers $\{k_i\}_{i=1}^{\infty}$ and an a.c. function $u : [0, \infty) \to R^n$ such that for each natural number m, we have

$$v_{k_i}(t) \to u(t) \text{ as } i \to \infty \text{ uniformly on } [0, m], \tag{3.115}$$

$$I^f(0, m, u) \le \liminf_{i \to \infty} I^f(0, m, v_{k_i}). \tag{3.116}$$

Relations (3.116) and (3.114) imply that for each natural number m, we have

$$I^f(0, m, u) \le mf(\bar{x}, 0) + c_0 + c_1. \tag{3.117}$$

Thus u is an (f)-good function and

$$\lim_{t \to \infty} |u(t) - \bar{x}| = 0. \tag{3.118}$$

We claim that u is an (f)-overtaking optimal function. Assume the contrary. Then there exists an a.c. function $w : [0, \infty) \to R^n$ such that

$$w(0) = u(0), \ \limsup_{T \to \infty}[I^f(0, T, u) - I^f(0, T, w)] > \Delta, \tag{3.119}$$

where Δ is a positive constant. Since u is and (f)-good function it follows from (3.119) and Proposition 3.1 that the function w is (f)-good. Thus

$$\lim_{t \to \infty} |w(t) - \bar{x}| = 0. \tag{3.120}$$

Assumption (A1) and Proposition 3.8 imply that there exists a positive number $\delta < 1$ such that

$$\{(y, z) \in R^n \times R^n : |y - \bar{x}| \le 4\delta, \ |z| \le 4\delta\} \subset \text{dom}(f), \tag{3.121}$$

$$|f(y, z) - f(\bar{x}, 0)| \le \Delta/16 \tag{3.122}$$

for each point $y \in R^n$ satisfying the inequality $|y - \bar{x}| \le 4\delta$ and each point $z \in R^n$ satisfying the inequality $|z| \le 4\delta$;

for each a.c. function $v : [0, 1] \to R^n$ satisfying the inequalities $|v(0) - \bar{x}|$, $|v(1) - \bar{x}| \le 4\delta$ we have

$$I^f(0, 1, v) \ge f(\bar{x}, 0) - \Delta/16. \tag{3.123}$$

By (3.118) and (3.120), there exists a number $\tau_0 \ge 4$ such that

$$|w(t) - \bar{x}|, \ |u(t) - \bar{x}| \le \delta/4 \text{ for all numbers } t \ge \tau_0. \tag{3.124}$$

In view of (3.119) there exists an integer $\tau_1 \ge 4(\tau_0 + 4)$ such that

$$I^f(0, \tau_1, u) - I^f(0, \tau_1, w) > \Delta. \tag{3.125}$$

It follows from (3.115) and (3.116) that there exists an integer $q \geq 1$ such that

$$T_q > 4(\tau_1 + 4), \tag{3.126}$$

$$|v_q(t) - u(t)| \leq \delta/16, \ t \in [0, 4\tau_1 + 4], \tag{3.127}$$

$$I^f(0, \tau_1, u) \leq I^f(0, \tau_1, v_q) + \Delta/64. \tag{3.128}$$

Define an a.c. function $\tilde{v} : [0, T_q] \to R^n$ by

$$\tilde{v}(t) = w(t), \ t \in [0, \tau_1], \tag{3.129}$$

$$\tilde{v}(t) = w(\tau_1) + (t - \tau_1)(v_q(\tau_1 + 1) - w(\tau_1)), \ t \in (\tau_1, \tau_1 + 1],$$
$$\tilde{v}(t) = v_q(t), \ t \in (\tau_1 + 1, T_q].$$

Relations (3.129), (3.128), and (3.125) imply that

$$\begin{aligned} I^f(0, T_q, \tilde{v}) - I^f(0, T_q, v_q) &= I^f(0, \tau_1 + 1, \tilde{v}) - I^f(0, \tau_1 + 1, v_q) \\ &= I^f(0, \tau_1, w) - I^f(0, \tau_1, v_q) \\ &\quad + I^f(\tau_1, \tau_1 + 1, \tilde{v}) - I^f(\tau_1, \tau_1 + 1, v_q) \\ &\leq I^f(0, \tau_1, w) - I^f(0, \tau_1, u) \\ &\quad + \Delta/64 + I^f(\tau_1, \tau_1 + 1, \tilde{v}) - I^f(\tau_1, \tau_1 + 1, v_q) \\ &\leq -\Delta + \Delta/64 + I^f(\tau_1, \tau_1 + 1, \tilde{v}) \\ &\quad - I^f(\tau_1, \tau_1 + 1, v_q). \end{aligned} \tag{3.130}$$

By (3.127) and (3.124), for $s = \tau_1, \ \tau_1 + 1$,

$$|v_q(s) - \bar{x}| \leq |v_q(s) - u(s)| + |u(s) - \bar{x}| \leq \delta/16 + \delta/4. \tag{3.131}$$

By (3.131) and (3.123),

$$I^f(\tau_1, \tau_1 + 1, v_q) \geq f(\bar{x}, 0) - \Delta/16. \tag{3.132}$$

It follows from (3.129), (3.131), and (3.124) that for all numbers $t \in (\tau_1, \tau_1 + 1)$

$$\begin{aligned} |\tilde{v}(t) - \bar{x}| &\leq (1 - t + \tau_1)|w(\tau_1) - \bar{x}| + (t - \tau_1)|v_q(\tau_1 + 1) - \bar{x}| \\ &\leq (1 - t + \tau_1)\delta/4 + (t - \tau_1)\delta/16 + (t - \tau_1)\delta/4 < \delta/2. \end{aligned} \tag{3.133}$$

By (3.129), (3.127), and (3.124), for all numbers $t \in (\tau_1, \tau_1 + 1)$,

$$|\tilde{v}'(t)| = |v_q(\tau_1 + 1) - w(\tau_1)| \le |v_q(\tau_1 + 1) - u(\tau_1 + 1)| + |u(\tau_1 + 1) - \bar{x}|$$
$$+ |\bar{x} - w(\tau_1)| \le \delta/16 + \delta/4 + \delta/4 < (3/4)\delta. \tag{3.134}$$

It follows from (3.133), (3.134), and (3.122) that for all numbers $t \in (\tau_1, \tau_1 + 1)$, $f(\tilde{v}(t), \tilde{v}'(t)) \le f(\bar{x}, 0) + \Delta/16$ and

$$I^f(\tau_1, \tau_1 + 1, \tilde{v}) \le f(\bar{x}, 0) + \Delta/16.$$

By this inequality, (3.130) and (3.132),

$$I^f(0, T_q, \tilde{v}) - I^(0, T_q, v_q) \le -\Delta + \Delta/64 + f(\bar{x}, 0) + \Delta/16$$
$$- f(\bar{x}, 0) + \Delta/16 < -\Delta/2.$$

Since $\tilde{v}(0) = w(0) = u(0) = x = v_q(0)$ the inequality above contradicts (3.109). The contradiction we have reached shows that u is an (f)-overtaking optimal function. Theorem 3.3 is proved.

3.7   Proof of Theorem 3.4

In view of Lemma 3.12 there exist positive numbers L_1 and M_3 such that the following property holds:

(P4)   for each integer $T \ge L_1$, each point $z_0 \in X_{M_0}$, and each point $z_1 \in Y_{M_1}$, we have

$$\sigma(f, T, z_0, z_1) \ge Tf(\bar{x}, 0) + M_3. \tag{3.135}$$

Let $\bar{r} \in (0, 1)$ be defined as in Sect. 3.4 (see (3.62), (3.63)). By Lemma 3.11 there exists a number $\delta \in (0, \bar{r}/2)$ such that the following property holds:

(P5)   for each number $T \ge 2$ and each a. c. function $v : [0, T] \to R^n$ which satisfies

$$|v(0) - \bar{x}|, \ |v(T) - \bar{x}| \le \delta,$$
$$I^f(0, T, v) \le \sigma(f, T, v(0), v(T)) + \delta$$

the inequality $|v(t) - \bar{x}| \le \epsilon$ is valid for all numbers $t \in [0, T]$.
Lemma 3.10 implies that there exists a positive number $\bar{L}_1$ such that the following property holds:

(P6)   for each integer $T \geq \bar{L}_1$, each a.c. function $v : [0, T] \to R^n$ which satisfies

$$|v(0)| \leq M_0, \ I^f(0, T, v) \geq Tf(\bar{x}, 0) + M_2 + M_3,$$

and each number $s \in [0, T - \bar{L}_1]$, the inequality

$$\min\{|v(t) - \bar{x}| : t \in [s, s + \bar{L}_1]\} \leq \delta$$

holds.
Set

$$L = 8(L_1 + \bar{L}_1 + 4) \tag{3.136}$$

and choose an integer

$$Q > 4 + \delta^{-1}M_2. \tag{3.137}$$

Assume that

$$T > L, \ z_0 \in X_{M_0}, \ z_1 \in Y_{M_1}. \tag{3.138}$$

It is clear that by property (P4), (3.138), and (3.136) the inequality (3.135) is valid.
Assume that an a. c. function $v : [0, T] \to R^n$ satisfies

$$v(0) = z_0, \ v(T) = z_1, \ I^f(0, T, v) \leq \sigma(f, T, z_0, z_1) + M_2. \tag{3.139}$$

Relations (3.135) and (3.139) imply that

$$I^f(0, T, v) \leq Tf(\bar{x}, 0) + M_2 + M_3. \tag{3.140}$$

It follows from (3.136), (3.138), (3.140), (3.139) and property (P6) that there exists a sequence of integers $\{S_i\}_{i=0}^q \subset [0, T]$ such that

$$S_0 \in [0, \bar{L}_1 + 2], \ S_{i+1} - S_i \in [2, 2 + \bar{L}_1], \ i = 0, \ldots, q - 1, \ T < S_q + 2\bar{L}_1, \tag{3.141}$$

$$|v(S_i) - \bar{x}| \leq \delta, \ i = 0, \ldots, q.$$

Put

$$E = \{i \in \{0, \ldots, q-1\} : I^f(S_i, S_{i+1}, v) \leq \sigma(f, S_{i+1} - S_i, v(S_i), v(S_{i+1})) + \delta\}. \tag{3.142}$$

It is easy to see that there exists an a.c. function $\bar{v} : [0, T] \to R^n$ such that

$$\bar{v}(t) = v(t), \ t \in [0, S_0] \cup [S_q, T] \cup \{S_i : i = 0, \ldots, q\} \cup \{[S_i, S_{i+1}] : i \in E\}, \tag{3.143}$$

$$I^f(S_i, S_{i+1}, \bar{v}) < I^f(S_i, S_{i+1}, v) + \delta, \ i \in \{0, \ldots, q-1\} \setminus E. \tag{3.144}$$

By (3.143), (3.144), and (3.139), we have

$$M_2 \geq I^f(0, T, v) - I^f(0, T, \bar{v}) \geq \delta \text{Card}(\{0, \ldots, q-1\} \setminus E)$$

and

$$\text{Card}(\{0, 1, \ldots, q-1\} \setminus E) \leq \delta^{-1} M_2. \tag{3.145}$$

It follows from (3.142), (3.141) and property (P5) that for each $i \in E$

$$|v(t) - \bar{x}| \leq \epsilon, \ t \in [S_i, S_{i+1}]$$

and

$$|v(t) - \bar{x}| \leq \epsilon$$

for all $t \in [0, T] \setminus ([0, S_0] \cup [S_q, T] \cup \{[S_i, S_{i+1}] : i \in [0, q-1] \setminus E\})$.

By (3.145) and (3.137),

$$\text{Card}(\{0, \ldots, q-1\} \setminus E) + 2 \leq \delta^{-1} M_2 + 2 < Q.$$

Theorem 3.4 is proved.

3.8   Proof of Theorem 3.5

Let $\bar{r} \in (0, 1)$ be defined as in Sect. 3.4 (see (3.62), (3.63)). Lemma 3.12 implies that there exist positive numbers L_1, M_2 such that the following property holds:

(P7)   for each number $T \geq L_1$, each point $z_0 \in X_{M_0}$, and each point $z_1 \in Y_{M_1}$,

$$\sigma(f, T, z_0, z_1) \leq Tf(\bar{x}, 0) + M_2 \tag{3.146}$$

In view of Lemma 3.11 there exists a number $\delta \in (0, \bar{r}/2)$ such that the following property holds:

(P8)   For each number $T \geq 2$ and each a.c. function $v : [0, T] \to R^n$ which satisfies the inequalities

$$|v(0) - \bar{x}|, \ |v(T) - \bar{x}| \leq \delta,$$

$$I^f(0, T, v) \leq \sigma(f, T, v(0), v(T)) + \delta$$

the inequality $|v(t) - \bar{x}| \leq \epsilon$ is true for all numbers $t \in [0, T]$.

Lemma 3.10 implies that there exists a positive number L_2 such that the following property holds:

(P9)   For each number $T \geq L_2$, each a.c. function $v : [0, T] \to R^n$ satisfying

$$|v(0)| \leq M, \ I^f(0, T, v) \leq Tf(\bar{x}, 0) + M_2 + 4,$$

and each number $s \in [0, T - L_2]$, we have

$$\min\{|v(t) - \bar{x}| : \ t \in [s, s + L_2]\} \leq \delta.$$

Set

$$L = 4(L_1 + L_2 + 1). \tag{3.147}$$

Assume that

$$T > 2L, \ z_0 \in X_{M_0}, \ z_1 \in Y_{M_1}. \tag{3.148}$$

By (3.148) and (3.147), inequality (3.146) holds.

Assume that an a.c. function $v : [0, T] \to R^n$ satisfies

$$v(0) = z_0, \ v(T) = z_1, \ I^f(0, T, v) \leq \sigma(f, T, z_0, z_1) + \delta. \tag{3.149}$$

Relations (3.146) and (3.149) imply that

$$I^f(0, T, v) \leq Tf(\bar{x}, 0) + M_2 + 1. \tag{3.150}$$

It follows from property (P9), (3.148), (3.147), (3.149), and (3.150) that there exist numbers τ_1 and τ_2 such that

$$\tau_1 \in [0, L_2], \ \tau_2 \in [T - L_2, T], \ |v(\tau_i) - \bar{x}| \leq \delta, \ i = 1, 2. \tag{3.151}$$

If $|v(t) - \bar{x}| \leq \delta$, then put $\tau_1 = 0$ and if $|v(T) - \bar{x}| \leq \delta$, the put $\tau_2 = T$. In view of (3.150), (3.151), (3.149) and property (P8), we have

$$|v(t) - \bar{x}| \leq \epsilon, \ t \in [\tau_1, \tau_2].$$

Theorem 3.5 is proved.

3.9 Examples

Example 3.13. Let a_0 be a positive number, $\psi_0 : [0, \infty) \to [0, \infty)$ be an increasing function satisfying

$$\lim_{t \to \infty} \psi_0(t) = \infty \tag{3.152}$$

and $L : R^n \times R^n \to [0, \infty]$ be a lower semicontinuous function such that

$$\operatorname{dom}(L) := \{(x, y) \in R^n \times R^n : L(x, y) < \infty\} \tag{3.153}$$

is nonempty, convex, and closed set and

$$L(x, y) \geq \max\{\psi_0(|x|), \ \psi_0(|y|)|y|\} - a_0 \text{ for each } x, y \in R^n. \tag{3.154}$$

Assume that for each point $x \in R^n$ the function $L(x, \cdot) : R^n \to R^1 \cup \{\infty\}$ is convex and that there exists a point $\bar{x} \in R^n$ such that

$$L(x, y) = 0 \text{ if and only if } (x, y) = (\bar{x}, 0), \tag{3.155}$$

$(\bar{x}, 0)$ is an interior point of $\operatorname{dom}(L)$, and that L is continuous at the point $(\bar{x}, 0)$.
    Let $\mu \in R^1$ and $l \in R^n$. Define

$$f(x, y) = L(x, y) + \mu + \langle l, y \rangle, \ x, y \in R^n. \tag{3.156}$$

We claim that all the assumptions introduced in Sect. 3.1 hold for f.
    It is clear that the function f is lower semicontinuous and that

$$\operatorname{dom}(f) = \operatorname{dom}(L).$$

Put

$$\psi(t) = (3/4)\psi_0(t), \ t \in [0, \infty).$$

It is clear that there exists a number $K_0 > 1$ such that

$$\psi_0(K_0) > 4|l| + 4. \tag{3.157}$$

Put

$$a = a_0 + |\mu| + |l|K_0. \tag{3.158}$$

We claim that relation (3.3) is true. Let $x, y \in R^n$. If $|y| \leq K_0$, then relations (3.156) and (3.154) imply that

$$f(x, y) = L(x, y) + \mu + \langle l, y \rangle \geq \max\{\psi_0(|x|), \ \psi_0(|y|)|y|\} - a_0 - |l||y| - |\mu|$$
$$\geq \max\{\psi_0(|x|), \psi_0(|y|)|y|\} - a_0 - |\mu| - |l|K_0 \geq \max\{\psi(|x|), \psi(|y|)|y|\} - a.$$

Hence relation (3.3) holds if $|y| \leq K_0$. Assume that

$$|y| > K_0. \tag{3.159}$$

There are two cases:

$$\psi_0(|x|) \geq \psi_0(|y|)|y|; \tag{3.160}$$
$$\psi_0(|x|) < \psi_0(|y|)|y|. \tag{3.161}$$

Assume that relation (3.160) holds. It follows from (3.159), (3.157) and (3.160) that

$$|\langle l, y \rangle| \leq |l||y| \leq 4^{-1}\psi_0(K_0)|y| \leq 4^{-1}\psi_0(|y|)|y| \leq 4^{-1}\psi_0|(x)|.$$

Combined with relations (3.156), (3.154), and (3.158) this inequality implies that

$$f(x, y) = L(x, y) + \mu + \langle l, y \rangle$$
$$\geq \max\{\psi_0(|x|), \ \psi_0(|y|)|y|\} - a_0 - |\mu| - |\langle l, y \rangle|$$
$$\geq \max\{\psi_0(|x|), \ \psi_0(|y|)|y|\} - a_0 - |\mu| - 4^{-1}\max\{\psi_0(|y|)|y|, \psi_0(|x|)\}$$
$$= (3/4)\max\{\psi_0(|x|), \ \psi_0(|y|)|y|\} - a_0 - |\mu|$$
$$\geq \max\{\psi(|x|), \ \psi(|y|)|y|\} - a.$$

Thus relation (3.3) holds if (3.160) is valid.

Assume that relation (3.161) holds. Then in view of (3.157) and (3.159)

$$|\langle l, y \rangle| \leq |l||y| \leq 4^{-1}\psi_0(K_0)|y| \leq 4^{-1}\psi_0(|y|)|y|.$$

When combined with (3.156), (3.154), and (3.161) this inequality implies that

$$f(x, y) = L(x, y) + \mu + \langle l, y \rangle \geq \psi_0(|y|)|y| - a_0 - |\mu| - 4^{-1}\psi_0(|y|)|y|$$
$$\geq -a_0 - |\mu| + (3/4)\psi_0(|y|)|y| \geq -a_0 - |\mu| + (3/4)\psi_0(|x|).$$

When combined with (3.158) and the definition of ψ this implies that

$$f(x, y) \geq (3/4)\max\{\psi_0(|x|), \ \psi_0(|y|)|y|\} - a_0 - |\mu| \geq \max\{\psi(|x|), \ \psi(|y|)|y|\} - a.$$

Thus relation (3.3) holds if (3.161) is valid and

$$f(x, y) \geq \max\{\psi(|x|), \ \psi(|y|)|y|\} - a \text{ for all } x, y \in R^n.$$

In view of (3.156) and (3.155), we have

$$\mu = f(\bar{x}, 0) \le f(x, 0) \text{ for each } x \in R^n.$$

It is clear that assumptions (A1) and (A3) hold.

Proposition 3.14. *Assumptions (A2) holds.*

Proof. Let $M > 0$ be given. In view of (3.152) there exists a number $M_0 > M + 1$ such that

$$\psi(M_0) > |\mu| + 1 + a. \tag{3.162}$$

Put

$$c_M = -(|l| + 1)(2M_0 + 1). \tag{3.163}$$

Let $T > 0$ be given and let a point $x \in R^n$ satisfy $|x| \le M$. We claim that

$$\sigma(f, T, x) \ge Tf(\bar{x}, 0) - c_M = T\mu - c_M. \tag{3.164}$$

We may assume without loss of generality that the value $\sigma(f, T, x)$ is finite. There exists an a.c. function $v : [0, T] \to R^n$ such that

$$v(0) = x, \quad \int_0^T f(v(t), v'(t))dt \le \sigma(f, T, x) + 1. \tag{3.165}$$

Inequality $|x| \le M$ and (3.165) imply that there exists a number $T_0 \in (0, T]$ such that

$$|v(T_0)| \le M_0, \ |v(t)| > M_0 \text{ if } t \text{ satisfies } T_0 < t \le T. \tag{3.166}$$

It follows from (3.166), (3.164), (3.162), (3.156), and (3.165) that

$$\int_0^T f(v(t), v'(t))dt = \int_0^{T_0} f(v(t), v'(t))dt + \int_{T_0}^T f(v(t), v'(t))dt$$

$$\ge \int_0^{T_0} f(v(t), v'(t))dt + (T - T_0)(\psi(M_0) - a)$$

$$\ge \int_0^{T_0} f(v(t), v'(t))dt + (T - T_0)|\mu|$$

$$\ge \int_0^{T_0} [\mu + \langle l, v'(t)\rangle]dt + (T - T_0)|\mu|$$

$$\ge T\mu + \langle l, v(T_0) - v(0)\rangle \ge T\mu - |l|2M_0,$$

$$\sigma(f, T, x) \ge T\mu - 2|l|M_0 - 1.$$

Proposition 3.14 is proved.   □

The next result shows that assumption (A4) holds for the integrand f.

Proposition 3.15. *Let* $v : [0, \infty) \to R^n$ *be an* (f)*-good function. Then*

$$\lim_{t \to \infty} |v(t) - \bar{x}| = 0.$$

Proof. Proposition 3.1 and (3.166) imply that

$$\sup\{|v(t)| : t \in [0, \infty)\} < \infty, \tag{3.167}$$

$$\lim_{T \to \infty} \int_0^T L(v(t), v'(t))dt < \infty. \tag{3.168}$$

For each nonnegative integer i put

$$v_i(t) = v(t + i), \ t \in [0, 1]. \tag{3.169}$$

Assume that the assertion of the proposition does not hold. Then there exist a positive number ϵ and a strictly increasing sequence of natural numbers $\{i_k\}_{k=1}^\infty$ such that for all natural numbers k, we have

$$\sup\{|v_{i_k}(t) - \bar{x}| : t \in [0, 1]\} \geq \epsilon. \tag{3.170}$$

In view of Proposition 3.7 and (3.167)-(3.169), extracting a sequence and re-indexing if necessary, we may assume without loss of generality that there exists an a.c. function $u : [0, 1] \to R^n$ such that

$$v_{i_k}(t) \to u(t) \text{ as } k \to \infty \text{ uniformly on } [0, 1], \tag{3.171}$$

$$I^f(0, 1, u) \leq \liminf_{k \to \infty} I^f(0, 1, v_{i_k}). \tag{3.172}$$

By (3.170) and (3.171),

$$\sup\{|u(t) - \bar{x}| : t \in [0, 1]\} \geq \epsilon/4. \tag{3.173}$$

It follows from (3.156), (3.172), (3.171). (3.169), and (3.168) that

$$\int_0^1 L(u(t), u'(t))dt = \int_0^1 f(u(t), u'(t))dt - \mu - \int_0^1 \langle l, u'(t) \rangle dt$$

$$\leq \liminf_{k \to \infty} \int_0^1 f(v_{i_k}(t), v'_{i_k}(t))dt - \mu - \lim_{k \to \infty} \int_0^1 \langle l, v'_{i_k}(t) \rangle dt$$

$$\leq \liminf_{k \to \infty} \int_0^1 L(v_{i_k}(t), v'_{i_k}(t))dt = 0.$$

Therefore $L(u(t), u'(t)) = 0$, $t \in [0, 1]$, (a.e.) and by relation (3.155) $u(t) = \bar{x}$ for all $t \in [0, 1]$. This contradicts (3.173). The contradiction we have reached proves Proposition 3.15. $\qquad\qquad\qquad\qquad\qquad\qquad\qquad\qquad\qquad\qquad\qquad\qquad\qquad \square$

Thus all the assumptions introduced in Sect. 3.1 hold for the function f.

Example 3.16. Let a be a positive number, $\psi : [0, \infty) \to [0, \infty)$ be an increasing function such that $\lim_{t \to \infty} \psi(t) = \infty$ and $f : R^n \times R^n \to R^1 \cup \{\infty\}$ be a convex lower semicontinuous function such that the set $\mathrm{dom}(f)$ is nonempty, convex, and closed and that

$$f(x, y) \geq \max\{\psi(|x|), \ \psi(|y|)|y|\} - a \text{ for each } x, y \in R^n.$$

We suppose that there exists a point $\bar{x} \in R^n$ such that

$$f(\bar{x}, 0) \leq f(x, 0) \text{ for each } x \in R^n$$

and that $(\bar{x}, 0)$ is an interior point of the set $\mathrm{dom}(f)$. It is known that the function f is continuous at the point $(\bar{x}, 0)$. It is well-known fact of convex analysis that there exists a point $l \in R^n$ such that

$$f(x, y) \geq f(\bar{x}, 0) + \langle l, y \rangle \text{ for each } x, y \in R^n.$$

We assume that for each pair of points (x_1, y_1), $(x_2, y_2) \in \mathrm{dom}(f)$ satisfying $(x_1, y_1) \neq (x_2, y_2)$ and each number $\alpha \in (0, 1)$, we have

$$f(\alpha(x_1, y_1) + (1 - \alpha)(x_2, y_2)) < \alpha f(x_1, y_1) + (1 - \alpha) f(x_2, y_2).$$

Put

$$L(x, y) = f(x, y) - f(\bar{x}, 0) - \langle l, y \rangle \text{ for each } x, y \in R^n.$$

It is not difficult to see that there exist a positive number a_0 and an increasing function $\psi_0 : [0, \infty) \to [0, \infty)$ such that

$$L(x, y) \geq \max\{\psi_0(|x|), \ \psi_0(|y|)|y|\} - a_0 \text{ for all } x, y \in R^n.$$

It is easy to see that L is a convex, lower semicontinuous function and that the equality $L(x, y) = 0$ holds if and only if $(x, y) = (\bar{x}, 0)$. Now it is easy to see that our example is a particular case of Example 3.13 and all the assumptions introduced in Sect. 3.1 hold for f.

3.10 Behavior of Solutions in the Regions Containing End Points

We continue to use the notation and definitions introduced in Sect. 3.1 and to study the structure of approximate solutions of problems (P_2). Our goal is to study their structure in the regions containing end points.

Let a be a positive number, $\psi : [0, \infty) \to [0, \infty)$ be an increasing function which satisfies

$$\lim_{t \to \infty} \psi(t) = \infty, \qquad (3.174)$$

and $f : R^n \times R^n \to R^1 \cup \{\infty\}$ be a convex lower semicontinuous function such that the set $\mathrm{dom}(f)$ is nonempty and closed and that

$$f(x, y) \ge \max\{\psi(|x|), \ \psi(|y|)|y|\} - a \text{ for each } x, y \in R^n. \qquad (3.175)$$

We suppose that there exists a point $\bar{x} \in R^n$ such that the following assumption holds:

(A5) $(\bar{x}, 0)$ is an interior point of the set $\mathrm{dom}(f)$ and

$$f(\bar{x}, 0) \le f(x, 0) \text{ for all } x \in R^n. \qquad (3.176)$$

Remark 3.17. It is easy to see that the existence of the point $\bar{x} \in R^n$ satisfying (3.176) follows from (3.174) and (3.175). Here we also assume in addition that $(\bar{x}, 0)$ is an interior point of the set $\mathrm{dom}(f)$.

They are well-known facts from convex analysis [45] that the function f is continuous at the point $(\bar{x}, 0)$ and that there exits a point $l \in R^n$ such that

$$f(x, y) \ge f(\bar{x}, 0) + \langle l, y \rangle \text{ for each } x, y \in R^n. \qquad (3.177)$$

We also assume that for each pair of points (x_1, y_1), $(x_2, y_2) \in \mathrm{dom}(f)$ such that $(x_1, y_1) \ne (x_2, y_2)$ and each number $\alpha \in (0, 1)$ the inequality

$$f(\alpha(x_1, y_1) + (1 - \alpha)(x_2, y_2)) < \alpha f(x_1, y_1) + (1 - \alpha)f(x_2, y_2) \qquad (3.178)$$

holds. This means that the function f is strictly convex. The integrand f was considered in Example 3.16. It was shown there that assumptions (A1)–(A4) and all the results of Sects. 3.1 and 3.9 hold for the integrand f.

In our study we will use an integrand L defined by

$$L(x, y) = f(x, y) - f(\bar{x}, 0) - \langle l, y \rangle \text{ for all } x, y \in R^n. \qquad (3.179)$$

We suppose that the following assumption holds:

(A6) For each pair of positive numbers M, ϵ there exists a positive number γ such that for each pair of points (ξ_1, ξ_2), $(\eta_1, \eta_2) \in \mathrm{dom}(f)$ which satisfies the inequalities $|\xi_i|, |\eta_i| \le M, i = 1, 2$ and $|\xi_1 - \xi_2| \ge \epsilon$, we have

$$2^{-1} f(\xi_1, \eta_1) + 2^{-1} f(\xi_2, \eta_2) - f(2^{-1}(\xi_1 + \xi_2), 2^{-1}(\eta_1 + \eta_2)) \ge \gamma.$$

Remark 3.18. It is clear that assumption (A6) follows from relation (3.178) if the restriction of the function f to the set dom(f) is continuous.

Since the restriction of the function f to the set dom(f) is strictly convex (see assumption (A6)), Theorem 3.3 implies the following result.

Theorem 3.19. *Assume that $x \in R^n$ and that there exists an (f)-good function $v : [0, \infty) \to R^n$ satisfying $v(0) = x$. Then there exists a unique (f)-overtaking optimal function $v_* : [0, \infty) \to R^n$ such that $v_*(0) = x$.*

Let $z \in R^n$ and there exists an (f)-good function $v : [0, \infty) \to R^n$ such that $v(0) = z$. Denote by $Y^{(f,z)} : [0, \infty) \to R^n$ a unique (f)-overtaking optimal function satisfying $Y^{(f,z)}(0) = z$ which exists by Theorem 3.19.

In this chapter we prove the following theorem obtained in [61] which describes the structure of approximate solutions of variational problems in the regions containing the left end point.

Theorem 3.20. *Let $M, \epsilon > 0$ be real numbers and let $L_0 \geq 1$ be an integer. Then there exist a positive number δ and an integer $L_1 > L_0$ such that for each number $T \geq L_1$, each point $z \in X_M$, and each a.c. function $v : [0, T] \to R^n$ which satisfies*

$$v(0) = z, \quad I^f(0, T, v) \leq \sigma(f, T, z) + \delta,$$

the inequality

$$|v(t) - Y^{(f,z)}(t)| \leq \epsilon, \; t \in [0, L_0]$$

holds.

We intend to describe the structure of approximate solutions of variational problems in the regions containing the right end point. In order to meet this goal define the functions $\bar{f}, \bar{L} : R^n \times R^n \to R^1 \cup \{\infty\}$ by

$$\bar{f}(x, y) = f(x, -y), \quad \bar{L}(x, y) = L(x, -y) \text{ for all } x, y \in R^n. \tag{3.180}$$

It is not difficult to see that

$$\text{dom}(\bar{f}) = \{(x, y) \in R^n \times R^n : (x, -y) \in \text{dom}(f)\}, \tag{3.181}$$

dom($\bar{f}$) is nonempty closed convex subset of $R^n \times R^n$,

$$\bar{f}(x, y) \geq \max\{\psi(|x|), \; \psi(|y|)|y|\} - a \text{ for each } x, y \in R^n \times R^n, \tag{3.182}$$

the point $(\bar{x}, 0)$ is an interior point of the set dom($\bar{f}$) and the function $\bar{f}$ is convex and lower semicontinuous.

It follows from relations (3.180), (3.177), and (3.179) that for each pair of points $x, y \in R^n$

$$\bar{f}(x, y) = f(x, -y) \geq f(\bar{x}, 0) + \langle l, -y \rangle = \bar{f}(\bar{x}, 0) + \langle -l, y \rangle, \tag{3.183}$$

$$\bar{L}(x, y) = L(x, -y) = f(x, -y) - f(\bar{x}, 0) - \langle l, -y \rangle$$
$$= \bar{f}(x, y) - \bar{f}(\bar{x}, 0) - \langle -l, y \rangle. \tag{3.184}$$

Relations (3.180), (3.181), and (3.178) imply that for each pair of points (x_1, y_1), $(x_2, y_2) \in \text{dom}(\bar{f})$ such that $(x_1, y_1) \neq (x_2, y_2)$ and each number $\alpha \in (0, 1)$ we have

$$\bar{f}(\alpha(x_1, y_1) + (1 - \alpha)(x_2, y_2)) < \alpha \bar{f}(x_1, y_1) + (1 - \alpha)\bar{f}(x_2, y_2). \tag{3.185}$$

Therefore all the assumptions posed in this section for the function f also hold for the function $\bar{f}$. Also all the results of this section and of Sect. 3.1 stated for the function f are valid for the function $\bar{f}$. In particular Theorems 3.2 and 3.3 hold for the integrand $\bar{f}$.

Assumption (A6) and relation (3.180) imply that the following assumption holds:

(A7) For each pair of numbers $M, \epsilon > 0$ there exists a positive number γ such that for each pair of points $(\xi_1, \xi_2), (\eta_1, \eta_2) \in \text{dom}(\bar{f})$ which satisfies

$$|\xi_i|, \ |\eta_i| \leq M, \ i = 1, 2 \text{ and } |\xi_1 - \xi_2| \geq \epsilon$$

the inequality

$$2^{-1} \bar{f}(\xi_1, \eta_1) + 2^{-1} \bar{f}(\xi_2, \eta_2) - \bar{f}(2^{-1}(\xi_1 + \xi_2), 2^{-1}(\eta_1 + \eta_2)) \geq \gamma_0$$

holds.

It is easy now to see that Theorems 3.19 and 3.20 hold for the integrand $\bar{f}$.

For each positive number M denote by $\bar{X}_M$ the set of all points $x \in R^n$ such that $|x| \leq M$ and that there exists an a.c. function $v : [0, \infty) \to R^n$ which satisfies

$$I^{\bar{f}}(0, T, v) - T\bar{f}(\bar{x}, 0) \leq M \text{ for each } T \in (0, \infty). \tag{3.186}$$

Set

$$\bar{X}_* = \cup\{\bar{X}_M : M \in (0, \infty)\}. \tag{3.187}$$

Since the function $\bar{f}$ is convex we conclude that the set $\bar{X}_M$ is convex for all positive numbers M. Proposition 3.7 implies that for each positive number M the set $\bar{X}_M$ is closed.

By Theorem 3.19, applied to the integrand $\bar{f}$, for each point $x \in \bar{X}_*$ there exists a unique $(\bar{f})$-overtaking optimal function $\Lambda^{(x)} : [0, \infty) \to R^n$ such that $\Lambda^{(x)}(0) = x$.

Proposition 3.1 implies that $\Lambda^{(x)}$ is $(\bar{f})$-good function for any point $x \in \bar{X}_*$. In view of Proposition 3.15, for each point $x \in \bar{X}_*$,

$$\lim_{t \to \infty} |\Lambda^{(x)}(t) - \bar{x}| = 0. \tag{3.188}$$

For each point $x \in \bar{X}_*$ set

$$\pi(x) = \lim_{T \to \infty} [I^{\bar{f}}(0, T, \Lambda^{(x)}) - T\bar{f}(\bar{x}, 0)]. \tag{3.189}$$

Let $x \in \bar{X}_*$ be given. We claim that the value $\pi(x)$ is well-defined and finite. Relations (3.184), (3.188), and (3.189) imply that

$$\pi(x) = \lim_{T \to \infty} \left[\int_0^T \bar{L}(\Lambda^{(x)}(t), (\Lambda^{(x)})'(t)) dt - \int_0^T \langle l, (\Lambda^{(x)})'(t) \rangle dt \right]$$

$$= \lim_{T \to \infty} \int_0^T \bar{L}(\Lambda^{(x)}(t), (\Lambda^{(x)})'(t)) dt - \lim_{T \to \infty} \langle l, \Lambda^{(x)}(T) - x \rangle$$

$$= \int_0^\infty \bar{L}(\Lambda^{(x)}(t), (\Lambda^{(x)})'(t)) dt - \langle l, \bar{x} - x \rangle. \tag{3.190}$$

Therefore the value $\pi(x)$ is well defined. Since the function $\Lambda^{(x)}$ is $(\bar{f})$-good, Proposition 3.1 implies that $\pi(x)$ is finite for each $x \in \bar{X}_*$.

The function π plays an important role in our study of the structure of approximate solutions of variational problems in the regions containing the right end point. We show that approximate solutions of the problem (P_2) are arbitrary close to the function $\Lambda^{(x_*)}(T - t)$ in a region which contains the right end point T, where x_* is a unique point of minimum of the function π.

In this chapter we prove the following result obtained in [61].

Proposition 3.21. *1. For each positive number M the function $\pi : \bar{X}_M \to R^1$ is lower semicontinuous.*
2. For all pairs of points $y, z \in \bar{X}_*$ satisfying $y \neq z$ and each number $\alpha \in (0, 1)$,

$$\pi(\alpha y + (1 - \alpha)z) < \alpha\pi(y) + (1 - \alpha)\pi(z).$$

3. $\pi(\bar{x}) = 0$.
4. There exists a number $\tilde{M} > |\bar{x}|$ such that $\pi(x) \geq 2$ for each point $x \in \bar{X}_* \setminus \bar{X}_{\tilde{M}}$.

Let a positive number $\tilde{M}$ be as guaranteed by Proposition 3.21. By Proposition 3.21, there exists a unique point $x_* \in \bar{X}_{\tilde{M}}$ such that

$$\pi(x_*) < \pi(x) \text{ for all points } x \in \bar{X}_{\tilde{M}} \setminus \{x_*\}. \tag{3.191}$$

By Proposition 3.21 if $x \in \bar{X}_* \setminus \bar{X}_{\tilde{M}}$, then

$$\pi(x) \geq 2 > \pi(\bar{x}) \geq \pi(x_*). \tag{3.192}$$

The following theorem obtained in [61] describes the structure of approximate solutions of variational problems in the regions containing the right end point.

Theorem 3.22. *Let $M, \epsilon > 0$ be real numbers and let $L_1 \geq 1$ be an integer. Then there exist a positive number δ and a natural number $L_2 > L_1$ such that if a number $T > 2L_2$ and if an a.c. function $v : [0, T] \to R^n$ satisfies*

$$v(0) \in X_M \text{ and } I^f(0, T, v) \leq \sigma(f, T, v(0)) + \delta,$$

then

$$|v(T - t) - \Lambda^{(x_*)}(t)| \leq \epsilon \text{ for all } t \in [0, L_1].$$

We can easily construct a broad class of integrands satisfying the assumptions posed in the section and for which our results hold. Namely, suppose that K is a closed convex subset of the space $R^n \times R^n$ with a nonempty interior and that $f : K \to R^1$ is a strictly convex continuous function for which the minimization problem $f(x, 0) \to \min$ subject to $(x, 0) \in K$ has a solution $\bar{x}$ such that the point $(\bar{x}, 0)$ belongs to the interior of the set K and such that

$$f(x, y) \geq c_1|x| + c_2|y|^p - c_3$$

for all points $(x, y) \in R^n \times R^n$, where $c_1, c_2, c_3 > 0$ and $p > 1$ are constants. We put $f(x, y) = \infty$ for all points $(x, y) \in R^{2n} \setminus K$. It is not difficult to see that the integrand f satisfies all the assumptions posed in Sects. 3.1 and 3.10 and Theorems 3.19, 3.20, and 3.22 hold for f.

The characterization of approximate solutions in the initial and final periods is implicit: it is in terms of unique (f)-overtaking functions satisfying certain boundary conditions. In order to obtain approximations of these (f)-overtaking functions we need to find a finite number of approximate solutions of the problem (P_2) with the same boundary condition x and with different large enough real numbers T. This information can be useful if we need to find an approximate solution of the problem (P_2) with the boundary condition x and with a new interval $[0, T]$ where T is large enough. This approximate solution is the concatenation of the approximation of $Y^{(f,x)}(t)$, the turnpike $\bar{x}$, and the approximation of $\Lambda^{(x_*)}(T - t)$.

3.11   Proof of Theorem 3.20

For simplicity we use the notation $Y^{(z)} = Y^{(f,z)}$ for each point $z \in \cup\{X_M : M \in (0, \infty)\}$.

Assume that the assertion of the theorem does not hold. Therefore for each integer k there exists a number

$$T_k \geq L_0 + 4k \tag{3.193}$$

and an a.c. function $v_k : [0, T_k] \to R^n$ such that

$$v_k(0) \in X_M, \quad I^f(0, T_k, v_k) \leq \sigma(f, T_k, v_k(0)) + k^{-1}, \tag{3.194}$$

$$\sup\{|v_k(t) - Y^{(v_k(0))}(t)| : t \in [0, L_0]\} > \epsilon. \tag{3.195}$$

We obtain some useful estimates for $|v_k(t)|, t \in [0, T_k]$ and $|Y^{(v_k(0))}(t)|, t \in [0, \infty)$ and for the integral functional with the integrand f and the functions v_k and $Y^{(v_k(0))}$, $k = 1, 2, \ldots$.

By relation (3.194) and the definition of the set X_M (see (3.9)), for each natural number k, we have

$$I^f(0, T_k, v_k) \leq \sigma(f, T_k, v_k(0)) + k^{-1} \leq M + T_k f(\bar{x}, 0) + k^{-1}, \tag{3.196}$$

$$|v_k(0)| \leq M. \tag{3.197}$$

In view of (3.196), (3.197) and Proposition 3.6, there exists a positive number M_0 such that for each natural number k

$$|v_k(t)| \leq M_0 \text{ for all numbers } t \in [0, T_k]. \tag{3.198}$$

It follows from (3.194) and the definition of the set X_M (see (3.9)) for each integer $k \geq 1$

$$I^f(0, T, Y^{(v_k(0))}) \leq Tf(\bar{x}, 0) + M + 1 \text{ for all large enough } T. \tag{3.199}$$

By (3.199), Proposition 3.6, and (3.197), there exists a number $M_1 > M_0$ such that for each integer $k \geq 1$

$$|Y^{(v_k(0))}(t)| \leq M_1 \text{ for all numbers } t \in [0, \infty). \tag{3.200}$$

Proposition 3.14 implies that there exists a positive number c_1 such that

$$\sigma(f, T, x) \geq Tf(\bar{x}, 0) - c_1 \text{ for each } T > 0$$

$$\text{and each } x \in R^n \text{ satisfying } |x| \leq M_1. \tag{3.201}$$

Now we show the existence of a subsequence $\{v_{k_i}\}_{i=1}^{\infty}$ and a subinterval $[a_0, b_0] \subset (0, L_0)$ such that

$$|v_{k_i}(t) - Y^{(v_{k_i}(0))}(t)| \geq \epsilon/4$$

for all numbers $t \in [a_0, b_0]$ and all large enough integers i. Moreover, we show that v_{k_i} (respectively, $Y^{(v_{k_i}(0))}$) converges to $\tilde{v}$ (respectively, $\tilde{y}$) as $i \to \infty$ uniformly on any bounded subinterval of the set $[0, \infty)$.

Fix a natural number j. By (3.193), (3.196), (3.198), (3.201) and the inequality $M_1 > M_0$, for each natural number $k \geq j$,

$$I^f(0, j, v_k) = I^f(0, T_k, v_k) - I^f(j, T_k, v_k)$$

$$\leq M + k^{-1} + T_k f(\bar{x}, 0) - \sigma(f, T_k - j, v_k(j))$$

$$\leq M + k^{-1} + T_k f(\bar{x}, 0) - (T_k - j) f(\bar{x}, 0) + c_1$$

$$\leq M + k^{-1} + j f(\bar{x}, 0) + c_1. \tag{3.202}$$

Let $k \geq 1$ be an integer. It follows from (3.199) that there exists a number $S_k > 2j + 2$ such that

$$I^f(0, S_k, Y^{(v_k(0))}) \leq S_k f(\bar{x}, 0) + M + 1. \tag{3.203}$$

By (3.203), (3.200), and (3.201),

$$I^f(0, j, Y^{(v_k(0))}) = I^f(0, S_k, Y^{(v_k(0))}) - I^f(j, S_k, Y^{(v_k(0))})$$

$$\leq S_k f(\bar{x}, 0) + M + 1 - \sigma(f, S_k - j, Y^{(v_k(0))}(j))$$

$$\leq S_k f(\bar{x}, 0) + M + 1 - (S_k - j) f(\bar{x}, 0) + c_1$$

$$= j f(\bar{x}, 0) + M + 1 + c_1. \tag{3.204}$$

In view of Proposition 3.7, (3.203), (3.204), and (3.197) extracting a subsequence and re-indexing we may assume without loss of generality that there exist a strictly increasing sequence of natural numbers $\{k_i\}_{i=1}^{\infty}$ and a pair of a.c. functions $\tilde{v} : [0, \infty) \to R^n$ and $\tilde{y} : [0, \infty) \to R^n$ such that for each natural number j, we have

$$v_{k_i}(t) \to \tilde{v}(t) \text{ as } i \to \infty \text{ uniformly on } [0, j], \tag{3.205}$$

$$Y^{(v_{k_i}(0))}(t) \to \tilde{y}(t) \text{ as } i \to \infty \text{ uniformly on } [0, j],$$

$$I^f(0, j, \tilde{v}) \leq \liminf_{i \to \infty} I^f(0, j, v_{k_i}),$$

$$I^f(0, j, \tilde{y}) \leq \liminf_{i \to \infty} I^f(0, j, Y^{(v_{k_i}(0))}).$$

By (3.202), (3.204), and (3.205), for each natural number j,

$$I^f(0, j, \tilde{v}) \leq M + j f(\bar{x}, 0) + c_1, \quad I^f(0, j, \tilde{y}) \leq j f(\bar{x}, 0) + M + 1 + c_1. \tag{3.206}$$

It follows from (3.206) and Proposition 3.1 that $\tilde{v}$ and $\tilde{y}$ are (f)-good functions. Together with Proposition 3.15 this implies that

$$\tilde{v}(t) \to \bar{x}, \ \tilde{y}(t) \to \bar{x} \text{ as } t \to \infty. \tag{3.207}$$

Relations (3.195) and (3.205) imply that

$$\sup\{|\tilde{v}(t) - \tilde{y}(t)| : t \in [0, L_0]\} \geq \epsilon/2. \tag{3.208}$$

In view of (3.208), there exists a pair of numbers $a_0, b_0 \in (0, L_0]$ such that

$$0 < b_0 - a_0 < 1 \text{ and } |\tilde{v}(t) - \tilde{y}(t)| \geq \epsilon/3 \text{ for all } t \in [a_0, b_0]. \tag{3.209}$$

By (3.205) and (3.209), there exists an integer $i_0 \geq 4 + L_0$ such that for each integer $i \geq i_0$, we have

$$|v_{k_i}(t) - Y^{(v_{k_i}(0))}(t)| \geq \epsilon/4 \text{ for all } t \in [a_0, b_0]. \tag{3.210}$$

Now we show that the values of the integral functional with the integrand f and with the functions v_{k_i} and $Y^{(v_{k_i}(0))}$, $i = 1, 2, \ldots$ are bounded by a constant which does not depend on i.

Assume that an integer i satisfies $i \geq i_0$. In view of (3.197), (3.198), (3.201), and the inequality $M_1 > M_0$,

$$
\begin{aligned}
I^f(a_0, b_0, v_{k_i}) &= I^f(0, T_{k_i}, v_{k_i}) - I^f(0, a_0, v_{k_i}) - I^f(b_0, T_{k_i}, v_{k_i}) \\
&\leq M + T_{k_i} f(\bar{x}, 0) + 1 - \sigma(f, a_0, v_{k_i}(0)) - \sigma(f, T_{k_i} - b_0, v_{k_i}(b_0)) \\
&\leq M + T_{k_i} f(\bar{x}, 0) + 1 - a_0 f(\bar{x}, 0) \\
&\quad + c_1 - (T_{k_i} - b_0) f(\bar{x}, 0) \\
&\quad + c_1 \leq M + (b_0 - a_0) f(\bar{x}, 0) + 2c_1 + 1. \tag{3.211}
\end{aligned}
$$

By (3.199), there exists a number $S_i > 4b_0 + 4$ such that

$$I^f(0, S_i, Y^{(v_{k_i}(0))}) \leq S_i f(\bar{x}, 0) + M + 1. \tag{3.212}$$

Relations (3.200), (3.201), and (3.212) imply that

$$
\begin{aligned}
I^f(a_0, b_0, Y^{(v_{k_i}(0))}) &= I^f(0, S_i, Y^{(v_{k_i}(0))}) \\
&\quad - I^f(0, a_0, Y^{(v_{k_i}(0))}) - I^f(b_0, S_i, Y^{(v_{k_i}(0))}) \\
&\leq S_i f(\bar{x}, 0) + M + 1 - \sigma(f, a_0, Y^{(v_{k_i}(0))}(0)) \\
&\quad - \sigma(f, S_i - b_0, Y^{(v_{k_i}(0))}(b_0))
\end{aligned}
$$

$$\leq S_i f(\bar{x}, 0) + M + 1 - a_0 f(\bar{x}, 0)$$
$$+ c_1 - (S_i - b_0) f(\bar{x}, 0) + c_1$$
$$= (b_0 - a_0) f(\bar{x}, 0) + 2c_1 + M + 1. \qquad (3.213)$$

It follows from (3.211) and (3.212) that for each integer $i \geq i_0$, we have

$$I^f(a_0, b_0, v_{k_i}), \ I^f(a_0, b_0, Y^{(v_{k_i}(0))})$$

$$\leq M + 2c_1 + 1 + (b_0 - a_0) f(\bar{x}, 0). \qquad (3.214)$$

Now we show that there exists a positive number γ_0 such that for each integer $i \geq i_0$ and each number $S \in [L_0, T_{k_i}]$, we have

$$I^f(0, S, 2^{-1}(v_{k_i} + Y^{(v_{k_i}(0))}))$$

$$\leq 2^{-1} I^f(0, S, v_{k_i}) + 2^{-1} I^f(0, S, Y^{(v_{k_i}(0))}) - \gamma_0(3/4)(b_0 - a_0).$$

By (3.174) there exists a real number $M_2 > M_1 + 1$ such that

$$\psi(M_2) > 4[2a + 2(M + 2c_1 + 1)(b_0 - a_0)^{-1} + |f(\bar{x}, 0)|]. \qquad (3.215)$$

For each integer $i \geq i_0$ put

$$E_i = \{t \in [a_0, b_0] : \ |v'_{k_i}(t)|, \ |(Y^{(v_{k_i}(0))})'(t)| \leq M_2\}$$

$$\cap \{t \in [a_0, b_0] : \ f(v_{k_i}(t), v'_{k_i}(t)), \ f(Y^{(v_{k_i}(0))}(t), (Y^{(v_{k_i}(0))})'(t)) < \infty\}. \qquad (3.216)$$

Let an integer $i \geq i_0$ be given. By (3.214), (3.175), (3.216), and the monotonicity of the function ψ,

$$2(M + 2c_1 + 1 + (b_0 - a_0) f(\bar{x}, 0)) \geq I^f(a_0, b_0, v_{k_i}) + I^f(a_0, b_0, Y^{(v_{k_i}(0))})$$

$$\geq \int_{a_0}^{b_0} (\psi(|v'_{k_i}(t)|)|v'_{i_k}(t)| - a) dt$$

$$+ \int_{a_0}^{b_0} (\psi(|(Y^{(v_{k_i}(0))})'(t)|)|(Y^{(v_{i_k}(0))})'(t)| - a) dt$$

$$\geq -2(b_0 - a_0)a + \text{mes}([a_0, b_0] \setminus E_i)\psi(M_2)M_2.$$

$$(3.217)$$

By (3.217), (3.215), and the inequality $M_2 > M_1 + 1$,

$$\text{mes}([a_0, b_0] \setminus E_i) \leq (\psi(M_2)M_2)^{-1}[2(b_0 - a_0)a$$

$$+ 2(M + 2c_1 + 1 + (b_0 - a_0)|f(\bar{x}, 0)|)]$$

$$\leq (\psi(M_2))^{-1}[2(b_0 - a_0)a$$

$$+ 2(M + 2c_1 + 1 + (b_0 - a_0)|f(\bar{x}, 0)|] \leq (b_0 - a_0)/4. \tag{3.218}$$

It follows from (3.216) and (3.218) that

$$\text{mes}(E_i) \geq (3/4)(b_0 - a_0) \text{ for each integer } i \geq i_0. \tag{3.219}$$

Assumption (A6) implies that there exists a number $\gamma_0 \in (0, 1)$ such that for each pair of points $(\xi_1, \xi_2), (\eta_1, \eta_2) \in \text{dom}(f)$ which satisfy

$$|\xi_i|, |\eta_i| \leq M_2, \ i = 1, 2, \ |\xi_1 - \xi_2| \geq \epsilon/8, \tag{3.220}$$

we have

$$- f(2^{-1}(\xi_1 + \xi_2), 2^{-1}(\eta_1 + \eta_2)) + 2^{-1} f(\xi_1, \eta_1) + 2^{-1} f(\xi_2, \eta_2) \geq \gamma_0. \tag{3.221}$$

For each integer $i \geq i_0$ set

$$u_i(t) = 2^{-1}(v_{k_i}(t) + Y^{(v_{k_i}(0))}(t)), \ t \in [0, T_{k_i}]. \tag{3.222}$$

Let an integer $i \geq i_0$ be given. In view of relation (3.222), for almost every number $t \in [0, T_{k_i}]$,

$$f(u_i(t), u_i'(t)) \leq 2^{-1} f(v_{k_i}(t), v_{k_i}'(t)) + 2^{-1} f(Y^{(v_{k_i}(0))}(t), (Y^{(v_{k_i}(0))})'(t)). \tag{3.223}$$

By (3.216), the inequality $M_2 > M_1 + 1 > M_0$, (3.200), (3.198), (3.210), the choice of γ_0 (see (3.220) and (3.221)), and (3.222), for almost every $t \in E_i$, we have

$$f(u_i(t), u_i'(t)) \leq 2^{-1} f(v_{i_k}(t), v_{k_i}'(t)) + 2^{-1} f(Y^{(v_{k_i}(0))}(t), (Y^{(v_{k_i}(0))})'(t)) - \gamma_0. \tag{3.224}$$

By (3.216), (3.219), (3.223), (3.224) and the inclusions $a_0, b_0 \in [0, L_0]$, for each number $S \in [L_0, T_{k_i}]$,

$$I^f(0, S, u_i) \leq 2^{-1} I^f(0, S, v_{k_i}) + 2^{-1} I^f(0, S, Y^{(v_{k_i}(0))}) - \gamma_0 \text{mes}(E_i)$$

$$\leq 2^{-1} I^f(0, S, v_{k_i}) + 2^{-1} I^f(0, S, Y^{(v_{k_i}(0))})$$

$$- \gamma_0(3/4)(b_0 - a_0). \tag{3.225}$$

Now we turn to the next step of our proof.

Put

$$\Delta = \gamma_0(b_0 - a_0)/16. \tag{3.226}$$

In view of assumption (A5), the continuity of the function f at the point $(\bar{x}, 0)$ and Proposition 3.8, there exists a number $r \in (0, 1)$ such that

$$\{(\xi_1, \xi_2) \in R^n \times R^n : |\xi_1 - \bar{x}| \le 4r \text{ and } |\xi_2| \le 4r\} \subset \text{dom}(f); \tag{3.227}$$

$$|f(\xi_1, \xi_2) - f(\bar{x}, 0)| \le 32^{-1}\Delta \tag{3.228}$$

for each $\xi_1, \xi_2 \in R^n$ satisfying $|\xi_1 - \bar{x}| \le 4r$, $|\xi_2| \le 4r$;

if an a.c. function $h : [0, 1] \to R^n$ satisfies the inequalities

$$|h(0) - \bar{x}|, \ |h(1) - \bar{x}| \le 4r,$$

then

$$I^f(0, 1, h) \ge f(\bar{x}, 0) - \Delta/16. \tag{3.229}$$

By (3.207) there exists an integer $L_2 \ge 1$ such that

$$|\tilde{v}(t) - \bar{x}|, \ |\tilde{y}(t) - \bar{x}| \le r/8 \text{ for all } t \ge L_2. \tag{3.230}$$

Relation (3.205) implies that there exists an integer $j \ge i_0 + 4L_2 + 4$ such that

$$k_j^{-1} < 16^{-1}\Delta, \tag{3.231}$$

$$|v_{k_j}(t) - \tilde{v}(t)|, \ |Y^{(v_{k_j}(0))}(t) - \tilde{y}(t)| \le r/32 \text{ for all } t \in [0, 4L_2 + 4L_0 + 4]. \tag{3.232}$$

We consider the function u_j defined by (3.222) and define a.c. functions $u_j^{(1)}, u_j^{(2)} :$ $[0, 4L_0 + 4L_2 + 4] \to R^n$ by

$$u_j^{(1)}(t) = u_j(t), \ t \in [0, 4L_0 + 4L_2 + 3], \tag{3.233}$$

$$u_j^{(1)}(t) = u_j(4L_0 + 4L_2 + 3)$$
$$\qquad + (t - (4L_0 + 4L_2 + 3))[v_{k_j}(4L_0 + 4L_2 + 4) - u_j(4L_0 + 4L_2 + 3)],$$
$$\qquad t \in [4L_0 + 4L_2 + 3, 4L_0 + 4L_2 + 4],$$

$$u_j^{(2)}(t) = u_j(t), \ t \in [0, 4L_0 + 4L_2 + 3], \tag{3.234}$$

$$u_j^{(2)}(t) = (t - (4L_0 + 4L_2 + 3))[Y^{(v_{k_j}(0))}(4L_0 + 4L_2 + 4) - u_j(4L_0 + 4L_2 + 3)]$$

$$+u_j(4L_0 + 4L_2 + 3), \ t \in [4L_0 + 4L_2 + 3, 4L_0 + 4L_2 + 4].$$

It is easy to see that

$$u_j^{(2)}(0) = u_j^{(1)}(0) = u_j(0) = v_{k_j}(0) = Y^{(v_{k_j}(0))}(0), \tag{3.235}$$

$$u_j^{(1)}(4L_0 + 4L_2 + 4) = v_{k_j}(4L_0 + 4L_2 + 4),$$

$$u_j^{(2)}(4L_0 + 4L_2 + 4) = Y^{(v_{k_j}(0))}(4L_0 + 4L_2 + 4).$$

Since the function $Y^{(v_{k_j}(0))}$ is (f)-overtaking optimal it follows from (3.235) that

$$I^f(0, 4L_0 + 4L_2 + 4, u_j^{(2)}) \geq I^f(0, 4L_0 + 4L_2 + 4, Y^{(v_{k_j}(0))}). \tag{3.236}$$

By (3.194) and (3.235), we have

$$I^f(0, 4L_0 + 4L_2 + 4, u_j^{(1)}) \geq I^f(0, 4L_0 + 4L_2 + 4, v_{k_j}) - k_j^{-1}. \tag{3.237}$$

By (3.230) and (3.232), for all numbers $t \in [L_2, 4L_2 + 4L_0 + 4]$,

$$|v_{k_j}(t) - \bar{x}| \leq r/32 + r/8, \ |Y^{(v_{k_j}(0))}(t) - \bar{x}| \leq r/32 + r/8. \tag{3.238}$$

It follows from (3.238) and the choice of r (see (3.229)) that

$$I^f(4L_2 + 4L_0 + 3, 4L_2 + 4L_0 + 4, v_{k_j}) \geq f(\bar{x}, 0) - \Delta/16, \tag{3.239}$$

$$I^f(4L_2 + 4L_0 + 3, 4L_2 + 4L_0 + 4, Y^{(v_{k_j}(0))}) \geq f(\bar{x}, 0) - \Delta/16.$$

Let

$$t \in [4L_2 + 4L_0 + 3, \ 4L_2 + 4L_0 + 4]$$

be given. We evaluate $f(u_j^{(p)}(t), (u_j^{(p)})'(t))$ for $p = 1, 2$. It follows from (3.222), (3.233), and (3.238) that

$$|u_j^{(1)}(t) - \bar{x}| \leq \max\{|u_j(4L_0 + 4L_2 + 3) - \bar{x}|, |v_{k_j}(4L_0 + 4L_2 + 4) - \bar{x}|\}$$

$$\leq \max\{|v_{k_j}(4L_0 + 4L_2 + 3) - \bar{x}|, |Y^{(v_{k_j}(0))}(4L_0 + 4L_2 + 3) - \bar{x}|,$$

$$|v_{k_j}(4L_0 + 4L_2 + 4) - \bar{x}|\}$$

$$\leq r/32 + r/8. \tag{3.240}$$

By (3.233) and (3.238),

$$|(u_j^{(1)})'(t)| \leq |v_{k_j}(4L_0 + 4L_2 + 4) - u_j(4L_0 + 4L_2 + 3)|$$
$$\leq |v_{k_j}(4L_0 + 4L_2 + 4) - \bar{x}| + |\bar{x} - u_j(4L_0 + 4L_2 + 3)|$$
$$\leq 2(r/32 + r/8). \tag{3.241}$$

Relations (3.222), (3.234), and (3.238) imply that

$$|u_j^{(2)}(t) - \bar{x}| \leq \max\{|Y^{(v_{k_j}(0))}(4L_0 + 4L_2 + 4) - \bar{x}|, \ |u_j(4L_0 + 4L_2 + 3) - \bar{x}|\}$$
$$\leq \max\{|Y^{(v_{k_j}(0))}(4L_0 + 4L_2 + 4) - \bar{x}|, \ |Y^{(v_{k_j}(0))}(4L_0 + 4L_2 + 3) - \bar{x}|,$$
$$|v_{k_j}(4L_0 + 4L_2 + 3) - \bar{x}|\} \leq r/32 + r/8. \tag{3.242}$$

By (3.234) and (3.238),

$$|(u_j^{(2)})'(t)| \leq |Y^{(v_{k_j}(0))}(4L_0 + 4L_2 + 4) - u_j(4L_0 + 4L_2 + 3)|$$
$$\leq |Y^{(v_{k_j}(0))}(4L_0 + 4L_2 + 4) - \bar{x}| + |\bar{x} - u_j(4L_0 + 4L_2 + 3)|$$
$$\leq 2(r/32 + r/8). \tag{3.243}$$

It follows from (3.228) and (3.240)–(3.243) that

$$|f(u_j^{(p)}(t), (u_j^{(p)})'(t)) - f(\bar{x}, 0)| \leq 32^{-1}\Delta$$

for $p = 1, 2$ and all $t \in [4L_2 + 4L_0 + 3, \ 4L_2 + 4L_0 + 4]$. \qquad (3.244)

By (3.233), (3.234), (3.244), (3.237), (3.236), (3.221), and (3.239),

$$2I^f(0, 4L_0 + 4L_2 + 3, u_j) = I^f(0, 4L_0 + 4L_2 + 4, u_j^{(1)})$$
$$+ I^f(0, 4L_0 + 4L_2 + 4, u_j^{(2)})$$
$$- I^f(4L_0 + 4L_2 + 3, 4L_0 + 4L_2 + 4, u_j^{(1)})$$
$$- I^f(4L_0 + 4L_2 + 3, 4L_0 + 4L_2 + 4, u_j^{(2)})$$
$$\geq I^f(0, 4L_0 + 4L_2 + 4, u_j^{(1)})$$
$$+ I^f(0, 4L_0 + 4L_2 + 4, u_j^{(2)})$$
$$- 16^{-1}\Delta - 2f(\bar{x}, 0)$$
$$\geq I^f(0, 4L_0 + 4L_2 + 4, v_{k_j}) - k_j^{-1}$$

$$+ I^f (0, 4L_0 + 4L_2 + 4, Y^{(v_{k_j}(0))})$$

$$- 16^{-1} \Delta - 2f(\bar{x}, 0)$$

$$\geq I^f (0, 4L_0 + 4L_2 + 4, v_{k_j})$$

$$+ I^f (0, 4L_0 + 4L_2 + 4, Y^{(v_{k_j}(0))})$$

$$- 8^{-1} \Delta - 2f(\bar{x}, 0)$$

$$\geq I^f (0, 4L_0 + 4L_2 + 3, v_{k_j}) + I^f (0, 4L_0 + 4L_2$$

$$+ 3, Y^{(v_{k_j}(0))}) + 2f(\bar{x}, 0) - \Delta/4 - 2f(\bar{x}, 0)$$

$$\geq 2I^f (0, 4L_0 + 4L_2 + 3, u_j) + \gamma_0(b_0 - a_0) - \Delta/4$$

$$\geq 2I^f (0, 4L_0 + 4L_2 + 3, u_j) + \gamma_0(b_0 - a_0)/2.$$

The contradiction we have reached proves Theorem 3.20.

3.12 Proof of Proposition 3.21

Relations (3.182) and (3.184) imply that there exists a positive number $a_0 > a$ such that for each pair of points $x, y \in R^n$,

$$\bar{L}(x, y) \geq (3/4) \max\{\psi(|x|), \ \psi(|y|)|y|\} - a_0. \tag{3.245}$$

Lemma 3.23. *Let $g \in \{\bar{L}, \bar{f}\}$ and let M be a positive number. Then there exists a positive number M_0 such that for each a.c. function $v : [0, 1] \to R^n$ satisfying $\int_0^1 g(v(t), v'(t))dt \leq M$ the inequality $|v(t)| \leq M_0$ holds for all numbers $t \in [0, 1]$.*

Proof. By (3.174) there exists a positive number M_1 such that

$$\psi(M_1) > (M + a_0)4 \tag{3.246}$$

and there exists a positive number a_1 such that

$$4^{-1}\psi(t)t \geq t - a_1 \text{ for all numbers } t \geq 0. \tag{3.247}$$

Choose a real number

$$M_0 > a_0 + a_1 + M + M_1. \tag{3.248}$$

Assume that an a.c. function $v : [0, 1] \to R^n$ satisfies

$$\int_0^1 g(v(t), v'(t))dt \leq M. \tag{3.249}$$

We claim that

$$|v(t)| \le M_0 \text{ for all numbers } t \in [0, 1].$$

Assume the contrary. Then there exists a number $t_0 \in [0, 1]$ such that

$$|v(t_0)| > M_0. \tag{3.250}$$

If $|v(t)| \ge M_1$ for all numbers $t \in [0, 1]$, then it follows from (3.182), (3.245), and (3.246) that

$$M \ge \int_0^1 g(v(t), v'(t)) dt \ge (3/4)\psi(M_1) - a_0.$$

This inequality contradicts (3.246). The contradiction we have reached proves that there exists a number $t_1 \in [0, 1]$ such that

$$|v(t_1)| < M_1.$$

It is clear that $t_1 \ne t_0$. In view of (3.250), the inequality $|v(t_1)| < M_1$, (3.182), (3.245), (3.247), and (3.249), we have

$$M_0 - M_1 \le |v(t_0)| - |v(t_1)| \le |\int_{t_1}^{t_0} |v'(t)| dt| \le |\int_{t_1}^{t_0} [a_1 + 4^{-1}\psi(|v'(t)|)|v'(t)|] dt|$$

$$\le a_1 + |\int_{t_1}^{t_0} [g(v(t), v'(t)) + a_0] dt| \le a_1 + a_0$$

$$+ \int_0^1 g(v(t), v'(t)) dt \le a_0 + a_1 + M.$$

This contradicts (3.248). The contradiction we have reached proves Lemma 3.23. $\square$

Proposition 3.24. *An a.c. function* $v : [0, \infty) \to R^n$ *is* $(\bar{f})$-*good if and only if*

$$\int_0^\infty \bar{L}(v(t), v'(t)) dt := \lim_{T \to \infty} \int_0^T \bar{L}(v(t), v'(t)) dt < \infty.$$

Proof. Let $v : [0, \infty) \to R^n$ be an a.c. function. Then by (3.184) for each positive number T,

$$\int_0^T \bar{f}(v(t), v'(t)) dt = \int_0^T \bar{L}(v(t), v'(t)) dt + T \bar{f}(\bar{x}, 0) - \int_0^T \langle l, v'(t) \rangle dt$$

$$= \int_0^T \bar{L}(v(t), v'(t)) dt + T \bar{f}(\bar{x}, 0) - \langle l, v(T) - v(0) \rangle. \tag{3.251}$$

If v is an $(\bar{f})$-good function, then Proposition 3.1 implies that the function v is bounded and that

$$\int_0^\infty \bar{L}(v(t), v'(t))dt < \infty. \tag{3.252}$$

If (3.252) holds, then Lemma 3.23 implies that the function v is bounded on $[0, \infty)$ and in view of (3.251) and Proposition 3.1, the function v is $(\bar{f})$-good. Proposition 3.24 is proved. $\qquad\square$

Proposition 3.25. *Let $x \in \bar{X}$ and let an $(\bar{f})$-good function $v : [0, \infty) \to R^n$ satisfy $v(0) = x$. Then*

$$\int_0^\infty \bar{L}(v(t), v'(t))dt - \langle l, \bar{x} - x\rangle \geq \pi(x).$$

Proof. Since the function $\Lambda^{(x)}$ is $(\bar{f})$-overtaking optimal it is also $(\bar{f})$-good and by (3.184), Propositions 3.24 and 3.15, we have

$$0 \leq \limsup_{T \to \infty}[\int_0^T \bar{f}(\Lambda^{(x)}(t), (\Lambda^{(x)})'(t))dt - \int_0^T \bar{f}(v(t), v'(t))dt]$$

$$= \limsup_{T \to \infty}[\int_0^T \bar{L}(\Lambda^{(x)}(t), (\Lambda^{(x)})'(t))dt - \int_0^T \bar{L}(v(t), v'(t))dt$$

$$- \langle l, -x + \Lambda^{(x)}(T)\rangle + \langle l, v(T) - x\rangle]$$

$$= \int_0^\infty \bar{L}(\Lambda^{(x)}(t), (\Lambda^{(x)})'(t))dt$$

$$- \langle l, \bar{x} - x\rangle - \int_0^\infty \bar{L}(v(t), v'(t))dt + \langle l, \bar{x} - x\rangle.$$

When combined with (3.190) this implies that

$$\int_0^\infty \bar{L}(v(t), v'(t))dt - \langle l, \bar{x} - x\rangle$$

$$\geq \int_0^\infty \bar{L}(\Lambda^{(x)}(t), (\Lambda^{(x)})'(t))dt - \langle l, \bar{x} - x\rangle = \pi(x).$$

This completes the proof of Proposition 3.25. $\qquad\square$

Corollary 3.26. $\pi(\bar{x}) = 0$.

Proposition 3.27. *There exists a positive number M_* such that for each point $x \in \bar{X}_*$ satisfying $|x| > M_*$ the inequality $\pi(x) \geq 2$ holds.*

Proof. In view of (3.174) there exists a positive number M_1 such that

$$\psi(M_1) > a + |\bar{f}(\bar{x}, 0)| + 4. \tag{3.253}$$

Lemma 3.23 implies that there exists a positive number M_* such that if an a.c. function $v : [0, 1] \to R^n$ satisfies

$$\int_0^1 \bar{f}(v(t), v'(t)) dt \le (|l| + 1)(|\bar{x}| + M_1 + 1) + |\bar{f}(\bar{x}, 0)| + 4,$$

then

$$|v(t)| \le M_* \text{ for all numbers } t \in [0, 1]. \tag{3.254}$$

Let

$$x \in \bar{X}_* \text{ and } |x| > M_*. \tag{3.255}$$

Consider an $(\bar{f})$-overtaking optimal function $\Lambda^{(x)} : [0, \infty) \to R^n$ which is also $(\bar{f})$-good. By (3.255) and the choice of M_* (see (3.254)), we have

$$\int_0^1 \bar{f}(\Lambda^{(x)}(t), (\Lambda^{(x)})'(t)) dt > (|l| + 1)(|\bar{x}| + M_1 + 1) + |\bar{f}(\bar{x}, 0)| + 4. \tag{3.256}$$

If $|\Lambda^{(x)}(t)| > M_1$ for each number $t \ge 1$, then by (3.182) and (3.253), for each number $t \ge 1$,

$$\bar{f}(\Lambda^{(x)}(t), (\Lambda^{(x)})'(t)) - \bar{f}(\bar{x}, 0) \ge \psi(|\Lambda^{(x)}(t)|) - a - \bar{f}(\bar{x}, 0) \ge 4$$

and

$$\int_0^T \bar{f}(\Lambda^{(x)}(t), (\Lambda^{(x)})'(t)) dt - T \bar{f}(\bar{x}, 0) \to \infty \text{ as } T \to \infty,$$

a contradiction.

Thus there exists a number $S_0 \ge 1$ such that

$$|\Lambda^{(x)}(S_0)| \le M_1, \ |\Lambda^{(x)}(t)| > M_1 \text{ for each } t \text{ satisfying } 1 \le t < S_0. \tag{3.257}$$

Relations (3.182), (3.253), and (3.257) imply that

$$\bar{f}(\Lambda^{(x)}(t), (\Lambda^{(x)})'(t)) \ge |\bar{f}(\bar{x}, 0)| \text{ for each } t \text{ such that } 1 \le t < S_0. \tag{3.258}$$

It follows from (3.190), (3.189), (3.258), (3.184), (3.257), (3.256), and Proposition 3.15 that

$$\pi(x) = \int_0^\infty \bar{L}(\Lambda^{(x)}(t), (\Lambda^{(x)})'(t))dt - \langle l, \bar{x} - x \rangle$$

$$= \lim_{T \to \infty} \int_0^T [\bar{f}(\Lambda^{(x)}(t), (\Lambda^{(x)})'(t)) - \bar{f}(\bar{x}, 0)]dt$$

$$= \int_0^1 [\bar{f}(\Lambda^{(x)}(t), (\Lambda^{(x)})'(t))]dt - \bar{f}(\bar{x}, 0)$$

$$+ \int_1^{S_0} [\bar{f}(\Lambda^{(x)}(t), (\Lambda^{(x)})'(t)) - \bar{f}(\bar{x}, 0)]dt$$

$$+ \lim_{T \to \infty} \int_{S_0}^T [\bar{f}(\Lambda^{(x)}(t), (\Lambda^{(x)})'(t)) - \bar{f}(\bar{x}, 0)]dt$$

$$\geq \int_0^1 \bar{f}(\Lambda^{(x)}(t), (\Lambda^{(x)})'(t))dt - \bar{f}(\bar{x}, 0)$$

$$+ \lim_{T \to \infty} \int_{S_0}^T \bar{L}(\Lambda^{(x)}(t), (\Lambda^{(x)})'(t))dt - \lim_{T \to \infty} \langle l, \Lambda^{(x)}(T) - \Lambda^{(x)}(S_0) \rangle$$

$$\geq \int_0^1 \bar{f}(\Lambda^{(x)}(t), (\Lambda^{(x)})'(t))dt - \bar{f}(\bar{x}, 0) - |l|(|\bar{x}| + M_1) > 4.$$

This completes the proof of Proposition 3.27. $\qquad\square$

Let $M_* > 4$ be as guaranteed by Proposition 3.27. Namely

$$\pi(x) \geq 2 \text{ for each point } x \in \bar{X}_* \text{ satisfying } |x| \geq M_*. \tag{3.259}$$

Proposition 3.28. *There exists a number $\tilde{M} > M_*$ such that for each point $x \in \bar{X}_* \setminus \bar{X}_{\tilde{M}}$ the inequality $\pi(x) \geq 2$ holds.*

Proof. It follows from Proposition 3.6 that there exists a number $\bar{M} > M_*$ such that for each positive number T and each a.c. function $v : [0, T] \to R^n$ which satisfies

$$|v(0)| \leq M_* + 1, \ I^{\bar{f}}(0, T, v) \leq T \bar{f}(\bar{x}, 0) + 4 \tag{3.260}$$

we have

$$|v(t)| \leq \bar{M} \text{ for all } t \in [0, T]. \tag{3.261}$$

Choose a number

$$\tilde{M} > M_* + \bar{M} + 2 + |l|(|\bar{x}| + M_*)2\bar{M}. \tag{3.262}$$

Let $x \in \bar{X}_* \setminus \bar{X}_{\tilde{M}}$ be given. If $|x| > M_*$, then relation (3.259) implies that $\pi(x) \geq 2$. Assume that

$$|x| \leq M_*. \tag{3.263}$$

In order to complete the proof of the proposition it is sufficient to show that $\pi(x) \geq 2$. Let us assume that

$$\pi(x) < 2. \tag{3.264}$$

By (3.189) and (3.190),

$$\pi(x) = \int_0^\infty \bar{L}(\Lambda^{(x)}(t), (\Lambda^{(x)})'(t))dt - \langle l, \bar{x} - x \rangle$$

$$= \lim_{T \to \infty} [I^{\bar{f}}(0, T, \Lambda^{(x)}) - Tf(\bar{x}, 0)]. \tag{3.265}$$

It follows from (3.264) and (3.265) that for all large enough numbers T,

$$I^{\bar{f}}(0, T, \Lambda^{(x)}) \leq Tf(\bar{x}, 0) + 2.$$

Together with (3.263) and the choice of $\bar{M}$ (see (3.260) and (3.261)) this inequality implies that

$$|\Lambda^{(x)}(t)| \leq \bar{M} \text{ for all } t \in [0, \infty). \tag{3.266}$$

By (3.263)–(3.265),

$$\int_0^\infty \bar{L}(\Lambda^{(x)}(t), (\Lambda^{(x)})'(t))dt < 2 + |l|(|\bar{x}| + M_*).$$

When combined with (3.184), (3.266), and (3.262) this implies that for each positive number T, we have

$$I^{\bar{f}}(0, T, \Lambda^{(x)}) - Tf(\bar{x}, 0) = \int_0^T \bar{L}(\Lambda^{(x)}(t), (\Lambda^{(x)})'(t))dt - \langle l, \Lambda^{(x)}(T) - x \rangle$$

$$< 2 + |l|(|\bar{x}| + M_*) + |l|2\bar{M} < \tilde{M}.$$

In view of this inequality and (3.262) and (3.263), $x \in \bar{X}_{\tilde{M}}$, a contradiction. The contradiction we have reached proves that $\pi(x) \geq 2$. Proposition 3.28 is proved. □

Proposition 3.29. *For any positive number M the function* $\pi : \bar{X}_M \to R^1$ *is lower semicontinuous.*

Proof. Let M be a positive number, $\{x_k\}_{k=1}^\infty \subset \bar{X}_M$, $x \in \bar{X}_M$, and let $\lim_{k \to \infty} x_k = x$. We show that $\pi(x) \leq \liminf_{k \to \infty} \pi(x_k)$. We may assume that there exists a finite limit $\lim_{k \to \infty} \pi(x_k)$. Relations (3.189) and (3.190) imply that for each natural number k,

$$\pi(x_k) = \int_0^\infty \bar{L}(\Lambda^{(x_k)}(t), (\Lambda^{(x_k)})'(t))dt - \langle l, \bar{x} - x_k \rangle$$

$$= \lim_{T \to \infty} [I^{\bar{f}}(0, T, \Lambda^{(x_k)}) - Tf(\bar{x}, 0)]. \tag{3.267}$$

Clearly,

$$|x| \le M \text{ and } |x_k| \le M \text{ for all integers } k \ge 1. \tag{3.268}$$

Let k be a natural number. Since the point $x_k \in \bar{X}_M$ there exists an a.c. function $v : [0, \infty) \to R^n$ such that

$$v(0) = x_k, \ I^{\bar{f}}(0, T, v) - Tf(\bar{x}, 0) \le M \text{ for all } T > 0. \tag{3.269}$$

Since the function $\Lambda^{(x_k)}$ is $(\bar{f})$-overtaking optimal it follows from (3.269) that

$$\lim_{T \to \infty} [I^{\bar{f}}(0, T, \Lambda^{(x_k)}) - Tf(\bar{x}, 0)] - M$$

$$\le \limsup_{T \to \infty} [I^{\bar{f}}(0, T, \Lambda^{(x_k)}) - Tf(\bar{x}, 0) - (I^f(0, T, v) - Tf(\bar{x}, 0))] \le 0.$$

Thus

$$\lim_{T \to \infty} (I^{\bar{f}}(0, T, \Lambda^{(x_k)}) - Tf(\bar{x}, 0)) \le M \text{ for all integers } k \ge 1. \tag{3.270}$$

By Proposition 3.6, (3.268), and (3.270), there exists a positive number M_0 such that

$$|\Lambda^{(x_k)}(t)| \le M_0 \text{ for all } t \in [0, \infty) \text{ and all integers } k \ge 1. \tag{3.271}$$

In view of (3.270), (3.271) and Proposition 3.14, there exists a positive number M_1 such that for each natural number k and each positive number T,

$$I^{\bar{f}}(0, T, \Lambda^{(x_k)}) \le Tf(\bar{x}, 0) + M_1. \tag{3.272}$$

It follows from (3.271), (3.272), and Proposition 3.7 that there exists a strictly increasing sequence of natural numbers $\{k_i\}_{i=1}^\infty$ and an a.c. function $u : [0, \infty) \to R^n$ such that for each natural number m, we have

$$\Lambda^{(x_{k_i})}(t) \to u(t) \text{ as } i \to \infty \text{ uniformly on } [0, m], \tag{3.273}$$

$$I^{\bar{f}}(0, m, u) \le \liminf_{i \to \infty} I^{\bar{f}}(0, m, \Lambda^{(x_{k_i})}).$$

In view of (3.273),

$$u(0) = \lim_{i \to \infty} \Lambda^{(x_{k_i})}(0) = \lim_{i \to \infty} x_{k_i} = x. \qquad (3.274)$$

By (3.184), (3.190), and (3.273), for each natural number m,

$$\int_0^m \bar{L}(u(t), u'(t))dt = \int_0^m \bar{f}(u(t), u'(t))dt - m\bar{f}(\bar{x}, 0) + \langle l, u(m) - u(0) \rangle$$

$$\leq \liminf_{i \to \infty} [I^{\bar{f}}(0, m, \Lambda^{(x_{k_i})}) - m\bar{f}(\bar{x}, 0)$$

$$+ \langle l, \Lambda^{(x_{k_i})}(m) - \Lambda^{(x_{k_i})}(0) \rangle]$$

$$= \liminf_{i \to \infty} \int_0^m \bar{L}(\Lambda^{(x_{k_i})}(t), (\Lambda^{(x_{k_i})})'(t))dt$$

$$\leq \liminf_{i \to \infty} \int_0^\infty \bar{L}(\Lambda^{(x_{k_i})}(t), (\Lambda^{(x_{k_i})})'(t))dt$$

$$\leq \liminf_{i \to \infty} [\pi(x_{k_i}) + \langle l, \bar{x} - x_{k_i} \rangle] = \liminf_{i \to \infty} \pi(x_{k_i}) + \langle l, \bar{x} - x \rangle.$$

This implies that

$$\int_0^\infty \bar{L}(u(t), u'(t))dt - \langle l, \bar{x} - x \rangle \leq \liminf_{i \to \infty} \pi(x_{k_i}).$$

By this inequality and Proposition 3.25, $\pi(x) \leq \liminf_{i \to \infty} \pi(x_{k_i})$. Proposition 3.29 is proved. $\qquad \square$

Proposition 3.30. *Let* $y, z \in \bar{X}_*$, $y \neq z$ *and let* $\alpha \in (0, 1)$. *Then*

$$\pi(\alpha y + (1 - \alpha)z) < \alpha\pi(y) + (1 - \alpha)\pi(z).$$

Proof. Since the points $y, z \in \bar{X}_*$ and the functions $\Lambda^{(y)}$ and $\Lambda^{(z)}$ are $(\bar{f})$-good Proposition 3.1 implies that the function $\alpha\Lambda^{(y)} + (1 - \alpha)\Lambda^{(z)}$ is also $\bar{f}$-good. Since the integrand $\bar{L}$ is convex we have that for all nonnegative numbers t,

$$\bar{L}(\alpha\Lambda^{(y)}(t) + (1 - \alpha)\Lambda^{(z)}(t), \alpha(\Lambda^{(y)})'(t) + (1 - \alpha)(\Lambda^{(z)})'(t))$$

$$\leq \alpha\bar{L}(\Lambda^{(y)}(t), (\Lambda^{(y)})'(t)) + (1 - \alpha)\bar{L}(\Lambda^{(z)}(t), (\Lambda^{(z)})'(t)).$$

The inequality $y \neq z$ implies that for all numbers $t > 0$ which are close enough to zero, we have $\Lambda^{(y)}(t) \neq \Lambda^{(z)}(t)$ and

$$\bar{L}(\alpha\Lambda^{(y)}(t) + (1 - \alpha)\Lambda^{(z)}(t), \alpha(\Lambda^{(y)})'(t) + (1 - \alpha)(\Lambda^{(z)})'(t))$$

$$< \alpha\bar{L}(\Lambda^{(y)}(t), (\Lambda^{(y)})'(t)) + (1 - \alpha)\bar{L}(\Lambda^{(z)}(t), (\Lambda^{(z)})'(t))].$$

By the inequalities above,

$$\int_0^\infty \bar{L}(\alpha\Lambda^{(y)}(t) + (1-\alpha)\Lambda^{(z)}(t)), \alpha(\Lambda^{(y)})'(t) + (1-\alpha)(\Lambda^{(z)})'(t))dt$$

$$< \alpha \int_0^\infty \bar{L}(\Lambda^{(y)}(t), (\Lambda^{(y)})'(t))dt + (1-\alpha)\int_0^\infty \bar{L}(\Lambda^{(z)}(t), (\Lambda^{(z)})'(t))dt.$$

$$(3.275)$$

It follows from (3.275), Proposition 3.25, and (3.190) that

$$\pi(\alpha y + (1-\alpha)z)$$

$$\leq \int_0^\infty \bar{L}(\alpha\Lambda^{(y)}(t) + (1-\alpha)\Lambda^{(z)}(t), \alpha(\Lambda^{(y)})'(t) + (1-\alpha)(\Lambda^{(z)})'(t))dt$$

$$- \langle l, \bar{x} - \alpha y - (1-\alpha)z \rangle < \alpha[\int_0^\infty \bar{L}(\Lambda^{(y)}(t), (\Lambda^{(y)})'(t))dt - \langle l, \bar{x} - y \rangle]$$

$$+ (1-\alpha)[\int_0^\infty \bar{L}(\Lambda^{(z)}(t), (\Lambda^{(z)})'(t))dt - \langle l, \bar{x} - z \rangle]$$

$$= \alpha\pi(y) + (1-\alpha)\pi(z).$$

Proposition 3.30 is proved. $\square$

Now Proposition 3.21 follows from Propositions 3.28–3.30 and Corollary 3.26.

3.13 Proof of Theorem 3.22

Let $v : [0, T] \to R^n$ be an a.c. function. Set

$$\bar{v}(t) = v(T - t), \ t \in [0, T].$$

Clearly,

$$\int_0^T \bar{f}(\bar{v}(t), \bar{v}'(t))dt = \int_0^T f(\bar{v}(t), -\bar{v}'(t))dt$$

$$= \int_0^T f(v(T - t), v'(T - t))dt$$

$$= \int_0^T f(v(t), v'(t))dt.$$

The following lemma is an important ingredient in the proof of Theorem 3.22.

Lemma 3.31. *Let $\epsilon, M > 0$ and let $L_1 \geq 1$ be an integer. Then there exist an integer $L_2 \geq 1$ and a positive number δ such that for each number $T \geq L_2$ and each a.c. function $v : [0, T] \to R^n$ which satisfies*

$$|v(0)| \leq M, \quad \int_0^T \bar{L}(v(t), v'(t))dt + <l, -\bar{x} + v(0) > \leq \pi(x_*) + \delta \qquad (3.276)$$

the inequality $|v(t) - \Lambda^{(x_*)}(t)| \leq \epsilon$ holds for all numbers $t \in [0, L_1]$.

Proof. Assume the contrary. Then for each integer $k \geq 1$, there exist a number $T_k \geq k$ and an a.c. function $v_k : [0, T_k] \to R^n$ such that

$$M \geq |v_k(0)|, \quad \int_0^{T_k} \bar{L}(v_k(t), v'_k(t))dt - \langle l, \bar{x} - v_k(0) \rangle \leq \pi(x_*) + k^{-1}, \qquad (3.277)$$

$$\sup\{|v_k(t) - \Lambda^{(x^*)}(t)| : t \in [0, L_1]\} > \epsilon. \qquad (3.278)$$

We claim that there exists a subsequence $\{v_{k_i}\}_{i=1}^{\infty}$ which converges uniformly on any bounded subinterval of $[0, \infty)$ to an $(\bar{f})$-overtaking optimal function v satisfying $v(0) = x_*$.

In view of Proposition 3.6 (with $f = \bar{L}$) and (3.277), there exists a positive number M_0 such that

$$|v_k(t)| \leq M_0 \text{ for all } t \in [0, T_k] \text{ and all integers } k \geq 1. \qquad (3.279)$$

By Proposition 3.6 (with $f = \bar{L}$), (3.277), and (3.279), there exist a strictly increasing sequence of natural numbers $\{k_i\}_{i=1}^{\infty}$ and an a.c. function $v : [0, \infty) \to R^n$ such that for each natural number m, we have

$$v_{k_i}(t) \to v(t) \text{ as } i \to \infty \text{ uniformly on } [0, m], \qquad (3.280)$$

$$\int_0^m \bar{L}(v(t), v'(t))dt \leq \liminf_{i \to \infty} \int_0^m \bar{L}(v_{k_i}(t), v'_{k_i}(t))dt.$$

By (3.277) and (3.280),

$$|v(0)| \leq M. \qquad (3.281)$$

In view of (3.277) and (3.280), for each natural number m,

$$\int_0^m \bar{L}(v(t), v'(t))dt \leq \liminf_{i \to \infty} \int_0^{T_{k_i}} \bar{L}(v_{k_i}(t), v'_{k_i}(t))dt$$

$$\leq \liminf_{i \to \infty}[\pi(x_*) + k_i^{-1} + \langle l, \bar{x} - v_{k_i}(0) \rangle]$$

$$= \pi(x_*) + \langle l, \bar{x} - v(0) \rangle.$$

This relation implies that

$$\int_0^\infty \bar{L}(v(t), v'(t))dt + \langle l, v(0) - \bar{x} \rangle \le \pi(x_*). \tag{3.282}$$

It follows from (3.282) and Proposition 3.24 that v is an $(\bar{f})$-good function. By Proposition 3.25 and (3.282),

$$\pi(v(0)) \le \int_0^\infty \bar{L}(v(t), v'(t))dt - \langle l, \bar{x} - v(0) \rangle \le \pi(x_*).$$

It follows from the choice of x_* (see (3.191) and (3.192)), (3.184), and Proposition 3.15 that

$$v(0) = x_*,$$

$$\pi(x_*) = \int_0^\infty \bar{L}(v(t), v'(t))dt - \langle l, \bar{x} - x_* \rangle = \lim_{T \to \infty} [I^{\bar{f}}(0, T, v) - Tf(\bar{x}, 0)]. \tag{3.283}$$

By (3.283), (3.189), and (3.190), v is an $(\bar{f})$-overtaking optimal function. When combined with Theorem 3.20 this implies that

$$v(t) = \Lambda^{(x_*)}(t) \text{ for all } t \in [0, \infty). \tag{3.284}$$

Relations (3.280) and (3.284) imply that for each sufficiently large natural numbers i

$$\sup\{|v_{k_i}(t) - \Lambda^{(x_*)}(t)| : t \in [0, L_1]\} < \epsilon/2.$$

This contradicts (3.278). The contradiction we have reached proves Lemma 3.31.

$\square$

Proof of Theorem 3.22. We will choose the constants δ and L_2. First we choose a positive number $\bar{r} < 4^{-1}$ such that

$$\{(x, y) \in R^n \times R^n : |x - \bar{x}| \le 4\bar{r}, |y| \le 4\bar{r}\} \subset \text{dom}(f). \tag{3.285}$$

Proposition 3.6 implies that there exists a number $M_0 > M$ such that for each positive number T and each a.c. function $u : [0, T] \to R^n$ which satisfies

$$|u(0)| \le M, \quad I^f(0, T, u) \le Tf(\bar{x}, 0) + M + 2, \tag{3.286}$$

we have

$$|u(t)| \le M_0, \quad t \in [0, T]. \tag{3.287}$$

By Lemma 3.31, there exist a natural number $L_{12} > L_1$ and a positive number δ_0 such that the following property holds:

(P10) For each number $T \geq L_{12}$ and each a.c. function $u : [0, T] \to R^n$ satisfying

$$|u(0)| \leq M_0, \quad \int_0^T \bar{L}(u(t), u'(t))dt + \langle l, -\bar{x} + u(0) \rangle \leq \pi(x_*) + \delta_0,$$

the inequality $|u(t) - \Lambda^{(x_*)}(t)| \leq \epsilon$ holds for all numbers $t \in [0, L_1]$.   □

Proposition 3.8 implies that there exists a number $\delta_1 \in (0, 1)$ such that for each a.c. function $u : [0, 1] \to R^n$ satisfying

$$|u(0) - \bar{x}|, \ |u(1) - \bar{x}| \leq \delta_1,$$

we have

$$I^f(0, 1, u) \geq f(\bar{x}, 0) - \delta_0/16. \tag{3.288}$$

Since the function f is continuous at the point $(\bar{x}, 0)$ there exists a number $\delta_2 > 0$ such that

$$|f(\xi_1, \xi_2) - f(\bar{x}, 0)| \leq \delta_0/16 \text{ for all } (\xi_1, \xi_2) \in R^n \times R^n \text{ satisfying}$$

$$|\xi_1 - \bar{x}| \leq 2\delta_2, \ |\xi_2| \leq 2\delta_2; \tag{3.289}$$

$$2(|l| + 1)\delta_2 < \delta_0/16 \text{ and } \delta_2 < \min\{\bar{r}, \delta_1\}. \tag{3.290}$$

In view of Theorem 3.3 there exist an integer $L_{13} > L_{12}$ and a number $\delta \in (0, \delta_2)$ such that the following property holds:

(P11)   If a number $T > 2L_{13}$ and if an a.c. function $u : [0, T] \to R^n$ satisfies

$$u(0) \in X_M, \ I^f(0, T, u) \leq \sigma(f, T, u(0)) + \delta,$$

then

$$|u(t) - \bar{x}| \leq \delta_2, \ t \in [L_{13}, T - L_{13}].$$

By Proposition 3.15 there exists a natural number $L_2 > L_{13}$ such that

$$|\Lambda^{(x_*)}(t) - \bar{x}| \leq \delta_2 \text{ for all numbers } t \geq L_2. \tag{3.291}$$

Therefore we have chosen the constants L_2 and δ.

Let $T > 2L_2$ be given and let an a.c. function $v : [0, T] \to R^n$ satisfy

$$v(0) \in X_M, \quad I^f(0, T, v) \le \sigma(f, T, v(0)) + \delta. \tag{3.292}$$

We claim that

$$|v(T - t) - \Lambda^{(*)}(t)| \le \epsilon \text{ for all } t \in [0, L_1]. \tag{3.293}$$

By (3.292) and the definition of the set X_M, there exists an a.c. function $u :$
$[0, \infty) \to R^n$ such that

$$u(0) = v(0) \text{ and } I^f(0, T, u) - Tf(\bar{x}, 0) \le M \text{ for all } T > 0. \tag{3.294}$$

It follows from (3.292) and (3.294) that

$$I^f(0, T, v) \le I^f(0, T, u) + \delta \le Tf(\bar{x}, 0) + M + 1. \tag{3.295}$$

When combined with the choice of M_0 (see (3.286) and (3.287)) and (3.292) this
inequality implies that

$$|v(t)| \le M_0 \text{ for all } t \in [0, T]. \tag{3.296}$$

Property (P11) and (3.292) imply that

$$|v(t) - \bar{x}| \le \delta_2 \text{ for all } t \in [L_{13}, T - L_{13}] \tag{3.297}$$

Set

$$y(t) = v(t), \ t \in [0, T - L_2 - 1], \ y(t) = \Lambda^{(x*)}(T - t), \ t \in [T - L_2, T], \tag{3.298}$$

$$y(t) = v(T - L_2 - 1) + (t - (T - L_2 - 1))[\Lambda^{(x*)}(L_2) - v(T - L_2 - 1)],$$
$$t \in (T - L_2 - 1, T - L_2).$$

It follows from (3.291), (3.297), and (3.298) that for all numbers $t \in [T - L_2 - 1, T - L_2]$, we have

$$|y(t) - \bar{x}| \le \max\{|v(T - L_2 - 1) - \bar{x}|, \ |\Lambda^{(x*)}(L_2) - \bar{x}|\} \le \delta_2, \ |y'(t)| \le 2\delta_2. \tag{3.299}$$

In view of (3.299) and the choice of δ_2 (see (3.289)), for all numbers $t \in [T - L_2 - 1, T - L_2]$,

$$|f(y(t), y'(t)) - f(\bar{x}, 0)| \le \delta_0/16, \tag{3.300}$$

$$|I^f(T - L_2 - 1, T - L_2, y) - f(\bar{x}, 0)| \le \delta_0/16.$$

By (3.297) and the choice of δ_1 (see (3.288)), we have $I^f(T - L_2 - 1, T - L_2, v) \geq f(\bar{x}, 0) - \delta_0/16$. This inequality, (3.292), and (3.298) imply that

$$\delta \geq I^f(0, T, v) - I^f(0, T, y)$$

$$= I^f(0, T - L_2 - 1, v) + I^f(T - L_2 - 1, T - L_2, v) + I^f(T - L_2, T, v)$$

$$- I^f(0, T - L_2 - 1, y) - I^f(T - L_2 - 1, T - L_2, y) - I^f(T - L_2, T, y)$$

$$\geq f(\bar{x}, 0) - \delta_0/16 - (f(\bar{x}, 0) + \delta_0/16)$$

$$+ I^f(T - L_2, T, v) - I^f(T - L_2, T, y)$$

$$\geq -\delta_0/8 + \int_{T-L_2}^{T} f(v(t), v'(t))dt - \int_0^{L_2} \bar{f}(\Lambda^{(x*)}(t), (\Lambda^{(x*)})'(t))dt.$$

$$\tag{3.301}$$

Put

$$\bar{v}(t) = v(T - t), \ t \in [0, L_2]. \tag{3.302}$$

It follows from (3.290), (3.301), (3.184), (3.291), and (3.297) that

$$\delta_0/4 \geq \delta + \delta_0/8 \geq \int_0^{L_2} \bar{f}(\bar{v}(t), \bar{v}'(t))dt - \int_0^{L_2} \bar{f}(\Lambda^{(x*)}(t), (\Lambda^{(x*)})'(t))dt$$

$$= \int_0^{L_2} \bar{L}(\bar{v}(t), \bar{v}'(t))dt - \langle l, \bar{v}(L_2) - \bar{v}(0)\rangle$$

$$- [\int_0^{L_2} \bar{L}(\Lambda^{(x*)}(t), (\Lambda^{(x*)})'(t))dt - \langle l, \Lambda^{(x*)}(L_2) - \Lambda^{(x*)}(0)\rangle]$$

$$\geq \int_0^{L_2} \bar{L}(\bar{v}(t), \bar{v}'(t))dt - \langle l, \bar{x} - \bar{v}(0)\rangle$$

$$- [\int_0^{L_2} \bar{L}(\Lambda^{(x*)}(t), (\Lambda^{(x*)})'(t))dt - \langle l, \bar{x} - \Lambda^{(x*)}(0)\rangle] - 2\delta_2|l|.$$

When combined with (3.290) and (3.190) this implies that

$$\int_0^{L_2} \bar{L}(\bar{v}(t), \bar{v}'(t))dt - \langle l, \bar{x} - \bar{v}(0)\rangle$$

$$\leq \delta_0/4 + \delta_0/8 + \int_0^{\infty} \bar{L}(\Lambda^{(x*)}(t), (\Lambda^{(x*)})'(t))dt - \langle l, \bar{x} - \Lambda^{(x*)}(0)\rangle]$$

$$\leq \delta_0/2 + \pi(x_*). \tag{3.303}$$

By property (P10), (3.296), (3.302), and (3.303),

$$|\bar{v}(t) - \Lambda^{(x*)}(t)| \leq \epsilon$$

for all numbers $t \in [0, L_1]$ and

$$|v(T - t) - \Lambda^{(x*)}(t)| \leq \epsilon$$

for all $t \in [0, L_1]$. This completes the proof of Theorem 3.22.

3.14   Optimal Solutions for Infinite Horizon Problems

In this section which is based on [63] we study the structure of optimal solutions
of infinite horizon autonomous variational problems with a lower semicontinuous
integrand $f : R^n \times R^n \to R^1 \cup \{\infty\}$ introduced in Sect. 3.1. We also show that all
the optimality notions used in the literature are equivalent for the problems with the
integrand f. We use the notation and definitions introduced in Sect. 3.1.

Let a be a positive number, $\psi : [0, \infty) \to [0, \infty)$ be an increasing function such
that

$$\lim_{t \to \infty} \psi(t) = \infty, \tag{3.304}$$

and $f : R^n \times R^n \to R^1 \cup \{\infty\}$ be a lower semicontinuous function such that the
set $\mathrm{dom}(f)$ is nonempty convex and closed and that

$$f(x, y) \geq \max\{\psi(|x|), \ \psi(|y|)|y|\} - a \text{ for each } x, y \in R^n. \tag{3.305}$$

We suppose that there exists a point $\bar{x} \in R^n$ such that

$$f(\bar{x}, 0) \leq f(x, 0) \text{ for each } x \in R^n \tag{3.306}$$

and that assumptions (A1)–(A4) introduced in Sect. 3.1 hold.

In this section we use the notion of an overtaking optimal function introduced in
Sect. 3.1. The following two optimality notions are also used in the infinite horizon
optimal control.

An a.c. function $v : [0, \infty) \to R^n$ is called (f)-weakly optimal [14, 56] if for
each a. c. function $u : [0, \infty) \to R^n$ satisfying $u(0) = v(0)$, we have

$$\liminf_{T \to \infty} [I^f(0, T, v) - I^f(0, T, u)] \leq 0.$$

An a. c. function $v : [0, \infty) \to R^n$ is called (f)-minimal [7, 56] if for each
pair of numbers $T_1 \geq 0$, each $T_2 > T_1$, and each a.c. function $u : [T_1, T_2] \to R^n$
satisfying $u(T_i) = v(T_i)$, $i = 1, 2$, we have

$$\int_{T_1}^{T_2} f(v(t), v'(t))dt \leq \int_{T_1}^{T_2} f(u(t), u'(t))dt.$$

We prove the following theorem obtained in [63] which shows that for the integrand considered in the section all the three optimality notions introduced before are equivalent.

Theorem 3.32. *Assume that $x \in R^n$ and that there exists an (f)-good function $\tilde{v} : [0, \infty) \to R^n$ satisfying $\tilde{v}(0) = x$. Let $v : [0, \infty) \to R^n$ be an a.c. function such that $v(0) = x$. Then the following conditions are equivalent:*

(i) *the function v is (f)-overtaking optimal; (ii) the function v is (f)-weakly optimal; (iii) the function v is (f)-good and (f)-minimal; (iv) the function v is (f)-minimal and $\lim_{t \to \infty} v(t) = \bar{x}$; (v) the function v is (f)-minimal and $\liminf_{t \to \infty} |v(t) - \bar{x}| = 0$.*

We prove the following two theorems obtained in [63] which describe the asymptotic behavior of overtaking optimal functions.

Theorem 3.33. *Let ϵ be a positive number. Then there exists a positive number δ such that:*

(i) *For each point $x \in R^n$ satisfying $|x - \bar{x}| \le \delta$ there exists an (f)-overtaking optimal and (f)-good function $v : [0, \infty) \to R^n$ such that $v(0) = x$.*
(ii) *If an (f)-overtaking optimal function $v : [0, \infty) \to R^n$ satisfies $|v(0) - \bar{x}| \le \delta$, then $|v(t) - \bar{x}| \le \epsilon$ for all numbers $t \in [0, \infty)$.*

Theorem 3.34. *Let ϵ, M be positive numbers. Then there exists a positive number L such that for each point $x \in X_M$ and each (f)-overtaking optimal function $v : [0, \infty) \to R^n$ satisfying $v(0) = x$, the following inequality holds:*

$$|v(t) - \bar{x}| \le \epsilon \text{ for all } t \in [L, \infty).$$

The next theorem obtained in [63] establishes a non-self-intersection property of overtaking optimal solutions analogous to the property established in [37, 57] for variational problems with finite-valued integrands.

Theorem 3.35. *Assume that $v : [0, \infty) \to R^n$ is an (f)-good (f)-overtaking optimal function and that $0 \le t_1 < t_2$ satisfy $v(t_1) = v(t_2)$. Then $v(t) = \bar{x}$ for all numbers $t \ge t_1$.*

3.15   Proof of Theorem 3.32

It is clear that (i) implies (ii). By Proposition 3.1, (ii) implies (iii). It follows from assumption (A4) that (iii) implies (iv). Evidently, (iv) implies (v).

We show that (v) implies (iii). Assume that v is an (f)-minimal function and that

$$\liminf_{t \to \infty} |v(t) - \bar{x}| = 0. \tag{3.307}$$

By assumption (A1) there exists a number $\delta \in (0, 1)$ such that

$$\{(z, y) \in R^n \times R^n : |\bar{x} - z| \leq 4\delta, \ |y| \leq 4\delta\} \subset \mathrm{dom}(f), \tag{3.308}$$

$$|f(z, y) - f(\bar{x}, 0)| \leq 1 \text{ for each } (z, y) \in R^n \times R^n \text{ satisfying } |z - \bar{x}| \leq 4\delta, \ |y| \leq 4\delta. \tag{3.309}$$

In view of (3.307), there exists a sequence of numbers $\{t_k\}_{k=1}^\infty$ such that

$$t_1 \geq 4, \ t_{k+1} - t_k \geq 4, \ k = 1, 2, \ldots \text{ and } |v(t_k) - \bar{x}| \leq \delta, \ k = 1, 2, \ldots \tag{3.310}$$

Since $\tilde{v}$ is an (f)-good function there exists a positive number M_1 such that

$$|I^f(0, T, \tilde{v}) - Tf(\bar{x}, 0)| < M_1 \text{ for all } T > 0. \tag{3.311}$$

Assumption (A4) implies that

$$\lim_{t \to \infty} \tilde{v}(t) = \bar{x}. \tag{3.312}$$

By (3.312), there exists a number $\tau > 1$ such that

$$|\tilde{v}(t) - \bar{x}| \leq \delta \text{ for all } t \in [\tau, \infty). \tag{3.313}$$

Let $k \geq 1$ be an integer such that

$$t_k \geq \tau + 8. \tag{3.314}$$

Define an a. c. function $u : [0, t_k] \to R^n$ by

$$u(t) = \tilde{v}(t), \ t \in [0, t_k - 1], \ u(t_k - 1 + t) = \tilde{v}(t_k - 1) + t(v(t_k) - \tilde{v}(t_k - 1)), \ t \in (0, 1] \tag{3.315}$$

It is clear that

$$u(0) = \tilde{v}(0) = x = v(0), \ u(t_k) = v(t_k). \tag{3.316}$$

We estimate $I^f(0, t_k, u)$. For a number $t \in (t_k - 1, t_k)$ relations (3.310) and (3.313)–(3.315) imply that

$$|u(t) - \bar{x}| \leq \max\{|\tilde{v}(t_k - 1) - \bar{x}|, \ |v(t_k) - \bar{x}|\} \leq \delta, \tag{3.317}$$

$$|u'(t)| = |v(t_k) - \tilde{v}(t_k - 1)| \leq |v(t_k) - \bar{x}| + |\bar{x} - \tilde{v}(t_k - 1)| \leq 2\delta. \tag{3.318}$$

By (3.308), (3.309), (3.317), and (3.318), for $t \in (t_k - 1, t_k)$, we have

$$|f(u(t), u'(t)) - f(\bar{x}, 0)| \leq 1. \tag{3.319}$$

It follows from (3.311), (3.314), (3.315), and (3.319) that

$$
\begin{aligned}
I^f(0, t_k, u) &= I^f(0, t_k - 1, u) + I^f(t_k - 1, t_k, u) \\
&\leq I^f(0, t_k - 1, \tilde{v}) + f(\bar{x}, 0) + 1 \\
&\leq (t_k - 1) f(\bar{x}, 0) + M_1 + f(\bar{x}, 0) + 1 = t_k f(\bar{x}, 0) + M_1 + 1.
\end{aligned}
\tag{3.320}
$$

Since the function v is (f)-minimal relations (3.316) and (3.320) imply that

$$
I^f(0, t_k, v) \leq I^f(0, t_k, u) \leq t_k f(\bar{x}, 0) + M_1 + 1.
$$

When combined with Proposition 3.1 the inequality above which holds for every integer $k \geq 1$ satisfying (3.314) implies that the function v is (f)-good. Therefore (v) implies (iii).

In order to complete the proof of the theorem it is sufficient to show that (iii) implies (i).

Assume that the function v is (f)-good and (f)-minimal. Assumption (A4) implies that

$$
\lim_{t \to \infty} v(t) = \bar{x}.
\tag{3.321}
$$

We claim that v is an (f)-overtaking optimal function. Assume the contrary. Then there exists an a.c. function $v_1 : [0, \infty) \to R^n$ such that

$$
v_1(0) = v(0), \; \limsup_{T \to \infty} [I^f(0, T, v) - I(0, T, v_1)] > 0.
\tag{3.322}
$$

In view of Theorem 3.3 and Proposition 3.1, there exists an (f)-good and (f)-overtaking optimal function $v_2 : [0, \infty) \to R^n$ such that

$$
v_2(0) = v(0).
\tag{3.323}
$$

Then

$$
\limsup_{T \to \infty} [I^f(0, T, v_2) - I^f(0, T, v_1)] \leq 0.
\tag{3.324}
$$

Relations (3.322) and (3.324) imply that

$$
\limsup_{T \to \infty} [I^f(0, T, v) - I^f(0, T, v_2)] > 0.
\tag{3.325}
$$

In view of assumption (A4), we have

$$
\lim_{t \to \infty} v_2(t) = \bar{x}.
\tag{3.326}
$$

By inequality (3.325) there exists a positive number ϵ such that

$$\limsup_{T \to \infty} [I^f(0, T, v) - I^f(0, T, v_2)] > 2\epsilon.$$

Therefore there exists a sequence of numbers $\{T_k\}_{k=1}^{\infty} \subset (4, \infty)$ such that for all natural numbers k, we have

$$T_{k+1} \geq T_k + 4, \quad I^f(0, T_k, v) - I^f(0, T_k, v_2) > \epsilon. \tag{3.327}$$

It follows from Proposition 3.8 and assumption (A1) that there exists a positive number δ such that

$$\{(y, z) \in R^n \times R^n : |y - \bar{x}| \leq 4\delta, |z| \leq 4\delta\} \subset \text{dom}(f), \tag{3.328}$$

$$|f(y, z) - f(\bar{x}, 0)| \leq \epsilon/8 \text{ for each } (y, z) \in R^n \times R^n$$

$$\text{satisfying } |y - \bar{x}| \leq 4\delta, |z| \leq 4\delta, \tag{3.329}$$

$$I^f(0, 1, h) \geq f(\bar{x}, 0) - \epsilon/8 \text{ for each a.c. function } h : [0, 1] \to R^n \tag{3.330}$$

$$\text{satisfying } |h(0) - \bar{x}|, |h(1) - \bar{x}| \leq 4\delta.$$

Since the functions v and v_2 are (f)-good, assumption (A4) implies that there exists a positive number τ such that

$$|v(t) - \bar{x}|, |v_2(t) - \bar{x}| \leq \delta \text{ for all } t \geq \tau. \tag{3.331}$$

Choose an integer $k \geq 1$ such that

$$T_k > \tau. \tag{3.332}$$

Define an a.c. function $u : [0, T_k + 1] \to R^n$ by

$$u(t) = v_2(t), \ t \in [0, T_k], \ u(T_k + t) = v_2(T_k) + t(v(T_k + 1) - v_2(T_k)), \ t \in (0, 1]. \tag{3.333}$$

Relations (3.323) and (3.333) imply that

$$u(0) = v_2(0) = v(0), \ u(T_k + 1) = v(T_k + 1). \tag{3.334}$$

We estimate $I^f(0, T_k+1, u) - I^f(0, T_k+1, v)$. By (3.331)–(3.333), for all numbers $t \in (T_k, T_k + 1)$, we have

$$|u(t) - \bar{x}| \leq \max\{|v_2(T_k) - \bar{x}|, |v(T_k + 1) - \bar{x}|\} \leq \delta, \tag{3.335}$$

$$|u'(t)| = |v(T_k + 1) - v_2(T_k)| \leq |v(T_k + 1) - \bar{x}| + |\bar{x} - v_2(T_k)| \leq 2\delta. \tag{3.336}$$

It follows from (3.328), (3.329), (3.335), and (3.336) that for $t \in (T_k, T_k + 1)$, we have

$$|f(u(t), u'(t)) - f(\bar{x}, 0)| \leq \epsilon/8. \tag{3.337}$$

By (3.330)–(3.332),

$$I^f(T_k, T_k + 1, v) \geq f(\bar{x}, 0) - \epsilon/8. \tag{3.338}$$

Relations (3.333), (3.337), and (3.338) imply that

$$\begin{aligned}
I^f(0, T_k + 1, u) - I^f(0, T_k + 1, v) &= I^f(0, T_k, u) + I^f(T_k, T_{k+1}, u) \\
&\quad - I^f(0, T_k, v) - I^f(T_k, T_k + 1, v) \\
&\leq I^f(0, T_k, v_2) - I^f(0, T_k, v) \\
&\quad + f(\bar{x}, 0) + \epsilon/8 - (f(\bar{x}, 0) - \epsilon/8) \\
&< -\epsilon + \epsilon/4 < 0.
\end{aligned}$$

When combined with (3.328) this contradicts our assumption that the function (v) is (f)-minimal. The contradiction we have reached proves that v is an (f)-overtaking optimal function. Therefore (iii) implies (i). Theorem 3.32 is proved.

3.16   Proofs of Theorems 3.33–3.35

Proof of Theorem 3.33. Assumption (A1) implies that there exists a number $\delta_0 \in (0, 1)$ such that

$$\{(y, z) \in R^n \times R^n : |y - \bar{x}| \leq 4\delta_0, \ |z| \leq 4\delta_0\} \subset \text{dom}(f), \tag{3.339}$$

$$|f(y, z) - f(\bar{x}, 0)| \leq 1 \text{ for all } (y, z) \in R^n \times R^n \text{ satisfying}$$

$$|y - \bar{x}| \leq 4\delta_0, \ |z| \leq 4\delta_0. \tag{3.340}$$

By Lemma 3.11, there exists a number

$$\delta \in (0, \min\{\epsilon/2, \ \delta_0/4\}) \tag{3.341}$$

such that the following property holds:

(P12)   For each number $T \geq 2$ and each a. c. function $w : [0, T] \to R^n$ which satisfies

$$|w(0) - \bar{x}|, \ |w(T) - \bar{x}| \leq \delta$$

and

$$I^f(0, T, w) \leq \sigma(f, T, w(0), w(T)) + \delta$$

the inequality $|w(t) - \bar{x}| \leq \epsilon$ holds for all numbers $t \in [0, T]$.

Assume that a point $x \in R^n$ satisfies

$$|x - \bar{x}| \leq \delta. \qquad (3.342)$$

Define an a. c. function $v : [0, \infty) \to R^n$ such that

$$v(t) = x + t(\bar{x} - x), \ t \in [0, 1], \ v(t) = \bar{x}, \ t \in (1, \infty). \qquad (3.343)$$

In view of (3.343) and the choice of δ_0, the function v is (f)-good. By Theorem 3.3 there exists an (f)-overtaking optimal function u such that $u(0) = x$. In view of Proposition 3.1, the function u is (f)-good and the assertion (i) is proved.

We now prove the assertion (ii). Assume that $v : [0, \infty) \to R^n$ is an (f)-overtaking optimal function and that

$$|v(0) - \bar{x}| \leq \delta. \qquad (3.344)$$

Assertion (i) implies that v is an (f)-good function. It follows from assumption (A4) that

$$\lim_{t \to \infty} |v(t) - \bar{x}| = 0. \qquad (3.345)$$

Let $s \in (0, \infty)$ be given. In view of (3.345), there exists a number $T > s + 4$ such that $|v(T) - \bar{x}| \leq \delta$. Since the function v is (f)-overtaking optimal, the inequality above, (3.344), and property (P12) imply that $|v(s) - \bar{x}| \leq \epsilon$. Thus (ii) is proved. This completes the proof of Theorem 3.33. $\qquad \square$

Proof of Theorem 3.34. It follows from Theorem 3.33 that there exists a number $\delta \in (0, \epsilon)$ such that for each (f)-overtaking optimal function $v : [0, \infty) \to R^n$ satisfying $|v(0) - \bar{x}| \leq \delta$, we have

$$|v(t) - \bar{x}| \leq \epsilon, \ t \in [0, \infty). \qquad (3.346)$$

Lemma 3.10 implies that there exists a positive number L_0 such that the following property holds:

(P13)   If a number $T \geq L_0$ and if an a.c. function $v : [0, T] \to R^n$ satisfies the inequalities $|v(0)| \leq M$ and

$$I^f(0, T, v) \leq Tf(\bar{x}, 0) + M + 2,$$

then for each number $s \in [0, T - L_0]$, we have

$$\min\{|v(t) - \bar{x}| : t \in [s, s + L_0]\} \leq \delta/2.$$

Assume that a point $x \in X_M$ and that an (f)-overtaking optimal function $v :$ $[0, \infty) \to R^n$ satisfies $v(0) = x$. In view of the definition of the set X_M,

$$|x| \leq M \qquad (3.347)$$

and there exists an a.c. function $u : [0, \infty) \to R^n$ such that

$$u(0) = x, \ I^f(0, T, u) - Tf(\bar{x}, 0) \leq M \text{ for each } T \in (0, \infty). \qquad (3.348)$$

Since the function v is (f)-overtaking optimal it follows from the equality $v(0) = x$ and (3.348) that for all large enough positive numbers T, we have

$$I^f(0, T, v) \leq I^f(0, T, u) + 1 \leq Tf(\bar{x}, 0) + M + 1.$$

Therefore there exists a number $T \geq L_0 + 1$ such that

$$I(0, T, v) \leq Tf(\bar{x}, 0) + M + 1. \qquad (3.349)$$

It follows from (3.347), (3.349) and property (P13) with $s = 0$ that there exists a number $\tau \in [0, L_0]$ such that $|v(\tau) - \bar{x}| < \delta$. By this inequality and the choice of δ, we have $|v(t) - \bar{x}| \leq \epsilon$ for all numbers $t \in [\tau, \infty)$. Since $\tau \in [0, L_0]$ we conclude that $|v(t) - \bar{x}| \leq \epsilon$ for all $t \in [L_0, \infty)$. Theorem 3.34 is proved. $\qquad \square$

Proof of Theorem 3.35. We may assume without loss of generality that $t_1 = 0$. It is clear that there exists an a.c. function $u : [0, \infty) \to R^n$ such that

$$u(t) = v(t), \ t \in [0, t_2], \ u(t + t_2) = u(t), \ t \in [0, \infty). \qquad (3.350)$$

We claim that $I^f(t_1, t_2, v) = (t_2 - t_1)f(\bar{x}, 0)$. Assume the contrary. Then

$$I^f(t_1, t_2, v) - (t_2 - t_1)f(\bar{x}, 0) \neq 0. \qquad (3.351)$$

Proposition 3.1 implies that the function

$$T \to I^f(0, T, u) - Tf(\bar{x}, 0), \ T \in (0, \infty)$$

is bounded from below. When combined with (3.350) this implies that

$$I^f(0, t_2, u) \geq t_2 f(\bar{x}, 0).$$

Together with inequality (3.351) this inequality implies that

$$\Delta := I^f(0, t_2, v) - t_2 f(\bar{x}, 0) > 0. \qquad (3.352)$$

Assumption (A1) implies that there exists $\delta \in (0,1)$ such that

$$\{(y,z) \in R^n \times R^n : |y - \bar{x}| \le 4\delta, \; |z| \le 4\delta\} \subset \text{dom}(f), \tag{3.353}$$

$$|f(y,z) - f(\bar{x},0)| \le (1+t_2)^{-1}4^{-1}\Delta \tag{3.354}$$

for each $(y,z) \in R^n \times R^n$ such that $|y - \bar{x}| \le 4\delta, \; |z| \le 4\delta$.

By assumption (A4), $\lim_{t \to \infty} v(t) = \bar{x}$. Then there exists a real number $\tau_0 > 4 + 4t_2$ such that

$$|v(t) - \bar{x}| \le \delta, \; t \in [\tau_0, \infty). \tag{3.355}$$

Define an a.c. function $w : [0, \infty) \to R^n$ by

$$w(t) = v(t + t_2), \; t \in [0, \tau_0], \; w(t) = v(\tau_0 + t_2), \; t \in (\tau_0, \tau_0 + t_2],$$

$$w(t) = v(t), \; t \in (\tau_0 + t_2, \infty). \tag{3.356}$$

It follows from the equality $t_1 = 0$ and (3.356) that

$$w(0) = v(t_2) = v(0), \; w(\tau_0 + t_2) = v(\tau_0 + t_2). \tag{3.357}$$

By (3.355) and (3.356), for all numbers $t \in (\tau_0, \tau_0 + t_2)$,

$$w'(t) = 0, \; |w(t) - \bar{x}| = |v(\tau_0 + t_2) - \bar{x}| \le \delta. \tag{3.358}$$

In view of (3.356), we have

$$
\begin{aligned}
I^f(0, \tau_0 + t_2, v) - I^f(0, \tau_0 + t_2, w) &= I^f(0, t_2, v) + I^f(t_2, t_2 + \tau_0, v) \\
&\quad - I^f(0, \tau_0, w) - I^f(\tau_0, \tau_0 + t_2, w) \\
&= I^f(0, t_2, v) - I^f(\tau_0, \tau_0 + t_2, w) \\
&= I^f(0, t_2, v) - t_2 f(v(\tau_0 + t_2), 0).
\end{aligned}
\tag{3.359}
$$

By (3.352), (3.354), (3.358), and (3.359),

$$I^f(0, \tau_0 + t_2, v) - I^f(0, \tau_0 + t_2, w) = \Delta - t_2 f(v(\tau_0 + t_2), 0) + t_2 f(\bar{x}, 0)$$

$$\ge \Delta - t_2(1 + t_2)^{-1}4^{-1}\Delta > 0.$$

Together with relation (3.357) this contradicts the assumption that the function v is (f)-overtaking optimal. The contradiction we have reached proves that

$I^f(0, t_2, v) = t_2 f(\bar{x}, 0)$. When combined with (3.359) and Proposition 3.1 this implies that the function u is (f)-good. In view of assumption (A4), we have $\lim_{t\to\infty} u(t) = \bar{x}$. When combined with (3.359) this implies that $u(t) = v(t) = \bar{x}$, $t \in [0, t_2]$. By this relation and Theorem 3.33 $v(t) = \bar{x}$ for all numbers $t \in [0, \infty)$. Theorem 3.35 is proved.   □

3.17   A Property of Overtaking Optimality Functions

We begin with the following auxiliary result.

Proposition 3.36. *Let ϵ be a positive number. Then there exists a positive number δ such that if a number $T \geq 1$ and if an a.c. function $v : [0, T] \to R^n$ satisfies the inequalities*

$$|v(0) - \bar{x}|, \ |v(T) - \bar{x}| \leq \delta,$$

then

$$I^f(0, T, v) \geq Tf(\bar{x}, 0) - \epsilon.$$

Proof. Assumption (A1) implies that there exists a positive number δ such that

$$\{(y, z) \in R^n \times R^n : |y - \bar{x}| \leq 4\delta, \ |z| \leq 4\delta\} \subset \text{dom}(f), \tag{3.360}$$

$$|f(y, z) - f(\bar{x}, 0)| \leq 8^{-1}\epsilon \text{ for each } (y, z) \in R^n \times R^n$$

$$\text{satisfying } |y - \bar{x}| \leq 4\delta, \ |z| \leq 4\delta. \tag{3.361}$$

Assume that a number $T \geq 1$ and that $v : [0, T] \to R^n$ is an a.c. function such that

$$|v(0) - \bar{x}|, \ |v(T) - \bar{x}| \leq \delta. \tag{3.362}$$

We claim that

$$I^f(0, T, v) \geq Tf(\bar{x}, 0) - \epsilon.$$

Assume the contrary. Then

$$I^f(0, T, v) < Tf(\bar{x}, 0) - \epsilon. \tag{3.363}$$

Define an a.c. function $u : [0, \infty) \to R^n$ as follows:

$$u(0) = \bar{x}, \; u(t) = \bar{x} + t(v(0) - \bar{x})), \; t \in (0, 1], \; u(t) = v(t - 1), \; t \in (1, T + 1], \tag{3.364}$$

$$u(t) = v(T) + (t - T - 1)(\bar{x} - v(T)), \; t \in (T + 1, T + 2],$$

$$u(t + T + 2) = u(t), \; \text{for all } t \geq 0.$$

By (3.362) and (3.364), for all numbers $t \in (0, 1) \cup (T + 1, T + 2)$,

$$|u(t) - \bar{x}| \leq \max\{|v(0) - \bar{x}|, |\bar{x} - v(T)|\} \leq \delta,$$

$$|u'(t)| \leq \max\{|v(0) - \bar{x}|, |v(T) - \bar{x}|\} \leq \delta. \tag{3.365}$$

It follows from (3.361) and (3.365) that for all numbers $t \in (0, 1) \cup (T + 1, T + 2)$,

$$f(u(t), u'(t)) \leq 8^{-1}\epsilon + f(\bar{x}, 0). \tag{3.366}$$

By (3.363), (3.364) and (3.366), we have

$$I^f(0, T + 2, u) = I^f(0, 1, u) + I^f(1, T + 1, u) + I^f(T + 1, T + 2, u)$$
$$\leq 2(f(\bar{x}, 0) + \epsilon/8) + I^f(0, T, v) \leq 2f(\bar{x}, 0) + \epsilon/4$$
$$+ Tf(\bar{x}, 0) - \epsilon = (T + 2)f(\bar{x}, 0) - \epsilon/2.$$

When combined with (3.364) this implies that

$$\lim_{p \to \infty} I^f(0, p(T + 2), u) - p(T + 2)f(\bar{x}, 0) = -\infty,$$

where $p \geq 1$ is an integer. This contradicts Proposition 3.1. The contradiction we have reached proves the proposition. □

Proposition 3.37. *Let ϵ be a positive number. Then there exists a positive number δ such that for each number $T \geq 2$ and each pair of points $y_1, y_2 \in R^n$ satisfying $|y_i - \bar{x}| \leq \delta$, $i = 1, 2$ the inequality $\sigma(T, y_1, y_2) \leq Tf(\bar{x}, 0) + \epsilon$ holds.*

Proof. Assumption (A1) implies that there exists a positive number δ such that

$$\{(y, z) \in R^n \times R^n : |y - \bar{x}| \leq 4\delta, |z| \leq 4\delta\} \subset \text{dom}(f), \tag{3.367}$$

$$|f(y, z) - f(\bar{x}, 0)| \leq 8^{-1}\epsilon \text{ for each } (y, z) \in R^n \times R^n$$

$$\text{satisfying } |y - \bar{x}| \leq 4\delta, |z| \leq \delta. \tag{3.368}$$

Assume that a pair of points $y_1, y_2 \in R^n$, a number $T \geq 2$, and that

$$|y_i - \bar{x}| \leq \delta, \; i = 1, 2. \tag{3.369}$$

Define an a .c. function $v : [0, T] \to R^n$ as follows:

$$v(t) = y_1 + t(\bar{x} - y_1), \; t \in [0, 1], \; v(t) = \bar{x}, \; t \in (1, T - 1],$$

$$v(t) = \bar{x} + (t - T + 1)(y_2 - \bar{x}), \; t \in (T - 1, T]. \tag{3.370}$$

It is clear that

$$v(0) = y_1, \; v(T) = y_2. \tag{3.371}$$

By (3.369) and (3.370), for all numbers $t \in (0, 1) \cup (T - 1, T)$,

$$|v(t) - \bar{x}| \leq \max\{|y_1 - \bar{x}|, \; |y_2 - \bar{x}|\} \leq \delta,$$

$$|v'(t)| \leq \max\{|y_1 - \bar{x}|, \; |y_2 - \bar{x}|\} \leq \delta. \tag{3.372}$$

By (3.368) and (3.372), for all numbers $t \in (0, 1) \cup (T - 1, T)$, we have

$$f(v(t), v'(t)) \leq f(\bar{x}, 0) + \epsilon/8. \tag{3.373}$$

In view of (3.370), (3.371), and (3.373),

$$\sigma(T, y_1, y_2) \leq I^f(0, T, v) = I^f(0, 1, v) + I^f(1, T - 1, v) + I^f(T - 1, T, v)$$
$$\leq 2f(\bar{x}, 0) + \epsilon/4 + (T - 2)f(\bar{x}, 0)$$
$$= Tf(\bar{x}, 0) + \epsilon/2.$$

Proposition 3.37 is proved.   □

The following theorem obtained in [63] is the main result of this section. It is an extension of an analogous result which was established in [37, 57] for variational problems with finite-valued integrands.

Theorem 3.38. *Assume that a function $v : [0, \infty) \to R^n$ is (f)-good and (f)-overtaking optimal. Then for all pairs of positive numbers T, S,*

$$\sigma(f, T, v(0), v(T)) - Tf(\bar{x}, 0) \leq \sigma(f, S, v(0), v(T)) - Sf(\bar{x}, 0).$$

Proof. Assume the contrary. Then there exist a pair of numbers $T_0, S_0 > 0$ such that

$$\lambda := \sigma(f, T_0, v(0), v(T_0)) - T_0 f(\bar{x}, 0) - (\sigma(f, S_0, v(0), v(T_0)) - S_0 f(\bar{x}, 0)) > 0. \tag{3.374}$$

Assumption (A4) implies that

$$\lim_{t \to \infty} v(t) = \bar{x}. \tag{3.375}$$

In view of Propositions 3.36 and 3.37 and assumption (A1) there exists $\delta \in (0, 1)$ such that the following properties hold:

for each number $T \geq 2$ and each pair of points $y_1, y_2 \in R^n$ satisfying $|y_i - \bar{x}| \leq \delta$, $i = 1, 2$, we have

$$\sigma(f, T, y_1, y_2) \leq T f(\bar{x}, 0) + \lambda/8; \tag{3.376}$$

$$\{(y, z) \in R^n \times R^n : |y - \bar{x}| \leq 4\delta, |z| \leq 4\delta\} \subset \text{dom}(f); \tag{3.377}$$

$$|f(y, z) - f(\bar{x}, 0)| \leq 8^{-1}\lambda(1 + (T_0 - S_0))^{-1},$$

$$\text{for each } (y, z) \in R^n \times R^n \text{ satisfying } |y - \bar{x}| \leq 4\delta, |z| \leq 4\delta; \tag{3.378}$$

for each number $T \geq 1$ and each a.c. function $v : [0, T] \to R^n$ which satisfies

$$|v(0) - \bar{x}|, \ |v(T) - \bar{x}| \leq \delta,$$

we have

$$I^f(0, T, v) \geq T f(\bar{x}, 0) - \lambda/8. \tag{3.379}$$

In view of (3.375) there exists a number τ_0 such that

$$\tau_0 > 2S_0 + 2T_0 + 4, \ |v(t) - \bar{x}| \leq \delta/4, \ t \in [\tau_0/2, \infty). \tag{3.380}$$

By (3.374) there exists an a.c. function $\tilde{v} : [0, S_0] \to R^n$ such that

$$\tilde{v}(0) = v(0), \ \tilde{v}(S_0) = v(T_0),$$

$$I^f(0, S_0, \tilde{v}) - S_0 f(\bar{x}, 0) < \sigma(T_0, v(0), v(T_0)) - T_0 f(\bar{x}, 0) - \lambda/2$$

$$\leq I^f(0, T_0, v) - T_0 f(\bar{x}, 0) - \lambda/2. \tag{3.381}$$

Define an a.c. function $u : [0, \infty) \to R^n$ by

$$u(t) = \tilde{v}(t), \ t \in [0, S_0], \ u(t) = v(t - S_0 + T_0), \ t \in (S_0, \tau_0], \tag{3.382}$$

$$u(t) = v(\tau_0 - S_0 + T_0) + (t - \tau_0)[v(\tau_0 + 1) - v(\tau_0 - S_0 + T_0)], \ t \in (\tau_0, \tau_0 + 1],$$

$$u(t) = v(t), \ t \in (\tau_0 + 1, \infty).$$

It follows from (3.381) and (3.382) that

$$u(0) = \tilde{v}(0) = v(0), \quad u(\tau_0 + 4 + T_0) = v(\tau_0 + 4 + T_0). \tag{3.383}$$

We estimate $I^f(0, \tau_0 + T_0 + 4, u) - I^f(0, \tau_0 + T_0 + 4, v)$. Since the function v is (f)-overtaking optimal relations (3.381) and (3.382) imply that

$$I^f(0, \tau_0 + 4 + T_0, u) - I^f(0, \tau_0 + 4 + T_0, v)$$

$$= [I^f(0, S_0, \tilde{v}) + I^f(S_0, \tau_0, u) + I^f(\tau_0 + 1, \tau_0 + 4 + T_0, u)$$

$$+ I^f(\tau_0, \tau_0 + 1, u)] - I^f(0, T_0 + \tau_0 + 4, v)$$

$$< [S_0 f(\bar{x}, 0) + I^f(0, T_0, v) - T_0 f(\bar{x}, 0) - \lambda 2^{-1}]$$

$$+ I^f(T_0, \tau_0 - S_0 + T_0, v) + I^f(\tau_0, \tau_0 + 1, u) + I^f(\tau_0 + 1, \tau_0 + 4 + T_0, v)$$

$$- I^f(0, T_0, v) - I^f(T_0, \tau_0 - S_0 + T_0, v) - I^f(\tau_0 - S_0 + T_0, \tau_0 + T_0 + 4, v)$$

$$= (S_0 - T_0) f(\bar{x}, 0) - \lambda/2 + I^f(\tau_0, \tau_0 + 1, u) + I^f(\tau_0 + 1, \tau_0 + 4 + T_0, v)$$

$$- I^f(\tau_0 - S_0 + T_0, \tau_0 + T_0 + 4, v)$$

$$= (S_0 - T_0) f(\bar{x}, 0) - \lambda/2 + I^f(\tau_0, \tau_0 + 4, u)$$

$$+ \sigma(f, 3 + T_0, v(\tau_0 + 1), v(\tau_0 + 4 + T_0)) - I^f(\tau_0 - S_0 + T_0, \tau_0 + T_0 + 1, v). \tag{3.384}$$

It follows from (3.378), (3.380), and (3.382) that for all numbers $t \in (\tau_0, \tau_0 + 1)$,

$$|u(t) - \bar{x}| \leq \max\{|v(\tau_0 - S_0 + T_0) - \bar{x}|, \ |v(\tau_0 + 1) - \bar{x}|\} \leq \delta,$$

$$|u'(t)| \leq 2 \max\{|v(\tau_0 - S_0 + T_0) - \bar{x}|, \ |v(\tau_0 + 1) - \bar{x}|\} \leq \delta \tag{3.385}$$

and

$$f(u(t), u'(t)) \leq f(\bar{x}, 0) + 8^{-1}\lambda, \tag{3.386}$$

$$I^f(\tau_0, \tau_0 + 1, u) \leq f(\bar{x}, 0) + 8^{-1}\lambda. \tag{3.387}$$

In view of (3.376) and (3.380),

$$\sigma(3 + T_0, v(\tau_0 + 1), v(\tau_0 + 4 + T_0)) \leq (T_0 + 3) f(\bar{x}, 0) + \lambda/8. \tag{3.388}$$

By (3.379) and (3.380),

$$I^f(\tau_0 - S_0 + T_0, \tau_0 + T_0 + 4, v) \geq (S_0 + 4) f(\bar{x}, 0) - \lambda/8. \tag{3.389}$$

It follows from (3.384) and (3.387)–(3.389) that

$$
\begin{aligned}
I^f(0, \tau_0 + 4 + T_0, u) - I^f(0, \tau_0 + 4 + T_0, v) \leq\ & (S_0 - T_0) f(\bar{x}, 0) - \lambda/2 \\
& + f(\bar{x}, 0) + 8^{-1}\lambda \\
& + (T_0 + 3) f(\bar{x}, 0) \\
& + \lambda/8 - (S_0 + 4) f(\bar{x}, 0) + \lambda/8 \\
=\ & -\lambda/8.
\end{aligned}
$$

When combined with (3.383) this contradicts the assumption that the function v is (f)-overtaking optimal. The contradiction we have reached proves the theorem.

$\square$

Chapter 4
Infinite Horizon Problems

In this chapter we establish the existence of solutions for classes of nonconvex (nonconcave) infinite horizon discrete-time optimal control problems. These classes contain optimal control problems arising in economic dynamics which describe general one-sector and two-sector models with nonconcave utility functions representing the preferences of the planner.

4.1   One-Dimensional Autonomous Problems

In this section we study the existence of solutions for a class of nonconvex infinite horizon autonomous discrete-time optimal control problems. This class contains optimal control problems arising in economic dynamics which describe a general one-sector model without discounting and with a nonconcave utility function representing the preferences of the planner.

Let

$$R_+^n = \{x \in R^n : x = (x_1, \dots, x_n), \ x_i \geq 0, \ i = 1, \dots, n\}$$

be the nonnegative orthant of the n-dimensional Euclidean space R^n,

$$v \in [0, 1), \tag{4.1}$$

$f : [0, \infty) \to [0, \infty)$ be an increasing continuous function such that

$$f(0) = 0, \ f(x) > 0 \text{ for all numbers } x > 0, \tag{4.2}$$

and $w : [0, \infty) \to [0, \infty)$ be an increasing continuous function such that

$$w(0) = 0, \ w(x) > 0 \text{ for all numbers } x > 0. \tag{4.3}$$

© Springer International Publishing Switzerland 2014
A.J. Zaslavski, *Turnpike Phenomenon and Infinite Horizon Optimal Control*,
Springer Optimization and Its Applications 99, DOI 10.1007/978-3-319-08828-0_4

A pair of sequences $(\{x_t\}_{t=0}^{\infty}, \{y_t\}_{t=0}^{\infty})$ is called a program if $x_t, y_t \in R_+^1$, $t = 0, 1 \ldots$ and for all nonnegative integers t,

$$x_{t+1} \geq v x_t, \tag{4.4}$$

$$x_{t+1} - v x_t + y_t \leq f(x_t). \tag{4.5}$$

Let $T_1 \geq 0$, $T_2 > T_1$ be a pair of integers. A pair of sequences

$$\left(\{x_t\}_{t=T_1}^{T_2}, \{y_t\}_{t=T_1}^{T_2-1}\right)$$

is called a program if

$$x_t \in R_+^1, \ t = T_1, \ldots, T_2, \ y_t \in R_+^1, \ t = T_1, \ldots, T_2 - 1$$

and for all integers $t = T_1, \ldots, T_2 - 1$, (4.4) and (4.5) hold.

We study an infinite horizon optimal control problem which corresponds to a finite horizon problem:

$$\sum_{t=0}^{T-1} w(y_t) \rightarrow \max, \ \left(\{x_t\}_{t=0}^{T}, \{y_t\}_{t=0}^{T-1}\right) \text{ is a program such that } x_0 = z,$$

where $T \geq 1$ is an integer and a point $z \in R_+^1$.

These optimal control systems describe a one-sector model of economic dynamics where x_t is funds at moment t, y_t is consumption at moment t and $w(y_t)$ evaluates consumption at moment t. The dynamics of the model is described by (4.4) and (4.5). It should be mentioned that this model was usually considered in the literature under an assumption that the functions w and f are concave. In this section we discuss the results obtained in [68] which establish the existence of optimal solutions without this assumption.

Assume that there exists a number $x^* > 0$ such that

$$f(x) > (1 - v)x \text{ for all numbers } x \in (0, x^*), \tag{4.6}$$

$$f(x) < (1 - v)x \text{ for all numbers } x \in (x^*, \infty). \tag{4.7}$$

It is clear that

$$f(x^*) = (1 - v)x^*. \tag{4.8}$$

It should be mentioned that the number x^* satisfying relations (4.6) and (4.7) exists for many one-sector models of economic dynamics. (For example, if $f(x) = x^{\alpha}$, $x \geq 0$ where a constant $\alpha \in (0, 1)$.)

In the sequel supremum over an empty set is $-\infty$.

For each number $z \in R^1_+$ and each pair of integers $T_2 > T_1 \geq 0$ define

$$U(z, T_1, T_2) = \sup \left\{ \sum_{t=T_1}^{T_2-1} w(y_t) : \left(\{x_t\}_{t=T_1}^{T_2}, \{y_t\}_{t=T_1}^{T_2-1} \right) \text{ is a program, } x_0 = z \right\}.$$

(4.9)

The following proposition easily follows from the continuity of the functions f, w and the compactness of the set of the programs over interval $[T_1, T_2]$ with the same initial state.

Proposition 4.1. *For each number $z \in R^1_+$ and each pair of integers $T_2 > T_1 \geq 0$ there exists a program $\left(\{x_t\}_{t=T_1}^{T_2}, \{y_t\}_{t=T_1}^{T_2-1} \right)$ such that $x(T_1) = z$ and $U(z, T_1, T_2) = \sum_{t=T_1}^{T_2-1} w(y_t)$.*

For each pair of numbers $m \geq 0$, $M > 0$ satisfying $m < x^* < M$ and each pair of integers $T_2 > T_1 \geq 0$ define

$$\hat{U}(m, M, T_1, T_2) = \sup\{U(z, T_1, T_2) : z \in [m, M]\}.$$

(4.10)

For each pair of numbers $z_1, z_2 \in R^1_+$ and each pair of integers $T_2 > T_1 \geq 0$ define

$$U(z_1, z_2, T_1, T_2) = \sup \left\{ \sum_{t=T_1}^{T_2-1} w(y_t) : \left(\{x_t\}_{t=T_1}^{T_2}, \{y_t\}_{t=T_1}^{T_2-1} \right) \text{ is a program,} \right.$$

$$\left. x_{T_1} = z_1, \ x_{T_2} \geq z_2 \right\}.$$

(4.11)

The following theorem obtained in [68] establishes for any initial state $x_0 > 0$ the existence of a solution of the corresponding infinite horizon optimal control problem.

Theorem 4.2. *Let $0 < m_0 < x^* < M_0$. Then there exists a positive number M_* such that for each number $x_0 \in [m_0, M_0]$ there exists a program $(\{x_t\}_{t=0}^{\infty}, \{y_t\}_{t=0}^{\infty})$ such that for each pair of integers $T_1, T_2 \geq 0$ satisfying $T_1 < T_2$,*

$$\left| \sum_{t=T_1}^{T_2-1} w(y_t) - \hat{U}(m_0, M_0, T_1, T_2) \right| \leq M_*$$

(4.12)

and that for each natural number T,

$$\sum_{t=0}^{T-1} w(y_t) = U(x_0, x_T, 0, T).$$

(4.13)

The following theorem obtained in [68] is the second main result of this section.

Theorem 4.3. *Let $0 < m_0 < x^* < M_0$. Then there exists a limit*

$$\mu = \lim_{p \to \infty} \hat{U}(m_0, M_0, 0, p)/p \tag{4.14}$$

and there exists a positive number M such that

$$|p^{-1}\hat{U}(m_0, M_0, 0, p) - \mu| \le 2M/p \text{ for all integers } p \ge 1. \tag{4.15}$$

Theorems 4.2 and 4.3 imply that the constant μ does not depend on the choice of m_0, M_0. It is not difficult to see that for each pair of nonnegative numbers m_0, M_0 satisfying $m_0 < x^* < M_0$ and each natural number p, we have

$$\hat{U}(m_0, M_0, 0, p) = \hat{U}(0, M_0, 0, p). \tag{4.16}$$

Theorems 4.2 and 4.3 imply the following result.

Theorem 4.4. *Let $0 < m_0 < x^* < M_0$. Then there exists a positive number M_* such that for each number $x_0 \in [m_0, M_0]$ there exists a program $(\{x_t\}_{t=0}^{\infty}, \{y_t\}_{t=0}^{\infty})$ such that for each pair of nonnegative integers T_1, T_2 satisfying $T_1 < T_2$, the inequality*

$$\left| \sum_{t=T_1}^{T_2-1} w(y_t) - (T_2 - T_1)\mu) \right| \le M_*$$

holds.

Theorem 4.5. *Let $(\{x_t\}_{t=0}^{\infty}, \{y_t\}_{t=0}^{\infty})$ be a program. Then either the sequence $\{\sum_{t=0}^{T-1} w(y_t) - T\mu\}_{T=1}^{\infty}$ is bounded or $\lim_{T \to \infty}[\sum_{t=0}^{T-1} w(y_t) - T\mu] = -\infty$.*

It should be mentioned that a program $(\{x_t\}_{t=0}^{\infty}, \{y_t\}_{t=0}^{\infty})$ such that the sequence

$$\left\{ \sum_{t=0}^{T-1} w(y_t) - T\mu \right\}_{T=1}^{\infty}$$

is bounded is called good in the literature [56].

Most results known in the literature which establish the existence of good programs were obtained for concave (convex) problems. For nonconcave (nonconvex) unconstrained problems existence of good programs was obtained in [28]. The problem considered here is constrained and nonconcave. This makes the situation more difficult and less understood.

In this section we prove Theorem 4.5.

Proof of Theorem 4.5. We may assume without loss of generality that x_0 is a positive number. Choose numbers $m_0, M_0 > 0$ such that

$$m_0 < x_0 < M_0, \ m_0 < x^* < M_0. \tag{4.17}$$

Let M be as guaranteed by Theorem 4.3 and assume that the sequence

$$\left\{ \sum_{t=0}^{T-1} w(y_t) - T\mu \right\}_{T=1}^{\infty}$$

is not bounded. Then it follows from (4.10), (4.17) and Theorem 4.3 that

$$\liminf_{T \to \infty} \left[\sum_{t=0}^{T-1} w(y_t) - T\mu \right] = -\infty. \tag{4.18}$$

Let $Q > 0$ be given. In view of (4.18), there exists a natural number T_0 such that

$$\sum_{t=0}^{T_0-1} w(y_t) - T_0\mu < -Q - 2M. \tag{4.19}$$

It follows from (4.7), (4.16), (4.17), (4.19) and the choice of M that for each integer $T > T_0$, we have

$$\sum_{t=0}^{T-1} w(y_t) - T\mu = \sum_{t=0}^{T_0-1} w(y_t) - T_0\mu + \sum_{t=T_0}^{T-1} w(y_t) - \hat{U}(0, M_0, T_0, T)$$

$$+ \hat{U}(m_0, M_0, T_0, T) - (T - T_0)\mu$$

$$< -Q - 2M + 2M < -Q.$$

Theorem 4.5 is proved.   □

4.2 Auxiliary Results

Put

$$g(z) = f(z) + vz, \ z \in R_+^1, \tag{4.20}$$

$$g^0 = g, \ g^{i+1} = g \circ g^i \text{ for all nonnegative integers } i. \tag{4.21}$$

Lemma 4.6. *Let x_0 be a positive number:*

$$x_{t+1} = vx_t + f(x_t) \text{ for all nonnegative integers } t. \tag{4.22}$$

Then $\lim_{t \to \infty} x_t = x^*$. Moreover, if the inequality $x_0 \geq x^*$ is true, then for all nonnegative integers t,

$$x_t \geq x^*, \ x_{t+1} \leq x_t \tag{4.23}$$

and if the inequality $x_0 \leq x^*$ is valid, then for all nonnegative integers t,

$$x_t \leq x^*, \; x_{t+1} \geq x_t. \tag{4.24}$$

Proof. Assume that t is a nonnegative integer and that $x_t \geq x^*$. It follows from (4.2), (4.7), (4.8), and monotonicity of the function f that $x_{t+1} \leq x_t$ and $x_{t+1} \geq x^*$. Therefore if $x_0 \geq x^*$, then (4.23) is valid for all nonnegative integers t.

Assume that t is a nonnegative integer and that $x_t \leq x^*$. In view of (4.22), (4.6), (4.8) and monotonicity of the function f, we have $x_{t+1} \geq x_t$ and $x_{t+1} \leq x^*$. Thus if $x_0 \leq x^*$, then (4.24) holds for all nonnegative integers t. It is clear that in both cases there exists $\lim_{t \to \infty} x_t > 0$.

It is not difficult to see that

$$f\left(\lim_{t \to \infty} x_t \right) + v \lim_{t \to \infty} x_t = \lim_{t \to \infty} (f(x_t) + v x_t) = \lim_{t \to \infty} x_{t+1}$$

and $f(\lim_{t \to \infty} x_t) = (1 - v) \lim_{t \to \infty} x_t$. When combined with relations (4.6)–(4.8) and the inequality $\lim_{t \to \infty} x_t > 0$ this implies that $\lim_{t \to \infty} x_t = x^*$. Lemma 4.6 is proved. $\qquad\square$

Lemma 4.6 implies the following result.

Lemma 4.7. *Let $M > x^*$ and ϵ be a positive number. Then there exists an integer $T_\epsilon \geq 1$ such that for each integer $T \geq T_\epsilon$ and each program $(\{x_t\}_{t=0}^T, \{y_t\}_{t=0}^{T-1})$ satisfying $x_0 \leq M$,*

$$x_t \leq x^* + \epsilon, \; t = T_\epsilon, \dots, T.$$

Assume that real positive numbers m_0, M_0 satisfy

$$2m_0 < x^* < M_0. \tag{4.25}$$

In view of (4.6) and (4.25), we have

$$f(2^{-1} x^*) > (1 - v) x^* / 2. \tag{4.26}$$

Put

$$\Lambda = f(x^*/2) - (1 - v) x^* / 2. \tag{4.27}$$

Choose a positive number δ_0 such that

$$\delta_0 < \min\{1, m_0/4\}, \; w(f(\delta_0)) < \Lambda/4. \tag{4.28}$$

Lemma 4.8. *Let M_1 be a positive number. Then there exists an integer $L \geq 1$ such that for each integer $T \geq 1$ and each program $(\{x_t\}_{t=0}^T, \{y_t\}_{t=0}^{T-1})$ which satisfies*

$$m_0 \leq x_0 \leq M_0,$$

$$\sum_{t=0,}^{T-1} w(y_t) \geq U(x_0, 0, T) - M_1 \tag{4.29}$$

the following property holds:
    For each integer $\tau \in \{0, \ldots, T\}$ there exists an integer $\tau_0 \in [\tau - L, \tau]$ such that
$x_{\tau_0} \geq \delta_0$.

Proof. Put

$$z_0^* = \delta, \; z_{t+1}^* = v z_t^* + f(z_t^*) \text{ for all integers } t \geq 0. \tag{4.30}$$

In view of (4.30) and Lemma 4.6, we have

$$\lim_{t \to \infty} z_t^* = x^*. \tag{4.31}$$

It follows from (4.31) that there exists an integer $L_0 > 4$ such that

$$z_{L_0}^* > x^*/2. \tag{4.32}$$

Choose a natural number:

$$L > 4L_0 + 4 + 4\Lambda^{-1}(M_1 + w(f(\delta_0))L_0 + w(M_0)). \tag{4.33}$$

Assume that $T \geq 1$ is an integer, a program $(\{x_t\}_{t=0}^T, \{y_t\}_{t=0}^{T-1})$ satisfies (4.29), and that $\tau \in \{0, \ldots, T\}$. Relations (4.28) and (4.29) imply that there exists

$$\tau_1 \in \{0, \ldots, \tau\} \tag{4.34}$$

such that

$$x_{\tau_1} \geq \delta_0 \text{ and } x_t < \delta_0 \text{ for all integers } t \text{ satisfying} \tag{4.35}$$

$$\tau_1 < t \leq \tau$$

(it may happen that $\tau_1 = \tau$). In order to complete the proof of the lemma it is sufficient to show that $\tau_1 \geq \tau - L$.
    Assume the contrary. Then

$$\tau_1 < \tau - L. \tag{4.36}$$

Define a program $(\{\tilde{x}_t\}_{t=0}^T, \{\tilde{y}_t\}_{t=0}^{T-1})$. Set

$$\tilde{x}_t = x_t, \; t = 0, \ldots, \tau_1, \tag{4.37}$$

if $\tau_1 \geq 1$ set

$$\tilde{y}_t = y_t, \ t = 0, \ldots, \tau_1 - 1 \tag{4.38}$$

and for $t = \tau_1, \ldots, \tau_1 + L_0 - 1$ set

$$\tilde{x}_{t+1} = v\tilde{x}_t + f(\tilde{x}_t), \ \tilde{y}_t = 0. \tag{4.39}$$

It follows from (4.30), (4.32), (4.35), (4.37), (4.39), and the monotonicity of the function f that

$$\tilde{x}_{\tau_1 + L_0} > x^*/2. \tag{4.40}$$

For all integers $t = \tau_1 + L_0, \ldots, \tau - 1$ we set

$$\tilde{y}_t = \Lambda, \ \tilde{x}_{t+1} = v\tilde{x}_t + f(\tilde{x}_t) - \Lambda. \tag{4.41}$$

By (4.27), (4.40), and (4.41), we can easily show using induction that

$$\tilde{x}_t \geq x^*/2, \ t = \tau_1 + L_0, \ldots, \tau \tag{4.42}$$

and that $(\{\tilde{x}_t\}_{t=0}^{\tau}, \{\tilde{y}_t\}_{t=0}^{\tau-1})$ is a program. It follows from (4.25), (4.28), (4.35), (4.36) and (4.44) that

$$\tilde{x}_\tau \geq x^*/2 > \delta_0 > x_\tau. \tag{4.43}$$

If $T > \tau$ set for all $t = \tau, \ldots, T - 1$,

$$\tilde{y}_t = y_t, \tag{4.44}$$

$$\tilde{x}_{t+1} = v\tilde{x}_t + f(\tilde{x}_t) - \tilde{y}_t. \tag{4.45}$$

In view of (4.43)–(4.45), we have

$$\tilde{x}_t \geq x_t, \ t = \tau, \ldots, T \tag{4.46}$$

and $(\{\tilde{x}_t\}_{t=0}^{T}, \{\tilde{y}_t\}_{t=0}^{T-1})$ is a program.

It follows from (4.28), (4.29), (4.33), (4.35)–(4.38), (4.41), and (4.44) that

$$M_1 \geq \sum_{t=0}^{T-1} w(\tilde{y}_t) - \sum_{t=0}^{T-1} w(y_t) = \sum_{t=0}^{\tau-1} w(\tilde{y}_t) - \sum_{t=0}^{\tau-1} w(y_t)$$

$$= \sum_{t=\tau_1}^{\tau-1} w(\tilde{y}_t) - \sum_{t=\tau_1}^{\tau-1} w(y_t) \geq \Lambda(\tau - \tau_1 - L_0) - w(M_0)$$

$$-(\tau - \tau_1)w(f(\delta_0)) = (\Lambda - w(f(\delta_0)))(\tau - \tau_1 - L_0) - w(f(\delta_0))L_0 - w(M_0)$$
$$\geq (3/4)\Lambda(\tau - \tau_1 - L_0) - w(f(\delta_0))L_0 - w(M_0)$$
$$\geq 4^{-1}L\Lambda - w(f(\delta_0))L_0 - w(M_0).$$

By the relation above,

$$L \leq (M_1 + w(f(\delta_0))L_0 + w(M_0))4\Lambda^{-1}.$$

This inequality contradicts (4.33). The contradiction we have reached proves Lemma 4.8. □

Lemma 4.9. *Let $M_1 > 0$ and let an integer $L \geq 1$ be as guaranteed by Lemma 4.8. Then there exists a number $\delta \in (0,1)$ such that for each integer $T \geq L$ and each program $(\{x_t\}_{t=0}^{T}, \{y_t\}_{t=0}^{T-1})$ which satisfies*

$$m_0 \leq x_0 \leq M_0,$$

$$\sum_{t=0}^{T-1} w(y_t) \geq U(x_0, 0, T) - M_1, \tag{4.47}$$

the following inequality holds:

$$x_t \geq \delta, \ t = 0, \ldots, T - L. \tag{4.48}$$

Proof. Choose a number $\delta \in (0, \delta_0)$ such that

$$g^i(\delta) < \delta_0, \ i = 1, \ldots, L \tag{4.49}$$

[see (4.20) and (4.21)].

Assume that an integer $T \geq L$ and that a program $(\{x_t\}_{t=0}^{T}, \{y_t\}_{t=0}^{T-1})$ satisfies (4.47). We claim that (4.48) holds.

Assume the contrary. Then there exists an integer

$$t_1 \in \{0, \ldots, T - L\} \tag{4.50}$$

such that

$$x_{t_1} < \delta. \tag{4.51}$$

In view of the choice of L, Lemma 4.8, (4.47), and (4.50) there exists an integer τ_0 such that

$$\tau_0 \leq t_1 + L, \ \tau_0 \geq t_1, \ x_{\tau_0} \geq \delta_0. \tag{4.52}$$

It follows from (4.52), (4.51), (4.6), (4.20), (4.21), and the choice of δ [see (4.49)] that

$$x_{\tau_0} \leq g^{\tau_0 - t_1}(\delta) < \delta_0.$$

This inequality contradicts (4.52). The contradiction we have reached proves (4.48). Lemma 4.9 is proved.   □

Choose a real number $\gamma_* > 0$ such that

$$\gamma_* < x^*/8, \ w(f(x^* + \gamma_*/8) - f(x^*) + 2\gamma_*) < w(\Lambda)/8. \tag{4.53}$$

Lemma 4.10. *Let M_1 be a positive number. Then there exists a pair of natural numbers L_1, L_2 such that for each natural number $T \geq L_1 + L_2$, each program $(\{x_t\}_{t=0}^T, \{y_t\}_{t=0}^{T-1})$ which satisfies*

$$m_0 \leq x_0 \leq M_0,$$

$$\sum_{t=0}^{T-1} w(y_t) \geq U(x_0, 0, T) - M_1, \tag{4.54}$$

and each integer $\tau \in \{L_1, \ldots, T - L_2\}$ the inequality

$$\min\{x_t : t = \tau, \ldots, \tau + L_2\} \leq x^* - \gamma_* \tag{4.55}$$

holds.

Proof. Lemma 4.7 implies that there exists an integer $L_1 \geq 4$ such that for each program $(\{x_t\}_{t=0}^{L_1}, \{y_t\}_{t=0}^{L_1-1})$ which satisfies $x_0 \leq M_0$ the inequality

$$x_{L_1} \leq x^* + \gamma_*/8 \tag{4.56}$$

holds. In view of (4.20), (4.21), and Lemma 4.6,

$$\lim_{i \to \infty} g^i(x^*/4) = x^*$$

and there exists an integer $L_0 \geq 4$ such that

$$g^{L_0}(x^*/4) \geq x^* - \gamma_*/8. \tag{4.57}$$

Choose a natural number:

$$L_2 > 4 + 8M_1 w(\Lambda)^{-1} + 8L_0 + 8w(\Lambda)^{-1} w(M_0). \tag{4.58}$$

Assume that an integer $T \geq L_1 + L_2$ is given, a program $(\{x_t\}_{t=0}^T, \{y_t\}_{t=0}^{T-1})$ satisfies (4.54), and that $\tau \in \{L_1, \ldots, T - L_2\}$.

In order to complete the proof of the lemma it is sufficient to show that (4.55) holds.

Assume the contrary. Then

$$x_t > x^* - \gamma_*, \ t = \tau, \ldots, \tau + L_2. \tag{4.59}$$

In view of (4.54), the choice of L_1 [see (4.56)], (4.7), and monotonicity of the function f, we have

$$x_t \le x^* + \gamma_*/8, \ t = L_1, \ldots, T. \tag{4.60}$$

There are two cases:

(1)

$$x_t > x^* - \gamma_*, \ t = \tau, \ldots, T; \tag{4.61}$$

(2) there is an integer $\tilde{\tau} > \tau + L_2$ such that

$$\tilde{\tau} \le T, \ x_{\tilde{\tau}} \le x^* - \gamma_*, \tag{4.62}$$

$$x_t > x^* - \gamma_* \text{ for all integers } t \text{ satisfying } \tau \le t < \tilde{\tau}. \tag{4.63}$$

Assume that the case (1) holds. Then inequality (4.61) holds for all integers $t = \tau + 1, \ldots, T$. It follows from (4.60) that

$$y_{t-1} + x_t - v x_{t-1} \le f(x_{t-1}) \le f(x^* + \gamma_*/8)$$

and when combined with (4.61), (4.8) and (4.60) this inequality implies that

$$
\begin{aligned}
y_{t-1} &\le f(x^* + \gamma_*/8) - x_t + v x_{t-1} \\
&\le f(x^* + \gamma_*/8) - x^* + \gamma_* + v(x^* + \gamma_*/8) \\
&\le f(x^* + \gamma_*/8) - f(x^*) + 2\gamma_*.
\end{aligned} \tag{4.64}
$$

Put

$$\tilde{x}_t = x_t, \ t = 0, \ldots, \tau, \ \tilde{y}_t = y_t, \ t = 0, \ldots, \tau - 1. \tag{4.65}$$

It follows from (4.53), (4.61), and (4.65) that

$$\tilde{x}_\tau > x^* - \gamma_* > x^*/2. \tag{4.66}$$

For all integers $t = \tau, \ldots, T - 1$ put

$$\tilde{y}_t = \Lambda, \ \tilde{x}_{t+1} = v \tilde{x}_t + f(\tilde{x}_t) - \Lambda. \tag{4.67}$$

By (4.66), (4.67), and (4.7), we show by induction that for all integers $t = \tau, \ldots, T$, we have

$$\tilde{x}_t > x^*/2 \tag{4.68}$$

and that $(\{\tilde{x}_t\}_{t=0}^{T}, \{\tilde{y}_t\}_{t=0}^{T-1})$ is a program. It follows from (4.65), (4.54), (4.61), (4.64), (4.53), and the choice of τ that

$$M_1 \geq \sum_{t=0}^{T-1} w(\tilde{y}_t) - \sum_{t=0}^{T-1} w(y_t) = \sum_{t=\tau}^{T-1} w(\tilde{y}_t) - \sum_{t=\tau}^{T-1} w(y_t)$$

$$\geq (T - \tau)w(\Lambda) - (T - \tau)w(f(x^* + \gamma_*/8) - f(x^*) + 2\gamma_*)$$

$$= (T - \tau)[w(\Lambda) - w(f(x^* + \gamma_*/8) - f(x^*) + 2\gamma_*)]$$

$$\geq (T - \tau)w(\Lambda)/2 \geq L_2 w(\Lambda)/2$$

and

$$L_2 \leq 2M_1 w(\Lambda)^{-1}.$$

This inequality contradicts (4.58). The contradiction we have reached proves that the case (1) does not holds. Thus the case (2) holds.

Set

$$\tilde{x}_t = x_t, \ t = 0, \ldots, \tau, \ \tilde{y}_t = y_t, \ t = 0, \ldots, \tau - 1 \tag{4.69}$$

and set for all integers $t = \tau, \ldots, \tilde{\tau} - L_0 - 1$,

$$\tilde{y}_t = \Lambda, \ \tilde{x}_{t+1} = v\tilde{x}_t + f(\tilde{x}_t) - \Lambda. \tag{4.70}$$

Relations (4.69), (4.59), and (4.53) imply that

$$\tilde{x}_\tau > x^*/2. \tag{4.71}$$

By (4.71), (4.70), (4.27), and monotonicity of the function f we can show using induction that

$$\tilde{x}_t \geq x^*/2, \ t = \tau, \ldots, \tilde{\tau} - L_0 \tag{4.72}$$

and that $(\{\tilde{x}_t\}_{t=0}^{\tilde{\tau}-L_0}, \{\tilde{y}_t\}_{t=0}^{\tilde{\tau}-L_0-1})$ is a program. Set for all integers $t = \tilde{\tau} - L_0, \ldots, \tilde{\tau} - 1$,

$$\tilde{y}_t = 0, \ \tilde{x}_{t+1} = v\tilde{x}_t + f(\tilde{x}_t). \tag{4.73}$$

It is clear that $(\{\tilde{x}_t\}_{t=0}^{\tilde{\tau}}, \{\tilde{y}_t\}_{t=0}^{\tilde{\tau}-1})$ is a program.

It follows from (4.72), (4.73), the choice of L_0 [see (4.57)], monotonicity of the function f, and (4.62) that

$$\tilde{x}_{\tilde{\tau}} \geq x^* - \gamma_*/8 \geq x_{\tilde{\tau}}. \tag{4.74}$$

For all integers t satisfying $\tilde{\tau} \leq t \leq T - 1$ set

$$\tilde{y}_t = y_t, \quad \tilde{x}_{t+1} = v\tilde{x}_t + x_{t+1} - vx_t. \tag{4.75}$$

Relations (4.74) and (4.75) imply that $(\{\tilde{x}_t\}_{t=0}^T, \{\tilde{y}_t\}_{t=0}^{T-1})$ is a program. By (4.60), (4.62), and the choice of τ, for all integers $t = \tau, \ldots, \tilde{\tau} - 1$, we have

$$x^* - \gamma_* < x_t \leq x^* + \gamma_*/8. \tag{4.76}$$

In view of (4.8) and (4.76), for all integers $t = \tau, \ldots, \tilde{\tau} - 2$,

$$y_t \leq f(x_t) - x_{t+1} + vx_t \leq f(x^* + \gamma_*/8) - x^* + \gamma_* + v(x^* + \gamma_*/8)$$
$$\leq f(x^* + \gamma_*/8) - f(x^*) + 2\gamma_*$$

and by (4.53), we have

$$w(y_t) \leq w(f(x^* + \gamma_*/8) - f(x^*) + 2\gamma_*) < w(\Lambda)/8. \tag{4.77}$$

It follows from (4.69), (4.54), (4.69), (4.75), (4.70), (4.77), (4.7), the inequality $\tilde{\tau} > \tau + L_2$, and (4.58) that

$$M_1 \geq \sum_{t=0}^{T-1} w(\tilde{y}_t) - \sum_{t=0}^{T-1} w(y_t) = \sum_{t=\tau}^{T-1} w(\tilde{y}_t) - \sum_{t=\tau}^{T-1} w(y_t)$$

$$= \sum_{t=\tau}^{\tilde{\tau}-1} w(\tilde{y}_t) - \sum_{t=\tau}^{\tilde{\tau}-1} w(y_t)$$

$$\geq (\tilde{\tau} - \tau - L_0)w(\Lambda) - (\tilde{\tau} - \tau - 1)w(\Lambda)/8 - w(M_0)$$

$$\geq (\tilde{\tau} - \tau - L_0)w(\Lambda)/2 - w(\Lambda)L_0 - w(M_0) \geq w(\Lambda)L_2/4 - w(\Lambda)L_0 - w(M_0)$$

and

$$L_2 \leq 4w(\Lambda)^{-1}(M_1 + w(\Lambda)L_0 + w(M_0)).$$

This inequality contradicts (4.58). The contradiction we have reached proves (4.55) and Lemma 4.10 itself. $\qquad\square$

Lemma 4.11. *Let M_1 be a positive number. Then there exists a pair of natural numbers $\tilde{L}_1, \tilde{L}_2$ and a real number $M_2 > 0$ such that for each natural number $T \geq \tilde{L}_1 + \tilde{L}_2$, each program $(\{x_t\}_{t=0}^T, \{y_t\}_{t=0}^{T-1})$ which satisfies*

$$m_0 \leq x_0 \leq M_0,$$

$$\sum_{t=0}^{T-1} w(y_t) \geq U(x_0, 0, T) - M_1, \tag{4.78}$$

and each pair of integers T_1, T_2 satisfying

$$0 \leq T_1 < T_2 \leq T - \tilde{L}_2, \; T_2 - T_1 \geq \tilde{L}_1 \tag{4.79}$$

the inequality

$$\sum_{t=T_1}^{T_2-1} w(y_t) \geq U(x_{T_1}, T_1, T_2) - M_2 \tag{4.80}$$

holds.

Proof. Lemma 4.9 implies that there exist an integer $L_1 \geq 1$ and a real number $\delta_1 \in (0, 1)$ such that for each natural number $T \geq L_1$ and each program $(\{x_t\}_{t=0}^{T}, \{y_t\}_{t=0}^{T-1})$ which satisfies (4.78), we have

$$x_t \geq \delta_1, \; t = 0, \ldots, T - L_1. \tag{4.81}$$

We may assume without loss of generality that

$$\delta_1 < m_0, \; \delta_1 < \delta_0. \tag{4.82}$$

It follows from Lemma 4.9 (applied with $m_0 = \delta_1$) that there exist an integer $L_2 \geq 1$ and a real number $\delta_2 \in (0, \delta_1)$ such that for each integer $T \geq L_2$ and each program $(\{x_t\}_{t=0}^{T}, \{y_t\}_{t=0}^{T-1})$ which satisfies

$$\delta_1 \leq x_0 \leq M_0,$$

$$\sum_{t=0}^{T-1} w(y_t) \geq U(x_0, 0, T) - M_1 - 1 \tag{4.83}$$

the following inequality holds:

$$x_t \geq \delta_2, \; t = 0, \ldots, T - L_2. \tag{4.84}$$

In view of Lemma 4.10, there exists a pair of natural numbers L_3, L_4 such that for each integer $T \geq L_3 + L_4$, each program $(\{x_t\}_{t=0}^{T}, \{y_t\}_{t=0}^{T-1})$ which satisfies

$$m_0 \leq x_0 \leq M_0,$$

$$\sum_{t=0}^{T-1} w(y_t) \geq U(x_0, 0, T) - M_1,$$

and each integer $\tau \in \{L_3, \ldots, T - L_4\}$, we have

$$\min\{x_t : t \in \{\tau, \ldots, \tau + L_4\}\} \leq x^* - \gamma_*. \tag{4.85}$$

It follows from (4.20), (4.21) and Lemma 4.6 that there exists an integer $L_5 \geq 1$ such that

$$g^t(\delta_2) > x^* - \gamma_*/8 \text{ for all integers } t \geq L_5. \tag{4.86}$$

Choose natural numbers

$$\tilde{L}_1 > L_1 + L_2 + L_3 + L_4 + L_5 + 4, \tag{4.87}$$

$$\tilde{L}_2 > 1 + L_1 + L_2 + 2L_4 + 4$$

and a real number

$$M_2 > M_1 + w(f(M_0))(L_2 + L_5 + 2L_4). \tag{4.88}$$

Assume that an integer $T \geq \tilde{L}_1 + \tilde{L}_2$ is given, a program $(\{x_t\}_{t=0}^T, \{y_t\}_{t=0}^{T-1})$ satisfies (4.78), and that a pair of integers T_1, T_2 satisfies (4.79). In view of the choice of L_1, δ_1 [see (4.81)], (4.87), and (4.78), we have

$$x_t \geq \delta_1, \ t = 0, \ldots, T - L_1. \tag{4.89}$$

Proposition 4.1 implies that there exists a program $(\{\tilde{x}_t\}_{t=T_1}^{T_2}, \{\tilde{y}_t\}_{t=T_1}^{T_2-1})$ such that

$$\tilde{x}_{T_1} = x_{T_1}, \ \sum_{t=T_1}^{T_2-1} w(\tilde{y}_t) = U(x_{T_1}, T_1, T_2). \tag{4.90}$$

In view of the choice of L_2 and δ_2 [see (4.83) and (4.84)], (4.90), (4.79), (4.89), (4.87), and (4.78), we have

$$\tilde{x}_t \geq \delta_2, \ t = T_1, \ldots, T_2 - L_2. \tag{4.91}$$

Relations (4.79) and (4.87) imply that

$$T_2 + 2L_4 \leq T - \tilde{L}_2 + 2L_4 \leq T, \tag{4.92}$$

$$L_3 < \tilde{L}_1 \leq T_2 < T_2 + L_4 \leq T - L_4.$$

It follows from the choice of L_3, L_4 (see (4.85) with $\tau = T_2 + L_4$), (4.78), (4.87), and (4.92) that there exists an integer t_0 such that

$$t_0 \in \{T_2 + L_4, \ldots, T_2 + 2L_4\},$$

$$x_{t_0} \leq x^* - \gamma_*. \tag{4.93}$$

Define a program $(\{\bar{x}_t\}_{t=0}^{T}, \{\bar{y}_t\}_{t=0}^{T-1})$ as follows. Put

$$\bar{x}_t = x_t, \; t = 0, \dots, T_1, \; \bar{y}_t = y_t, \; t = 0, \dots, T_1 - 1 \text{ if } T_1 > 0, \tag{4.94}$$

$$\bar{x}_t = \tilde{x}_t, \; t = T_1 + 1, \dots, T_2 - L_2 - L_5, \; \bar{y}_t = \tilde{y}_t, \; t = T_1, \dots, T_2 - L_2 - L_5 - 1.$$

In view of (4.94), (4.90), (4.79), and (4.87), $(\{\bar{x}_t\}_{t=0}^{T_2-L_2-L_5}, \{\bar{y}_t\}_{t=0}^{T_2-L_2-L_5-1})$ is a program. For all integers $t = T_2 - L_2 - L_5, \dots, t_0 - 1$ put

$$\bar{x}_{t+1} = v\bar{x}_t + f(\bar{x}_t), \; \bar{y}_t = 0. \tag{4.95}$$

It is clear that $(\{\bar{x}_t\}_{t=0}^{t_0}, \{\bar{y}_t\}_{t=0}^{t_0-1})$ is a program. It follows from (4.94), (4.91), and (4.79) that

$$\bar{x}_{T_2-L_2-L_5} = \tilde{x}_{T_2-L_2-L_5} \geq \delta_2. \tag{4.96}$$

By (4.95), (4.96), (4.93), (4.86), and (4.97), we have

$$\tilde{x}_{t_0} \geq x^* - \gamma_*/8 > x_{t_0}. \tag{4.97}$$

For all integers $t = t_0, \dots, T - 1$ put

$$\bar{x}_{t+1} = v\bar{x}_t + x_{t+1} - vx_t, \; \bar{y}_t = y_t. \tag{4.98}$$

By (4.97) and (4.98),

$$\bar{x}_t \geq x_t \text{ for all integers } t = t_0, \dots, T$$

and $(\{\bar{x}_t\}_{t=0}^{T}, \{\bar{y}_t\}_{t=0}^{T-1})$ is a program. It follows from (4.94), (4.78), (4.98), (4.93), (4.79), (4.87), (4.90), (4.78), and (4.7) that

$$M_1 \geq \sum_{t=0}^{T-1} w(\bar{y}_t) - \sum_{t=0}^{T-1} w(y_t) = \sum_{t=0}^{t_0-1} w(\bar{y}_t) - \sum_{t=0}^{t_0-1} w(y_t)$$

$$= \sum_{t=T_1}^{t_0-1} w(\bar{y}_t) - \sum_{t=T_1}^{t_0-1} w(y_t) \geq \sum_{t=T_1}^{T_2-L_2-L_5-1} w(\tilde{y}_t) - \sum_{t=T_1}^{t_0-1} w(y_t)$$

$$\geq U(x_{T_1}, T_1, T_2) - (L_2 + L_5)w(f(M_0)) - \sum_{t=T_1}^{T_2-1} w(y_t) - 2L_4 w(f(M_0))$$

and in view of (4.88)

$$\sum_{t=T_1}^{T_2-1} w(y_t) \geq U(x_{T_1}, T_1, T_2) - M_1 - w(f(M_0))(L_2 + L_5 + 2L_4)$$

$$\geq U(x_{T_1}, T_1, T_2) - M_2.$$

Lemma 4.11 is proved.   □

Lemma 4.12. *There exist an integer $L \geq 1$ and a positive number $\tilde{M}$ such that for each pair of real numbers $x_0, \tilde{x}_0 \in [m_0, M_0]$ and each integer $T \geq L$*

$$|U(x_0, 0, T) - U(\tilde{x}_0, 0, T)| \leq \tilde{M}.$$

Proof. Let natural numbers L_1, L_2 be as guaranteed by Lemma 4.10 with $M_1 = 1$. It follows from (4.20), (4.21) and Lemma 4.6 that there exists an integer $L_3 \geq 1$ such that

$$g^L(m_0) > x^* - \gamma_* \text{ for all integers } t \geq L_3. \tag{4.99}$$

Choose a natural number

$$L > L_1 + L_2 + L_3 \tag{4.100}$$

and set $\tilde{M} = Lw(f(M_0))$.

Assume that an integer

$$T \geq L, \ x_0, \tilde{x}_0 \in [m_0, M_0]. \tag{4.101}$$

Proposition 4.1 implies that there exists a program $(\{x_t\}_{t=0}^T, \{y_t\}_{t=0}^{T-1})$ such that

$$\sum_{t=0}^{T-1} w(y_t) = U(x_0, 0, T). \tag{4.102}$$

In view of (4.100) and (4.101), we have

$$L_1 + L_3 > L_1, \ L_1 + L_3 < T - L_2. \tag{4.103}$$

It follows from (4.101), (4.102), (4.103), the choice of L_1, L_2 and Lemma 4.10 (applied with $\tau = L_1 + L_3$) that there exists an integer

$$t_0 \in [L_1 + L_3, L_1 + L_2 + L_3] \tag{4.104}$$

such that

$$x_{t_0} < x^* - \gamma_*. \tag{4.105}$$

For all integers $t = 0, \dots, t_0 - 1$ put

$$\tilde{x}_{t+1} = v\tilde{x}_t + f(\tilde{x}_t), \ \tilde{y}_t = 0. \tag{4.106}$$

It is clear that $(\{\tilde{x}_t\}_{t=0}^{t_0}, \{\tilde{y}_t\}_{t=0}^{t_0-1})$ is a program. In view of (4.106), (4.101), (4.99), (4.104), and (4.105),

$$\tilde{x}_{t_0} > x^* - \gamma_* \geq x_{t_0}. \tag{4.107}$$

For all integers $t = t_0, \ldots, T - 1$ put

$$\tilde{x}_{t+1} = v\tilde{x}_t + x_{t+1} - vx_t, \quad \tilde{y}_t = y_t. \tag{4.108}$$

Relations (4.107) and (4.108) imply that $\tilde{x}_t \geq x_t$ for all integers $t = t_0, \ldots, T$ and that $(\{\tilde{x}_t\}_{t=0}^T, \{\tilde{y}_t\}_{t=0}^{T-1})$ is a program. It follows from (4.108), (4.101), (4.7), (4.102), (4.104), (4.100) and the choice of M that

$$\tilde{U}(\tilde{x}_0, 0, T) \geq \sum_{t=0}^{T-1} w(\tilde{y}_t) \geq \sum_{t=0}^{T-1} w(y_t) - t_0 w(f(M_0))$$

$$\geq U(x_0, 0, T) - Lw(f(M_0)) = U(x_0, 0, T) - \tilde{M}.$$

Lemma 4.12 is proved.   □

Lemma 4.12 implies the following result.

Lemma 4.13. *Let a natural number L and $\tilde{M} > 0$ be as guaranteed by Lemma 4.12. Then for each real number $x_0 \in [m_0, M_0]$ and each integer $T \geq L$ the inequality*

$$|U(x_0, 0, T) - \hat{U}(m_0, M_0, 0, T)| \leq \tilde{M}$$

holds.

Lemma 4.14. *Let M_1 be a positive number. Then there exists a pair of natural numbers $\tilde{L}_1, \tilde{L}_2$ and a real number $\tilde{M}_2 > 0$ such that for each integer $T \geq \tilde{L}_1 + \tilde{L}_2$, each program $(\{x_t\}_{t=0}^T, \{y_t\}_{t=0}^{T-1})$ which satisfies*

$$m_0 \leq x_0 \leq M_0,$$

$$\sum_{t=0}^{T-1} w(y_t) \geq U(x_0, 0, T) - M_1, \tag{4.109}$$

and each pair of integers T_1, T_2 satisfying

$$0 \leq T_1 < T_2 \leq T - \tilde{L}_2, \quad T_2 - T_1 \geq \tilde{L}_1,$$

the inequality

$$\sum_{t=T_1}^{T_2-1} w(y_t) \geq \hat{U}(m_0, M_0, T_1, T_2) - \tilde{M}_2$$

holds.

Proof. Lemma 4.9 implies that there exist an integer $L_1 \geq 1$ and a real number $\delta \in (0, 1)$ such that for each integer $T \geq L_1$ and each program $(\{x_t\}_{t=0}^T, \{y_t\}_{t=0}^{T-1})$ which satisfies (4.109), the following inequality holds:

$$x_t \geq \delta, \quad t = 0, \ldots, T - L_1. \tag{4.110}$$

We may assume that

$$\delta < m_0/4. \tag{4.111}$$

It follows from Lemma 4.13 that there exist an integer $L_2 \geq 1$ and a positive number $\tilde{M}$ such that for each number $x_0 \in [\delta, M_0]$ and each integer $T \geq L_2$, we have

$$|U(x_0, 0, T) - \hat{U}(\delta, M_0, 0, T)| \leq \tilde{M}. \tag{4.112}$$

Let natural numbers $\tilde{L}_1, \tilde{L}_2$ and a number $M_2 > 0$ be as guaranteed by Lemma 4.11. We may assume without loss of generality that

$$\tilde{L}_1, \tilde{L}_2 > L_1 + L_2. \tag{4.113}$$

Choose

$$\bar{M}_2 \geq \tilde{M} + M_2. \tag{4.114}$$

Assume that an integer $T \geq \tilde{L}_1 + \tilde{L}_2$, a program $(\{x_t\}_{t=0}^T, \{y_t\}_{t=0}^{T-1})$ satisfies (4.109), and a pair of integers T_1, T_2 satisfies

$$0 \leq T_1 < T_2 \leq T - \tilde{L}_2, \ T_2 - T_1 \geq \tilde{L}_1. \tag{4.115}$$

In view of the choice of $\tilde{L}_1, \tilde{L}_2, M_2$, and Lemma 4.11, we have

$$\sum_{t=T_1}^{T_2-1} w(y_t) \geq U(x_{T_1}, T_1, T_2) - M_2. \tag{4.116}$$

It follows from the choice of δ, (4.113), and (4.109) that (4.110) holds. By (4.110) and (4.113),

$$x_{T_1} \geq \delta. \tag{4.117}$$

Relation (4.109) implies that

$$x_{T_1} \leq M_0. \tag{4.118}$$

In view of the choice of L_2 and $\tilde{M}$, (4.117), (4.118), (4.115), and (4.113), we have

$$|U(x_{T_1}, T_1, T_2) - \hat{U}(\delta, M_0, T_1, T_2)| \leq \tilde{M}. \tag{4.119}$$

It follows from (4.111) and monotonicity of the function f that

$$\hat{U}(\delta, M_0, T_1, T_2) = \hat{U}(m_0, M_0, T_1, T_2). \tag{4.120}$$

By (4.116), (4.119), and (4.120),

$$\sum_{t=T_1}^{T_2-1} w(y_t) \geq \hat{U}(\delta, M_0, T_1, T_2) - \tilde{M} - M_2 = \hat{U}(m_0, M_0, T_1, T_2) - \tilde{M}_2.$$

Lemma 4.14 is proved. □

Lemma 4.15. *For each integer* $T \geq 1$,

$$\hat{U}(m_0, M_0, 0, T) \geq Tw(f(2^{-1}x^*) - (1 - v)x^*/2).$$

Proof. Put

$$x_0 = x^*/2, \ x_t = x^*/2 \text{ for all integers } t > 0$$

and for all nonnegative integers t, $y_t = f(2^{-1}x^*) - (1 - v)x^*/2$. It is not difficult to see that $(\{x_t\}_{t=0}^T, \{y_t\}_{t=0}^{T-1})$ is a program and that for all integers $T \geq 1$, we have

$$\hat{U}(m_0, M_0, 0, T) \geq \sum_{t=0}^{T-1} w(y_t) = Tw(f(2^{-1}x^*) - (1 - v)x^*/2).$$

Lemma 4.15 is proved. □

4.3 Proof of Theorem 4.2

We may assume without loss of generality that

$$2m_0 < x^*. \tag{4.121}$$

Let $M_1 = 1$ and let pair of natural numbers $\tilde{L}_1, \tilde{L}_2$ and $\tilde{M}_2 > 0$ be as guaranteed by Lemma 4.14. Let

$$x_0 \in [m_0, M_0] \tag{4.122}$$

be given. Proposition 4.1 implies that for each integer $k \geq 1$ there exists a program

$$(\{x_t^{(k)}\}_{t=0}^k, \{y_t^{(k)}\}_{t=0}^{k-1})$$

such that

$$x_0^{(k)} = x_0, \ \sum_{t=0}^{k-1} w(y_t^{(k)}) = U(x_0, 0, k). \tag{4.123}$$

By (4.122), (4.123), the choice of $\tilde{L}_1, \tilde{L}_2$ and $\tilde{M}_2$, and Lemma 4.14, the following property holds:

P(i) for each integer $k \geq \tilde{L}_1 + \tilde{L}_2$ and each pair of integers $T_1, T_2 \in [0, k - \tilde{L}_2]$ satisfying $T_2 - T_1 \geq \tilde{L}_1$, we have

$$\sum_{t=T_1}^{T_2-1} w(y_t^{(k)}) \geq \hat{U}(m_0, M_0, T_1, T_2) - \tilde{M}_2.$$

It is clear that there exists a strictly increasing sequence of natural numbers $\{k_j\}_{j=1}^{\infty}$ such that for each nonnegative integer t there exist limits

$$x_t = \lim_{j \to \infty} x_t^{(k_j)}, \quad y_t = \lim_{j \to \infty} y_t^{(k_j)}. \tag{4.124}$$

Evidently, $(\{x_t\}_{t=0}^{\infty}, \{y_t\}_{t=0}^{\infty})$ is a program.

By (4.124) and property P(i), for each pair of integers $T_1, T_2 \geq 0$ satisfying $T_2 - T_1 \geq \tilde{L}_1$, we have

$$\left| \sum_{t=T_1}^{T_2-1} w(y_t) - \hat{U}(m_0, M_0, T_1, T_2) \right| \leq \tilde{M}_2. \tag{4.125}$$

In order to complete the proof of Theorem 4.2 it is sufficient to show that for each natural number T,

$$\sum_{t=0}^{T-1} w(y_t) = U(x_0, x_T, 0, T). \tag{4.126}$$

Assume the contrary. Then there exists an integer $T \geq 1$ such that

$$\Delta_0 := U(x_0, x_T, 0, T) - \sum_{t=0}^{T-1} w(y_t) > 0. \tag{4.127}$$

In view of (4.11) there exists a program $(\{\tilde{x}_t\}_{t=0}^{T}, \{\tilde{y}_t\}_{t=0}^{T-1})$ such that

$$\tilde{x}_T \geq x_T, \quad \tilde{x}_0 = x_0, \quad \sum_{t=0}^{T-1} w(\tilde{y}_t) = U(x_0, x_T, 0, T). \tag{4.128}$$

It follows from (4.125) and Lemma 4.15 that there exists a natural number $S > T + 8$ such that

$$w(y_S) > w(f(2^{-1}x^*)) - (1 - v)x^*/2)/2. \tag{4.129}$$

Choose a real number $\Delta_1 > 0$ such that

$$y_S > 8\Delta_1, \tag{4.130}$$

$$|w(z_1) - w(z_2)| \leq \Delta_0/8 \text{ for each } z_1, z_2 \in [0, f(M_0)]$$

$$\text{satisfying } |z_1 - z_2| \leq 2\Delta_1. \tag{4.131}$$

In view of (4.124) and continuity of the function f there exists a natural number $k > S + 4$ such that

$$|x_{S+1} - x_{S+1}^{(k)}| \leq \Delta_1/2, \tag{4.132}$$

$$|w(y_t) - w(y_t^{(k)})| \leq (\Delta_0/8)(2S + 2)^{-1} \text{ for all integers } t = 0, \ldots, S + 1. \tag{4.133}$$

For all integers t satisfying $T \leq t < S$ set

$$\tilde{x}_{t+1} = v\tilde{x}_t + x_{t+1} - vx_t, \quad \tilde{y}_t = y_t. \tag{4.134}$$

Relations (4.128) and (4.134) imply that

$$\tilde{x}_t \geq x_t, \quad t = T, \ldots, S \tag{4.135}$$

and that $(\{\tilde{x}_t\}_{t=0}^S, \{\tilde{y}_t\}_{t=0}^{S-1})$ is a program. Put

$$\tilde{y}_S = y_S - \Delta_1, \tag{4.136}$$

$$\tilde{x}_{S+1} = v\tilde{x}_S + f(\tilde{x}_S) - y_S + \Delta_1.$$

By (4.135) and (4.136), we have

$$\tilde{x}_{S+1} \geq vx_S + f(x_S) - y_S + \Delta_1 \tag{4.137}$$

and

$$\tilde{x}_{S+1} \geq v\tilde{x}_S. \tag{4.138}$$

Thus relations (4.136), (4.130), and (4.138) imply that $(\{\tilde{x}_t\}_{t=0}^{S+1}, \{\tilde{y}_t\}_{t=0}^S)$ is a program. It follows from (4.136), (4.130), (4.131), (4.124), (4.123), and (4.122) that

$$|w(\tilde{y}_S) - w(y_S)| \leq \Delta_0/8. \tag{4.139}$$

Relations (4.132) and (4.137) imply that

$$\tilde{x}_{S+1} \geq x_{S+1} + \Delta_1 \geq x_{S+1}^{(k)}. \tag{4.140}$$

In view of (4.123), (4.128), and (4.140), we have

$$\tilde{x}_0 = x_0^{(k)}, \quad \tilde{x}_{S+1} \geq x_{S+1}^{(k)}. \tag{4.141}$$

It follows from (4.133), (4.134), (4.128), (4.127), and (4.139) that

$$\sum_{t=0}^{S} w(\tilde{y}_t) - \sum_{t=0}^{S} w(y_t^{(k)})$$

$$= \sum_{t=0}^{S} w(\tilde{y}_t) - \sum_{t=0}^{S} w(y_t) + \sum_{t=0}^{S} w(y_t) - \sum_{t=0}^{S} w(y_t^{(k)})$$

$$\geq \sum_{t=0}^{S} w(\tilde{y}_t) - \sum_{t=0}^{S} w(y_t) - \Delta_0/8$$

$$= \sum_{t=0}^{T-1} w(\tilde{y}_t) - \sum_{t=0}^{T-1} w(y(t) + w(\tilde{y}_S) - w(y_S) - \Delta_0/8$$

$$\geq \Delta_0 - \Delta_0/4 = (3/4)\Delta_0 > 0.$$

When combined with (4.141) this contradicts (4.123). The contradiction we have reached proves Theorem 4.2.

Proof of Theorem 4.3. Let $x_0 \in [m_0.M_0]$ and let M_* be as guaranteed by Theorem 4.2. Theorem 4.2 implies that there exists a program $(\{x_t\}_{t=0}^{\infty}, \{y_t\}_{t=0}^{\infty})$ such that for each pair of integers $T_1, T_2 \geq 0$ satisfying $T_1 < T_2$, we have

$$\left| \sum_{t=T_1}^{T_2} w(y_t) - \hat{U}(m_0, M_0, T_1, T_2) \right| \leq M_*. \tag{4.142}$$

Let $p \geq 1$ be an integer. We claim that for all sufficiently large integers $T \geq 1$,

$$\left| p^{-1}\hat{U}(m_0, M_0, 0, p) - T^{-1}\sum_{t=0}^{T-1} w(y_t) \right| \leq 2M_* p^{-1}. \tag{4.143}$$

Assume that an integer $T \geq p$. Then there exists a pair of integers q, s such that

$$q \geq 1, \ 0 \leq s < p, \ T = pq + s. \tag{4.144}$$

By (4.144),

$$T^{-1}\sum_{t=0}^{T-1} w(y_t) - p^{-1}\hat{U}(m_0, M_0, 0, p)$$

$$= T^{-1}\left(\sum_{t=0}^{pq-1} w(y_t) + \sum \{w(y_t) : \ t \text{ is an integer such that}\right.$$

$$\left. pq \leq t \leq T - 1\} \right) - p^{-1}\hat{U}(m_0, M_0, 0, p)$$

$$= T^{-1} \sum \{w(y_t) : t \text{ is an integer such that } pq \le t \le T - 1\}$$

$$+ (T^{-1}pq)(pq)^{-1} \sum_{i=0}^{q-1} \sum_{t=ip}^{(i+1)p-1} w(y_t) - p^{-1}\hat{U}(m_0, M_0, 0, p)$$

$$= T^{-1} \sum \{w(y_t) : t \text{ is an integer such that } pq \le t \le T - 1\}$$

$$+ (T^{-1}pq)(pq)^{-1} \left[\sum_{i=0}^{q-1} \left(\sum_{t=ip}^{(i+1)p-1} w(y_t) \right. \right.$$

$$\left. \left. -\hat{U}(m_0, M_0, 0, p) \right) + q\hat{U}(m_0, M_0, 0, p) \right] - p^{-1}\hat{U}(m_0, M_0, 0, p).$$

$$\tag{4.145}$$

It follows from (4.145), the inclusion $x_0 \in [m_0, M_0]$, (4.142), and (4.144) that

$$\left| T^{-1} \sum_{t=0}^{T-1} w(y_t) - p^{-1}\hat{U}(m_0, M_0, 0, p) \right| \le T^{-1}pw(M_0) + (pq)^{-1}qM_*$$

$$+ \hat{U}(m_0, M_0, 0, p)|q/T - 1/p| \le T^{-1}pw(M_0) + M_*/p$$

$$+ \hat{U}(m_0, M_0, 0, p)s(pT)^{-1} \to M_*/p \text{ as } T \to \infty.$$

Thus (4.143) is valid for all sufficiently large integers $T \ge 1$, as claimed.

Since p is any natural number we conclude that $\{T^{-1}\sum_{t=0}^{T-1} w(y_t)\}_{T=1}^{\infty}$ is a Cauchy sequence. It is clear that there exists

$$\lim_{T \to \infty} T^{-1} \sum_{t=0}^{T-1} w(y_t)$$

and for each natural number p,

$$\left| p^{-1}\hat{U}(m_0, M_0, 0, p) - \lim_{T \to \infty} T^{-1} \sum_{t=0}^{T-1} w(y_t) \right| \le p^{-1}(2M_*). \tag{4.146}$$

Since (4.146) holds for each natural number p we obtain that

$$\lim_{T \to \infty} T^{-1} \sum_{t=0}^{T-1} w(y_t) = \lim_{p \to \infty} \hat{U}(m_0, M_0, 0, p)/p. \tag{4.147}$$

Set

$$\mu = \lim_{T \to \infty} T^{-1} \sum_{t=0}^{T-1} w(y_t). \tag{4.148}$$

In view of (4.146) and (4.148), for all integers $p \geq 1$, we have

$$|\hat{U}(m_0, M_0, 0, p)/p - \mu| \leq 2M_*/p.$$

Theorem 4.3 is proved.   □

4.4 One-Dimensional Nonautonomous Problems

Let $v \in [0, 1)$ and let $f : [0, \infty) \to [0, \infty)$ be a monotone increasing continuous function such that

$$f(0) = 0 \text{ and } f(z) > 0 \text{ for all numbers } z > 0. \tag{4.149}$$

Assume that $w_t : [0, \infty) \to [0, \infty)$, $t = 0, 1, \dots$ are monotone increasing continuous functions and that for all nonnegative integers t,

$$w_t(0) = 0 \text{ and } w_t(x) > 0 \text{ for all numbers } x > 0. \tag{4.150}$$

We suppose that the following assumption holds:

(A1) for each positive number M, $\lim_{t \to \infty} w_t(M) = 0$.

Recall that a sequence $\{x_t, y_t\}_{t=0}^{\infty}$ is called a program if for all integers $t \geq 0$, $x_t, y_t \in R_+^1$ and

$$x_{t+1} \geq v x_t, \quad x_{t+1} - v x_t + y_t \leq f(x_t). \tag{4.151}$$

Let integers T_1, T_2 satisfy $T_2 > T_1 \geq 0$. A sequence $(\{x_t\}_{t=T_1}^{T_2}, \{y_t\}_{t=T_1}^{T_2-1})$ is called a program if $x_t \in R_+^1$, $t = T_1, \dots, T_2$, $y_t \in R_+^1$, $t = T_1, \dots, T_2 - 1$ and if for all integers $t = T_1, \dots, T_2 - 1$ the inequalities (4.151) hold.

We study an infinite horizon optimal control problem which corresponds to a finite horizon problem

$$\sum_{t=0}^{T-1} w_t(y_t) \to \max, \quad (\{x_t\}_{t=0}^{T}, \{y_t\}_{t=0}^{T-1}) \text{ is a program such that } x_0 = z,$$

where T is a natural number and $z \in R_+^1$.

These optimal control systems describe a one-sector model of economic dynamics where x_t, is funds at moment t, y_t is consumption at moment t and $w_t(y_t)$ evaluates consumption at moment t. The dynamics of the model is described by (4.151).

Assume that there exists a real number $x^* > 0$ such that

$$f(x) > (1 - v)x \text{ for all numbers } x \in (0, x^*), \tag{4.152}$$

$$f(x) < (1 - v)x \text{ for all numbers } x \in (x^*, \infty). \tag{4.153}$$

It follows from (4.152) and (4.153) that

$$f(x^*) = (1 - v)x^*.$$ (4.154)

Put

$$g(z) = f(z) + vz, \ z \in R^1_+, \ g^0 = g, \ g^{i+1} = g \circ g^i$$ (4.155)

for all nonnegative integers i. In view of (4.152) and (4.153), we have

$$\lim_{i \to \infty} g^i(z) = x^* \text{ for all } z > 0.$$ (4.156)

A program $\{x_t, y_t\}_{t=0}^\infty$ is called overtaking optimal if for each program $\{x_t', y_t'\}_{t=0}^\infty$ satisfying $x_0' = x_0$ the inequality

$$\limsup_{T \to \infty} \sum_{t=0}^{T-1} [w_t(y_t') - w_t(y_t)] \le 0$$

holds.

In this section we use the following assumptions:

(A2) For each positive number M, $\sum_{t=0}^\infty w_t(M) < \infty$.
(A3) For each pair of real numbers M_1, M_2 which satisfy $M_2 > M_1 > 0$ the following equation holds:

$$\sum_{t=0}^\infty (w_t(M_2) - w_t(M_1)) = \infty.$$

We prove the results obtained in [74] which show the existence of an overtaking optimal program for any initial state if at least one of the Assumptions (A2) and (A3) holds. In the case of Assumption (A2) the proof of the existence result (Theorem 4.18) is not difficult and standard while in the case of Assumption (A3) (see Theorem 4.19) the situation is more difficult and less understood.

Remark 4.16. If $w : [0, \infty) \to [0, \infty)$ is a continuous increasing function which satisfies $w(0) = 0$, $w(x) > 0$ for all positive numbers x, $\{\lambda_t\}_{t=0}^\infty \subset (0, \infty)$ satisfies $\lim_{t \to \infty} \lambda_t = 0$ and $w_t = \lambda_t w, t = 0, 1, \ldots$, then Assumption (A1) holds. It is clear that Assumption (A2) holds if $\sum_{t=0}^\infty \lambda_t < \infty$ and that Assumption (A3) holds if $\sum_{t=0}^\infty \lambda_t = \infty$ and the function w is strictly increasing. The case with $\sum_{t=0}^\infty \lambda_t < \infty$ is studied in the economic literature. Here our main interest is in the case when $\sum_{t=0}^\infty \lambda_t = \infty$.

For each real number $x_0 \in R^1_+$ and each natural number T put

$$U(x_0, T) = \sup \left\{ \sum_{t=0}^{T-1} w_t(y_t) : (\{x_t\}_{t=0}^T, \{y_t\}_{t=0}^{T-1}) \text{ is a program} \right\}.$$ (4.157)

The following proposition follows immediately from the continuity of f and w_t, $t = 0, 1, \ldots$.

Proposition 4.17. *For each real number $x_0 \in R^1_+$ and each integer $T \geq 1$ there exists a program $(\{x_t\}_{t=0}^T, \{y_t\}_{t=0}^{T-1})$ such that $\sum_{t=0}^{T-1} w_t(y_t) = U(x_0, T)$.*

We prove the following two theorems obtained in [74].

Theorem 4.18. *Assume that Assumption (A2) holds and let $M_0 > x^*$. Then for each number $z \in [0, M_0]$ there exists a program $(\{x_t^{(z)}\}_{t=0}^\infty, \{y_t^{(z)}\}_{t=0}^\infty)$ such that $x_0^{(z)} = z$ and that the following assertion holds.*

Let δ be a positive number. Then there exists an integer $L^{(\delta)} \geq 1$ such that for each integer $S \geq L^{(\delta)}$ and each real number $z \in [0, M_0]$ the inequality

$$\sum_{t=0}^{S-1} w_t(y_t^{(z)}) \geq U(z, S) - \delta$$

holds.

Theorem 4.19. *Assume that Assumption (A3) holds and let $0 < m_0 < x^* < M_0$. Then for each real number $z \in [m_0, M_0]$ there exists a program $(\{x_t^{(z)}\}_{t=0}^\infty, \{y_t^{(z)}\}_{t=0}^\infty)$ such that $x_0^{(z)} = z$ and that the following assertion holds.*

Let δ be a positive number. Then there exists an integer $L^{(\delta)} \geq 1$ such that for each integer $S \geq L^{(\delta)}$ and each real number $z \in [m_0, M_0]$ the inequality

$$\sum_{t=0}^{S-1} w_t(y_t^{(z)}) \geq U(z, S) - \delta$$

holds.

It is clear that the program $(\{x_t^{(z)}\}_{t=0}^\infty, \{y_t^{(z)}\}_{t=0}^\infty)$ in the statement of Theorem 4.18 (Theorem 4.19 respectively) is overtaking optimal.

4.5 Proof of Theorem 4.18

Let $z \in [0, M_0]$ be given. Proposition 4.17 implies that for each natural number k there exists a program $(\{x_t^{(z,k)}\}_{t=0}^k, \{y_t^{(z,k)}\}_{t=0}^{k-1})$ such that

$$x_0^{(z,k)} = z, \quad \sum_{t=0}^{k-1} w_t(y_t^{(z,k)}) = U(z, k). \tag{4.158}$$

It is clear that there exists a strictly increasing sequence of natural numbers $\{k_i\}_{i=1}^\infty$ such that for each nonnegative integer t there exists

$$x_t^{(z)} = \lim_{i \to \infty} x_t^{(z,k_i)}, \quad y_t^{(z)} = \lim_{i \to \infty} y_t^{(z,k_i)}. \tag{4.159}$$

It is easy to see that $(\{x_t^{(z)}\}_{t=0}^\infty, \{y_t^{(z)}\}_{t=0}^\infty)$ is a program and

$$x_0^{(z)} = z. \tag{4.160}$$

Let $\delta > 0$ be given. Assumption (A2) implies that there exists an integer $L^{(\delta)} \geq 1$ such that

$$\sum_{t=L^{(\delta)}}^\infty w_t(f(M_0)) < \delta. \tag{4.161}$$

Assume that an integer $S \geq L^{(\delta)}$ and that a number $z \in [0, M_0]$. Proposition 4.17 implies that there exists a program $(\{x_t\}_{t=0}^S, \{y_t\}_{t=0}^{S-1})$ such that

$$U(z, S) = \sum_{t=0}^{S-1} w_t(y_t), \quad x_0 = z. \tag{4.162}$$

For all integers $t \geq S$ put

$$y_t = 0, \quad x_{t+1} = vx_t + f(x_t). \tag{4.163}$$

It is clear that $(\{x_t\}_{t=0}^\infty, \{y_t\}_{t=0}^\infty)$ is a program. In view of (4.158) and the inequalities $z \leq M_0, x^* < M_0$, we have

$$y_t^{(z,k)} \leq f(M_0) \text{ for all integers } k \geq 0 \text{ and } t = 0, \dots, k - 1. \tag{4.164}$$

It follows from (4.164), (4.158), (4.162), and (4.161) that for each natural number i satisfying $k_i > S$,

$$\sum_{t=0}^{S-1} w_t(y_t^{(z,k_i)}) \geq \sum_{t=0}^{k_i-1} w_t(y_t^{(z,k_i)}) - \sum_{t=S}^{k_i-1} w_t(f(M_0))$$

$$\geq \sum_{t=0}^{k_i-1} w_t(y_t) - \sum_{t=S}^\infty w_t(f(M_0)) \geq \sum_{t=0}^{S-1} w_t(y_t)$$

$$- \sum_{t=L^{(\delta)}}^\infty w_t(f(M_0)) \geq U(z, S) - \delta.$$

When combined with (4.159) this implies that

$$\sum_{t=0}^{S-1} w_t(y_t^{(z)}) \geq U(z, S) - \delta.$$

Theorem 4.18 is proved.

4.6   Auxiliary Results for Theorem 4.19

Let

$$0 < m_0 < x^* < M_0. \qquad (4.165)$$

Relations (4.152) and (4.165) imply that

$$f(m_0) > (1 - v)m_0. \qquad (4.166)$$

Choose a real number

$$\gamma_0 \in (0, m_0/4) \qquad (4.167)$$

such that

$$f(\gamma_0) < 4^{-1}[f(m_0) - (1 - v)m_0]. \qquad (4.168)$$

It follows from (4.167), (4.165), and (4.149) that

$$f(\gamma_0) > (1 - v)\gamma_0. \qquad (4.169)$$

Lemma 4.20. *Let $p \geq 1$ be an integer. Then there exists an integer $L_0 \geq 1$ such that for each integer $T \geq L_0 + p$, each program $(\{x_t\}_{t=0}^{T}, \{y_t\}_{t=0}^{T-1})$ satisfying*

$$x_0 \in [m_0, M_0], \ \sum_{t=0}^{T-1} w_t(y_t) = U(x_0, T), \qquad (4.170)$$

and each integer $S \in [0, p]$ the inequality

$$\max\{x_t : t \in \{S, \dots, S + L_0 - 1\}\} \geq \gamma_0 \qquad (4.171)$$

holds.

Proof. In view of (4.165) and (4.156), there exists an integer $L_1 \geq 4$ such that

$$g^{L_1}(\gamma_0) > m_0. \qquad (4.172)$$

It follows from Assumption (A1), (4.168), and Assumption (A3) that there exists a natural number

$$L_0 > 4L_1 + 4 + 4p \qquad (4.173)$$

such that

$$\sum_{t=p+L_1}^{L_0-2} [w_t(f(m_0)) - (1-v)m_0) - w_t(f(\gamma_0))]$$

$$> \sum_{t=0}^{p+L_1-1} w_t(f(M_0)) + 1 + \sup\{w_t(f(M_0)) : t = 0, 1, \dots\}. \qquad (4.174)$$

Assume that an integer $T \geq L_0 + p$, a program $(\{x_t\}_{t=0}^T, \{y_t\}_{t=0}^{T-1})$ satisfies (4.170), and an integer $S \in [0, p]$. We claim that (4.171) holds. Assume the contrary. Then

$$x_t < \gamma_0 \text{ for all integers } t = S, \dots, S + L_0 - 1. \qquad (4.175)$$

Relations (4.167), (4.170), and (4.175) imply that there exists an integer $\tau_0 \geq 0$ such that

$$\tau_0 < S, \ x_{\tau_0} \geq \gamma_0, \ x_t < \gamma_0, \ t = \tau_0 + 1, \dots, S + L_0 - 1. \qquad (4.176)$$

There are two cases:

$$\max\{x_t : t = S, \dots, T - 1\} < \gamma_0; \qquad (4.177)$$

$$\max\{x_t : t = S, \dots, T - 1\} \geq \gamma_0. \qquad (4.178)$$

Assume that (4.177) holds. Set

$$\bar{x}_t = x_t, \ t = 0, \dots, \tau_0, \ \bar{y}_t = y_t \text{ for all integers } t \text{ satisfying } 0 \leq t < \tau_0. \qquad (4.179)$$

For all integers t satisfying $\tau_0 \leq t < \tau_0 + L_1$ set

$$\bar{y}_t = 0, \ \bar{x}_{t+1} = v\bar{x}_t + f(\bar{x}_t). \qquad (4.180)$$

It is clear that

$$(\{\bar{x}_t\}_{t=0}^{\tau_0+L_1}, \{\bar{y}_t\}_{t=0}^{\tau_0+L_1-1})$$

is a program. It follows from (4.180), (4.149), (4.179), (4.176), and (4.172) that

$$\bar{x}_{\tau_0+L_1} = g^{L_1}(\bar{x}_{\tau_0}) = g^{L_1}(x_{\tau_0}) \geq g^{L_1}(\gamma_0) > m_0. \qquad (4.181)$$

For all integers t satisfying $\tau_0 + L_1 \leq t \leq T - 1$ set

$$\bar{y}_t = f(m_0) - (1-v)m_0,$$

$$\bar{x}_{t+1} = v\bar{x}_t + f(\bar{x}_t) - f(m_0) + (1-v)m_0. \qquad (4.182)$$

By (4.181), (4.182), and (4.166),

$$\bar{x}_t \geq m_0, \ t = \tau_0 + L_1, \ldots, T \tag{4.183}$$

and $(\{\bar{x}_t\}_{t=0}^T, \{\bar{y}_t\}_{t=0}^{T-1})$ is a program. In view of (4.179), (4.157), (4.170), (4.182), (4.165), (4.177), (4.176), (4.174), and (4.168),

$$U(x_0, T) \geq \sum_{t=0}^{T-1} w_t(\bar{y}_t) = \sum_{t=0}^{T-1} w_t(y_t) + \sum_{t=0}^{T-1} w_t(\bar{y}_t) - \sum_{t=0}^{T-1} w_t(y_t)$$

$$= U(x_0, T) + \sum_{t=\tau_0}^{T-1} w_t(\bar{y}_t) - \sum_{t=\tau_0}^{T-1} w_t(y_t) \geq U(x_0, T)$$

$$+ \sum_{t=\tau_0+L_1}^{T-1} w_t(f(m_0) - (1-v)m_0) - \sum_{t=\tau_0}^{S+L_1-1} w_t(f(M_0))$$

$$- \sum_{t=S+L_1}^{T-1} w_t(f(\gamma_0))$$

$$\geq U(x_0, T) + \sum_{t=S+L_1}^{T-1} [w_t(f(m_0) - (1-v)m_0) - w_t(f(\gamma_0))]$$

$$- \sum_{t=0}^{p+L_1-1} w_t(f(M_0))$$

$$\geq U(x_0, T) + \sum_{t=p+L_1}^{p+L_0-1} w_t(f(m_0) - (1-v)m_0) - w_t(f(\gamma_0))]$$

$$- \sum_{t=0}^{p+L_1-1} w_t(f(M_0)) \geq U(x_0, T) + 1,$$

a contradiction. The contradiction we have reached shows that (4.177) does not hold. Hence (4.178) holds.

In view of (4.178) there exists an integer τ_1 such that

$$S + L_0 \leq \tau_1 < T, \ x_{\tau_1} \geq \gamma_0, \ x_t < \gamma_0, \ t = S, \ldots, \tau_1 - 1. \tag{4.184}$$

Set

$$\bar{x}_t = x_t, \ t = 0, \ldots, \tau_0, \ \bar{y}_t = y_t \text{ for all integers } t \text{ satisfying } 0 \leq t < \tau_0 \tag{4.185}$$

and for all integers t satisfying $\tau_0 \leq t < \tau_0 + L_1$ put

$$\bar{y}_t = 0, \ \bar{x}_{t+1} = v\bar{x}_t + f(\bar{x}_t). \tag{4.186}$$

It is not difficult to see that $(\{\bar{x}_t\}_{t=0}^{\tau_0+L_1}, \{\bar{y}_t\}_{t=0}^{\tau_0+L_1-1})$ is a program and that (4.181) holds. For all integers t satisfying $\tau_0 + L_1 \le t \le \tau_1 - 2$ set

$$y_t = f(m_0) - (1-v)m_0, \; \bar{x}_{t+1} = v\bar{x}_t + f(\bar{x}_t) - f(m_0) + (1-v)m_0. \quad (4.187)$$

Relations (4.181), (4.187) and (4.166) imply that

$$\bar{x}_t \ge m_0, \; t = \tau_0 + L_1, \ldots, \tau_1 - 1 \quad (4.188)$$

and that $(\{\bar{x}_t\}_{t=0}^{\tau_1-1}, \{\bar{y}_t\}_{t=0}^{\tau_1-2})$ is a program. Set

$$\bar{y}_{\tau_1-1} = 0, \; \bar{x}_{\tau_1} = v\bar{x}_{\tau_1-1} + f(\bar{x}_{\tau_1-1}). \quad (4.189)$$

By (4.189), we have $(\{\bar{x}_t\}_{t=0}^{\tau_1}, \{\bar{y}_t\}_{t=0}^{\tau_1-1})$ is a program and in view of (4.189), (4.188), (4.184), and (4.167),

$$\bar{x}_{\tau_1} \ge vm_0 + f(m_0) \ge vx_{\tau_1-1} + f(x_{\tau_1-1}) \ge x_{\tau_1}. \quad (4.190)$$

For all integers t satisfying $\tau_1 \le t < T$ set

$$\bar{y}_t = y_t, \; \bar{x}_{t+1} = v\bar{x}_t + x_{t+1} - vx_t. \quad (4.191)$$

Relations (4.190) and (4.191) imply that

$$\bar{x}_t \ge x_t, \; t = \tau_1, \ldots, T. \quad (4.192)$$

It follows from (4.191) and (4.192) that $(\{\bar{x}_t\}_{t=0}^{T}, \{\bar{y}_t\}_{t=0}^{T-1})$ is a program. By (4.185), (4.157), (4.170), (4.191), (4.187), (4.184), (4.165), (4.167), (4.176), (4.168), and (4.174),

$$U(x_0, T) \ge \sum_{t=0}^{T-1} w_t(\bar{y}_t) = \sum_{t=0}^{T-1} w_t(y_t) + \sum_{t=0}^{T-1} w_t(\bar{y}_t) - \sum_{t=0}^{T-1} w_t(y_t)$$

$$= U(x_0, T) + \sum_{t=\tau_0}^{\tau_1-1} w_t(\bar{y}_t) - \sum_{t=\tau_0}^{\tau_1-1} w_t(y_t)$$

$$\ge U(x_0, T) + \sum_{t=\tau_0+L_1}^{\tau_1-2} w_t(f(m_0) - (1-v)m_0) - \sum_{t=\tau_0}^{S+L_1-1} w_t(f(M_0))$$

$$- \sum_{t=S+L_1}^{\tau_1-2} w_t(f(\gamma_0)) - w_{\tau_1-1}(f(\gamma_0))$$

$$\geq U(x_0, T) - \sum_{t=0}^{p+L_1-1} w_t(f(M_0)) - \sup\{w_t(f(M_0)) : t = 0, 1, \dots\}$$

$$+ \sum_{t=p+L_1}^{L_0-2} [w_t(f(m_0) - (1-v)m_0) - w_t(f(\gamma_0))] > U(x_0, T) + 1,$$

a contradiction. The contradiction we have reached proves (4.171) and Lemma 4.20 itself. □

Choose a real number $\gamma_1 > 0$ such that

$$\gamma_1 < [f(\gamma_0) - (1-v)\gamma_0]/4. \tag{4.193}$$

In view of (4.156) and (4.165) there exists an integer $L_* \geq 4$ such that for all integers $i \geq L_*$, we have

$$g^i(\gamma_0) > x^* - \gamma_1/4, \ g^i(\gamma_0) > m_0, \ g^i(M_0) < x^* + \gamma_1/4. \tag{4.194}$$

Lemma 4.21. *Let δ be a positive number. Then there exists an integer $\bar{L} \geq 1$ such that for each integer $L \geq \bar{L}$ there exists an integer $\tau \geq L$ such that the following assertion holds:*

For each integer $T \geq \tau$ and each program $(\{x_t\}_{t=0}^T, \{y_t\}_{t=0}^{T-1})$ satisfying

$$x_0 \in [m_0, M_0], \ \sum_{t=0}^{T-1} w_t(y_t) = U(x_0, T) \tag{4.195}$$

the inequality

$$\sum_{t=0}^{L-1} w_t(y_t) \geq U(x_0, L) - \delta \tag{4.196}$$

holds.

Proof. Assumption (A1) implies that there exists a natural number $p \geq 4$ such that

$$w_t(f(M_0)) < (\delta/8)(2L_* + 2)^{-1} \text{ for all integers } j \geq p. \tag{4.197}$$

In view of Lemma 4.20 there exists an integer $L_0 \geq 1$ such that the following property holds:

(P1) for each integer $T \geq L_0 + p$, each program $(\{x_t\}_{t=0}^T, \{y_t\}_{t=0}^{T-1})$ satisfying (4.195), and each integer $S \in [0, p]$, we have

$$\max\{x_t : t = S, \dots, S + L_0 - 1\} \geq \gamma_0.$$

Choose an integer

$$\bar{L} > L_0 + p + 4L_*. \tag{4.198}$$

Assume that a natural number $L \geq \bar{L}$ and choose an integer

$$\tau > L + 4 + L_*. \tag{4.199}$$

Assume that an integer $T \geq \tau$ and that a program $(\{x_t\}_{t=0}^T, \{y_t\}_{t=0}^{T-1})$ satisfies (4.195). We claim that (4.196) holds. Assume the contrary. Then

$$\sum_{t=0}^{L-1} w_t(y_t) < U(x_0, L) - \delta. \tag{4.200}$$

In view of Proposition 4.17 there exists a program $(\{\tilde{x}_t\}_{t=0}^L, \{\tilde{y}_t\}_{t=0}^{L-1})$ such that

$$\tilde{x}_0 = x_0, \ \sum_{t=0}^{L-1} w_t(\tilde{y}_t) = U(x_0, L). \tag{4.201}$$

Relations (4.201), (4.195), and (4.167) imply that there exists an integer $\tau_0 \geq 0$ such that

$$\tau_0 \leq L, \ \tilde{x}(\tau_0) \geq y_0, \ \text{if an integer } t \text{ satisfies } \tau_0 < t \leq L, \text{ then } \tilde{x}_{\tau_0} < y_0. \tag{4.202}$$

It follows from (4.202), (4.195), (4.199), (4.198), and property (P1) that

$$\tau_0 \geq p. \tag{4.203}$$

Define a program $(\{\bar{x}_t\}_{t=0}^L, \{\bar{y}_t\}_{t=0}^{L-1})$. There are two cases:

$$L - \tau_0 \leq L_*, \tag{4.204}$$

$$L - \tau_0 > L_*. \tag{4.205}$$

If inequality (4.204) is true, put

$$\bar{x}_t = \tilde{x}_t, \ t = 0, \dots, L, \ \bar{y}_t = \tilde{y}_t, \ t = 0, \dots, L - 1. \tag{4.206}$$

Assume that inequality (4.205) is valid. Put

$$\bar{x}_t = \tilde{x}_t, \ t = 0, \dots, \tau_0, \ \bar{y}_t = \tilde{y}_t, \ t = 0, \dots, \tau_0 - 1 \tag{4.207}$$

and for all integers $t = \tau_0, \dots, \tau_0 + L_* - 1$ set

$$\bar{y}_t = 0, \ \bar{x}_{t+1} = v\bar{x}_t + f(\bar{x}_t). \tag{4.208}$$

It is not difficult to see that $(\{\bar{x}_t\}_{t=0}^{\tau_0+L_*}, \{\bar{y}_t\}_{t=0}^{\tau_0+L_*-1})$ is a program. Relations (4.208), (4.155), (4.202), and (4.194) imply that

$$\bar{x}_{\tau_0+L_*} = g^{L_*}(\bar{x}(\tau_0)) \geq g^{L_*}(\gamma_0) > m_0. \tag{4.209}$$

For all integers t satisfying $\tau_0 + L_* \leq t < L$ set

$$\bar{y}_t = f(m_0) - (1 - v)m_0, \tag{4.210}$$
$$\bar{x}_{t+1} = v\bar{x}_t + (1 - v)m_0.$$

In view of (4.210) and (4.209), we have

$$\bar{x}_t \geq m_0, \ t = \tau_0 + L_*, \ldots, L. \tag{4.211}$$

In follows from (4.166), (4.210), and (4.211) that $(\{\bar{x}_t\}_{t=0}^{L}, \{\bar{y}_t\}_{t=0}^{L-1})$ is a program. By (4.207) and (4.211),

$$\bar{x}_0 = x_0. \tag{4.212}$$

By (4.210), (4.205), (4.202), (4.194), and (4.168), we have

$$\bar{y}_t \geq \tilde{y}_t \text{ for all integers } t = \tau_0 + L_*, \ldots, L - 1. \tag{4.213}$$

It follows from (4.201), (4.207), (4.213), (4.195), (4.165), (4.203), and (4.197) that

$$U(x_0, L) - \sum_{t=0}^{L-1} w_t(\bar{y}_t)$$

$$= U(x_0, L) - \sum_{t=0}^{L-1} w_t(\tilde{y}_t) + \sum_{t=0}^{L-1} w_t(\tilde{y}_t) - \sum_{t=0}^{L-1} w_t(\bar{y}_t)$$

$$\leq \sum_{t=\tau_0}^{\tau_0+L_*-1} w_t(\tilde{y}_t) - \sum_{t=\tau_0}^{\tau_0+L_*-1} w_t(\bar{y}_t)$$

$$\leq \sum_{t=\tau_0}^{\tau_0+L_*-1} w_t(f(M_0)) < \delta/8.$$

Thus in both cases we have constructed a program $(\{\bar{x}_t\}_{t=0}^{L}, \{\bar{y}_t\}_{t=0}^{L-1})$ such that

$$\bar{x}_0 = x_0, \ U(x_0, L) - \delta/8 < \sum_{t=0}^{L-1} w_t(\bar{y}_t) \tag{4.214}$$

and there exists an integer

$$S_0 \in [L - L_*, L] \text{ such that } \bar{x}_{S_0} \geq \gamma_0. \tag{4.215}$$

There are two cases:

(1) there exists an integer

$$S_1 \in [S_0 + L_*, T - 2] \text{ such that } y_{S_1+1} \geq \gamma_1; \tag{4.216}$$

(2)

$$y_t < \gamma_1 \text{ for all integers } t \in [S_0 + L_* + 1, T - 1]. \tag{4.217}$$

Assume that the case (2) holds. Set

$$x'_t = \bar{x}_t, \ t = 0, \ldots, S_0, \ y'_t = \bar{y}_t, \ t = 0, \ldots, S_0 - 1. \tag{4.218}$$

For all integers t satisfying $S_0 \leq t < T$ put

$$y'_t = f(\gamma_0) - (1 - v)\gamma_0, \tag{4.219}$$
$$x'_{t+1} = vx'_t + (1 - v)\gamma_0.$$

By (4.219), (4.218), and (4.215),

$$x'_t \geq \gamma_0, \ t = S_0, \ldots, T. \tag{4.220}$$

In view of (4.218), (4.219), (4.169), and (4.220), $(\{x'_t\}_{t=0}^{T}, \{y'_t\}_{t=0}^{T-1})$ is a program. It follows from (4.219), (4.217), and (4.193) that for all integers t satisfying $1 + L_* + S_0 \leq t < T$ we have

$$y'_t \geq y_t. \tag{4.221}$$

By (4.218) and (4.214),

$$x'_0 = x_0. \tag{4.222}$$

It follows from (4.222), (4.157), (4.195), (4.221), (4.198), (4.199), (4.165), (4.214), (4.215), (4.200), and (4.197) that

$$0 \leq U(x_0, T) - \sum_{t=0}^{T-1} w_t(y'_t)$$

$$= \sum_{t=0}^{T-1} w_t(y_t) - \sum_{t=0}^{T-1} w_t(y'_t) \leq \sum_{t=0}^{L_*+S_0} w_t(y_t) - \sum_{t=0}^{L_*+S_0} w_t(y'_t)$$

$$\leq \sum_{t=0}^{S_0-1} w_t(y_t) - \sum_{t=0}^{S_0-1} w_t(y_t') + \sum_{t=S_0}^{L_*+S_0} w_t(f(M_0))$$

$$= \sum_{t=0}^{S_0-1} w_t(y_t) - \sum_{t=0}^{S_0-1} w_t(\bar{y}_t) + \sum_{t=S_0}^{L_*+S_0} w_t(f(M_0))$$

$$\leq \sum_{t=0}^{L-1} w_t(y_t) - \sum_{t=0}^{L-1} w_t(\bar{y}_t)$$

$$+ (L - S_0 + 1) \max\{w_t(f(M_0)) : t = S_0, \dots, L\} + \sum_{t=S_0}^{L_*+S_0} w_t(f(M_0))$$

$$<U(x_0, L) - \delta - (U(x_0, L) - \delta/8) + (L_* + 1) \max\{w_t(f(M_0)) : t = S_0, \dots, L\}$$

$$+ \sum_{t=S_0}^{L_*+S_0} w_t(f(M_0)) \leq -(7/8)\delta + \delta/4 < -\delta/2,$$

a contradiction. The contradiction we have reached proves that the case (2) does not hold. Thus the case (1) holds and there is an integer S_1 satisfying (4.216). We may assume without loss of generality that

$$\text{for each integer } t \text{ satisfying } 1 + S_0 + L_* \leq t \leq S_1, y_t < \gamma_1. \tag{4.223}$$

Set

$$x_t' = \bar{x}_t, \ t = 0, \dots, S_0, \ y_t' = \bar{y}_t, \ t = 0, \dots, S_0 - 1. \tag{4.224}$$

For all integers t satisfying $S_0 \leq t < S_1 - L_*$ set

$$y_t' = f(\gamma_0) - (1-v)\gamma_0, \ x_{t+1}' = vx_t' + (1-v)\gamma_0. \tag{4.225}$$

Relations (4.224), (4.225), and (4.215) imply that

$$x_t' \geq \gamma_0, \ t = S_0, \dots, S_1 - L_*. \tag{4.226}$$

In view of (4.224) and (4.169), $(\{x_t'\}_{t=0}^{S_1-L_*}, \{y_t'\}_{t=0}^{S_1-L_*-1})$ is a program. For all integers t satisfying $S_1 - L_* \leq t \leq S_1 + 1$ put

$$y_t' = 0, \ x_{t+1}' = vx_t' + f(x_t'). \tag{4.227}$$

Clearly, $(\{x_t'\}_{t=0}^{S_1+2}, \{y_t'\}_{t=0}^{S_1+1})$ is a program. Relations (4.227), (4.155), (4.226) and (4.194), imply that

$$x_{S_1+2}' = g^{L_*+2}(x_{S_1-L_*}') \geq g^{L_*+2}(\gamma_0) \geq x^* - \gamma_1/4. \tag{4.228}$$

In view of (4.155), (4.151), (4.195), (4.216), and (4.194), we have

$$x_{S_1+2} + y_{S_1+1} = vx_{S_1+1} + (x_{S_1+2} - vx_{S_1+1}) + y_{S_1+1} \leq vx_{S_1+1} + f(x_{S_1+1})$$
$$= g(x_{S_1+1}) \leq g^{S_1+2}(x_0) \leq g^{S_1+2}(M_0) \leq x^* + \gamma_1/4. \qquad (4.229)$$

By (4.229), (4.216), and (4.228),

$$x_{S_1+2} \leq x^* + \gamma_1/4 - y_{S_1+1} \leq x'_{S_1+2} - \gamma_1/2 < x'_{S_1+2}. \qquad (4.230)$$

For all integers t satisfying $S_1 + 2 \leq t < T$ set

$$y'_t = y_t, \ x'_{t+1} = x_{t+1} - vx_t + vx'_t. \qquad (4.231)$$

By (4.230) and (4.231),

$$x'_t \geq x_t, \ t = S_1 + 2, \ldots, T. \qquad (4.232)$$

Relations (4.231) and (4.232) imply that $(\{x'_t\}_{t=0}^T, \{y'_t\}_{t=0}^{T-1})$ is a program. By (4.224), (4.214), (4.157), (4.195), (4.231), (4.216), (4.215), (4.169), and (4.198), we have

$$U(x_0, T) - \sum_{t=0}^{T-1} w_t(y'_t) = \sum_{t=0}^{T-1} w_t(y_t) - \sum_{t=0}^{T-1} w_t(y'_t)$$
$$\leq - \sum_{t=0}^{S_1+1} w_t(y'_t) + \sum_{t=0}^{S_1+1} w_t(y_t)$$
$$\leq \sum_{t=0}^{S_1-L_*-1} w_t(y_t) - \sum_{t=0}^{S_1-L_*-1} w_t(y'_t)$$
$$+ (L_* + 2) \max\{w_t(f(M_0)) : \text{an integer } t \geq p\}.$$
$$(4.233)$$

We suppose that the sum over empty set is zero. It follows from (4.233), (4.197), (4.195), (4.165), (4.223), (4.224), (4.226), (4.193), (4.215), (4.198), (4.214), (4.197), and (4.200) that

$$0 \leq \sum_{t=0}^{S_1-L_*-1} w_t(y_t) - \sum_{t=0}^{S_1-L_*-1} w_t(y'_t) + \delta/8$$
$$\leq \sum_{t=0}^{S_0-1} w_t(y_t) + \sum_{t=S_0}^{S_0+L_*} w_t(f(M_0))$$

$$+ \sum \{w_t(\gamma_1) : \text{an integer } t \text{ such that } 1 + L_* + S_0 \le t \le S_1 - L_* - 1\}$$

$$- \sum_{t=0}^{S_0-1} w_t(\bar{y}_t) + \delta/8$$

$$- \sum \{w_t(f(\gamma_0)) - (1-v)\gamma_0) : \text{an integer } t \text{ such that } S_0 \le t < S_1 - L_* - 1\}$$

$$\le \sum_{t=S_0}^{S_0+L_*} w_t(f(M_0)) + \sum_{t=0}^{S_0-1} w_t(y_t) + \delta/8 - \sum_{t=0}^{S_0-1} w_t(\bar{y}_t)$$

$$\le \sum_{t=0}^{L-1} w_t(y_t) + \delta/8 + \sum_{t=S_0}^{S_0+L_*} w_t(f(M_0)) - \sum_{t=0}^{L-1} w_t(\bar{y}_t)$$

$$+ (L - S_0) \max\{w_t(f(M_0)) : \text{an integer } t \text{ such that } S_0 \le t \le L - 1\}$$

$$\le \sum_{t=0}^{L-1} w_t(y_t) - \sum_{t=0}^{L-1} w_t(\bar{y}_t) + \delta/8 + 2(L_*+1) \max\{w_t(f(M_0)) : \text{an integer } t \ge p\}$$

$$\le U(x_0, L) - \delta - (U(x_0, L) - \delta/8) + \delta/4 < -\delta/2,$$

a contradiction. The contradiction we have reached proves (4.196). Lemma 4.21 is proved.   □

4.7 Proof of Theorem 4.19

Proposition 4.17 implies that for each $z \in R_+^1$ and each natural number T there exists a program $(\{x_t^{(z,T)}\}_{t=0}^T, \{y_t^{(z,T)}\}_{t=0}^{T-1})$ such that

$$x_0^{(z,T)} = z, \quad \sum_{t=0}^{T-1} w_t(y_t^{(z,T)}) = U(z, T). \tag{4.234}$$

Let

$$z \in [m_0, M_0] \tag{4.235}$$

be given. In view of (4.234) and (4.153) there exist a strictly increasing sequence of natural numbers $\{T_k\}_{k=1}^\infty$ and a program $(\{x_t^{(z)}\}_{t=0}^\infty, \{y_t^{(z)}\}_{t=0}^\infty)$ such that for each nonnegative integer t,

$$x_t^{(z)} = \lim_{k \to \infty} x_t^{(z,T_k)}, \quad y_t^{(z)} = \lim_{k \to \infty} y_t^{(z,T_k)}. \tag{4.236}$$

It is clear that

$$x_0^{(z)} = z. \tag{4.237}$$

Let $\delta > 0$ be given. Lemma 4.21 and (4.234) imply that there exists an integer $L_\delta \geq 1$ such that the following property holds:

(P2) For each integer $L \geq L_\delta$ there exists a natural number $\tau_L \geq L$ such that for each integer $T \geq \tau_L$ and each number $z \in [m_0, M_0]$ we have

$$\sum_{t=0}^{T-1} w_t(y_t^{(z,T)}) \geq U(z, L) - \delta/4.$$

Let an integer L satisfy $L \geq L_\delta$ and let an integer $\tau_L \geq L$. be as guaranteed by property (P2). Let

$$z \in [m_0, M_0]$$

be given. By (4.236) there exists an integer $k \geq 1$ such that

$$T_k > \tau_L,$$

$$\left| \sum_{t=0}^{L-1} w_t(y_t^{(z)}) - \sum_{t=0}^{L-1} w_t(y_t^{(z,T_k)}) \right| \leq \delta/4. \tag{4.238}$$

It follows from (4.234) and property (P2) that

$$\sum_{t=0}^{L-1} w_t(y_t^{(z,T_k)}) \geq U(z, L) - \delta/4.$$

Together with (4.238) this inequality implies that

$$\sum_{t=0}^{L-1} w_t(y_t^{(z)}) \geq U(z, L) - \delta.$$

Theorem 4.19 is proved.

4.8 Two-Dimensional Autonomous Problems

We study a large class of nonconvex infinite horizon discrete-time optimal control problems. This class contains optimal control problems arising in economic dynamics which describe a general two-sector model without discounting and with nonconcave utility functions representing the preferences of the planner.

Let

$$v_1, v_2 \in [0, 1), \tag{4.239}$$

let $f_1, f_2 : [0, \infty) \to [0, \infty)$ be strictly increasing continuous functions such that for $i = 1, 2$,

$$f_i(0) = 0 \text{ and } f_i(z) > 0 \text{ for all numbers } z > 0, \tag{4.240}$$

and let $w : [0, \infty) \to [0, \infty)$ be a strictly increasing continuous function such that

$$w(0) = 0 \text{ and } w(x) > 0 \text{ for all numbers } x > 0. \tag{4.241}$$

A sequence $\{x_{1,t}, x_{2,t}, y_t\}_{t=0}^{\infty}$ is called a program if for all nonnegative integers t,

$$x_{1,t}, x_{2,t}, y_t \in R_+^1, \ x_{1,t+1} \geq v_1 x_{1,t}, \ x_{2,t+1} \geq v_2 x_{2,t}, \tag{4.242}$$

$$x_{1,t+1} - v_1 x_{1,t} + x_{2,t+1} - v_2 x_{2,t} \leq f_1(x_{1,t}), \tag{4.243}$$

$$0 \leq y_t \leq f_2(x_{2,t}). \tag{4.244}$$

Let integers T_1, T_2 satisfy $T_2 > T_1 \geq 0$. A sequence

$$\left(\{x_{1,t}\}_{t=T_1}^{T_2}, \{x_{2,t}\}_{t=T_1}^{T_2}, \{y_t\}_{t=T_1}^{T_2-1} \right)$$

is called a program if

$$x_{1,t}, \ x_{2,t} \in R_+^1, \ t = T_1, \ldots, T_2, \ y_t \in R_+^1, \ t = T_1, \ldots, T_2 - 1$$

and if for all integers $t = T_1, \ldots, T_2 - 1$ inequalities (4.242)–(4.244) hold.

We study an infinite horizon optimal control problem which corresponds to a finite horizon problem:

$$\sum_{t=0}^{T-1} w(y_t) \to \max, \ \left(\{x_{1,t}\}_{t=T_1}^{T_2}, \{x_{2,t}\}_{t=T_1}^{T_2}, \{y_t\}_{t=T_1}^{T_2-1} \right)$$

is a program such that $x_{1,0} = z_1$, $x_{2,0} = z_2$,

where T is a natural number and $z_1, z_2 \in R_+^1$.

These optimal control systems describe a two-sector model of economic dynamics where the first sector produces funds, the second sector produces consumption, $x_{1,t}$ is funds of the first sector at moment t, $x_{2,t}$ is funds of the second sector at moment t, y_t is consumption at moment t and $w(y_t)$ evaluates consumption at moment t. The dynamics of the model is described by (4.242)–(4.244). It should be mentioned that this model was usually considered in the literature under an assumption that the functions f_1, f_2 and w are concave. In this section we present the results obtained in [75] which establish the existence of optimal solutions without this assumption.

Assume that there exists a real number $x^* > 0$ such that

$$f_1(x) > (1 - v_1)x \text{ for all numbers } x \in (0, x^*),\qquad (4.245)$$

$$f_1(x) < (1 - v_1)x \text{ for all numbers } x \in (x^*, \infty).\qquad (4.246)$$

By (4.245) and (4.246), we have

$$f_1(x^*) = (1 - v_1)x^*.\qquad (4.247)$$

Put

$$g(z) = f_1(z) + v_1 z, \ z \in R^1_+, \ g^0 = g, \ g^{i+1} = g \circ g^i \qquad (4.248)$$

for all nonnegative integers i. It follows from (4.245) and (4.246) that

$$\lim_{i \to \infty} g^i(z) = x^* \text{ for all } z > 0.\qquad (4.249)$$

In the sequel supremum over an empty set is $-\infty$ and the sum over an empty set is zero. For each pair of real numbers $z_1, z_2 \in R^1_+$ and each pair of integers $T_2 > T_1 \geq 0$ define

$$U(z_1, z_2, T_1, T_2) = \sup \left\{ \sum_{t=T_1}^{T_2-1} w(y_t) : (\{x_{1,t}\}_{t=T_1}^{T_2}, \{x_{2,t}\}_{t=T_1}^{T_2}, \{y_t\}_{t=T_1}^{T_2-1}) \text{is a program,} \right.$$

$$\left. x_{1,0} = z_1, \ x_{2,0} = z_2 \right\} .\qquad (4.250)$$

The following proposition follows immediately from the continuity of f_1, f_2, w.

Proposition 4.22. *For each pair of real numbers $z_1, z_2 \in R^1_+$ and each pair of integers $T_2 > T_1 \geq 0$ there exists a program*

$$(\{x_{1,t}\}_{t=T_1}^{T_2}, \{x_{2,t}\}_{t=T_1}^{T_2}, \{y_t\}_{t=T_1}^{T_2-1})$$

such that $x_{1,0} = z_1$, $x_{2,0} = z_2$ and that $\sum_{t=T_1}^{T_2-1} w(y_t) = U(z_1, z_2, T_1, T_2)$.

For each pair of real numbers $m \geq 0$, $M > 0$ satisfying $m < x^* < M$ and each pair of integers $T_2 > T_1 \geq 0$ define

$$\hat{U}(m, M, T_1, T_2) = \sup\{U(z_1, z_2, T_1, T_2) : z_1 \in [m, M], \ z_2 \in [0, M]\}.\qquad (4.251)$$

For each $z_1, z_2, z_3, z_4 \in R_+^1$ and each pair of integers $T_2 > T_1 \geq 0$, define

$$U(z_1, z_2, z_3, z_4, T_1, T_2) = \sup \left\{ \sum_{t=T_1}^{T_2-1} w(y_t) : \ (\{x_{1,t}\}_{t=T_1}^{T_2}, \ \{x_{2,t}\}_{t=T_1}^{T_2}, \{y_t\}_{t=T_1}^{T_2-1}) \right.$$

$$\left. \text{is a program}, \ x_{1,T_1} = z_1, \ x_{2,T_1} = z_2, \ x_{1,T_2} \geq z_3, \ x_{2,T_2} \geq z_4 \right\}. \qquad (4.252)$$

We suppose that the following assumption holds:

(A) For each positive number M there exists a positive number c_M such that for all pairs of real numbers $z_1, z_2 \in [0, M]$,

$$|f_2(z_1) - f_2(z_2)| \leq c_M |z_1 - z_2| \text{ and } |w(z_1) - w(z_2)| \leq c_M |z_1 - z_2|.$$

The following theorem which was obtained in [75] establishes for any initial state $(x_{1,0}, x_{2,0}) \in (0, \infty) \times [0, \infty)$ the existence of a solution of the corresponding infinite horizon optimal control problem.

Theorem 4.23. *Let* $0 < m_0 < x^* < M_0$. *Then there exists a positive number* M_* *such that for each real number* $x_{1,0} \in [m_0, M_0]$ *and each real number* $x_{2,0} \in [0, M_0]$ *there exists a program* $\{x_{1,t}, x_{2,t}, y_t\}_{t=0}^{\infty}$ *such that for each pair of integers* $T_1, T_2 \geq 0$ *satisfying* $T_1 < T_2$,

$$\left| \sum_{t=T_1}^{T_2-1} w(y_t) - \hat{U}(m_0, M_0, T_1, T_2) \right| \leq M_*$$

and that for each natural number T,

$$\sum_{t=0}^{T-1} w(y_t) = U(x_{1,0}, x_{2,0}, x_{1,T}, x_{2,T}, 0, T).$$

The following theorem obtained in [75] is our second main result in this section.

Theorem 4.24. *Let* $0 < m_0 < x^* < M_0$. *Then there exists*

$$\mu := \lim_{p \to \infty} \hat{U}(m_0, M_0, 0, p)/p, \qquad (4.253)$$

and there exists a positive number M *such that*

$$|p^{-1}\hat{U}(m_0, M_0, 0, p) - \mu| \leq 2M/p \text{ for all natural numbers } p. \qquad (4.254)$$

It follows from Theorems 4.23 and 4.24 that the constant μ does not depend on the choice of m_0, M_0. It is not difficult to see that for each pair of positive numbers m_0, M_0 satisfying $m_0 < x^* < M_0$ and each pair of integers $q > p > 0$, we have

$$\hat{U}(m_0, M_0, p, q) = \hat{U}(0, M_0, p, q). \tag{4.255}$$

Theorems 4.23 and 4.24 imply the following result.

Theorem 4.25. *Let* $0 < m_0 < x^* < M_0$. *Then there exists a positive number* M_* *such that for each real number* $x_{1,0} \in [m_0, M_0]$ *and each real number* $x_{2,0} \in [0, M_0]$ *there exists a program* $\{x_{1,t}, x_{2,t}, y_t\}_{t=0}^{\infty}$ *such that for each pair of integers* $T_1, T_2 \geq 0$ *satisfying* $T_1 < T_2$ *the inequality*

$$\left| \sum_{t=T_1}^{T_2-1} w(y_t) - \mu(T_2 - T_1) \right| \leq M_*$$

holds.

We will show (see Lemma 4.35) that $\mu > 0$.

Theorem 4.26. *Assume that* $\{x_{1,t}, x_{2,t}, y_t\}_{t=0}^{\infty}$ *is a program. Then either the sequence* $\{\sum_{t=0}^{T-1} w(y_t) - T\mu\}_{T=1}^{\infty}$ *is bounded or*

$$\lim_{T \to \infty} \left[\sum_{t=0}^{T-1} w(y_t) - T\mu \right] = -\infty.$$

A program $\{x_{1,t}, x_{2,t}, y_t\}_{t=0}^{\infty}$ such that the sequence

$$\left\{ \sum_{t=0}^{T-1} w(y_t) - T\mu \right\}_{T=1}^{\infty}$$

is bounded is called good.

It should be mentioned that most results known in the literature which establish the existence of good programs were obtained for concave (convex) problems. The problem considered in this section is constrained and nonconcave.

Proof of Theorem 4.26. We may assume without loss of generality that $x_{1,0} > 0$. Choose real numbers $m_0, M_0 > 0$ such that

$$m_0 < x^* < M_0, \ m_0 < x_{1,0} < M_0, x_{2,0} \leq M_0. \tag{4.256}$$

Let $M > 0$ be as guaranteed by Theorem 4.24 and assume that the sequence $\{\sum_{t=0}^{T-1} w(y_t) - T\mu\}_{T=1}^{\infty}$ is not bounded. Then it follows from (4.256) and Theorem 4.24 that

$$\liminf_{T \to \infty} \left[\sum_{t=0}^{T-1} w(y_t) - T\mu \right] = -\infty. \tag{4.257}$$

Let $Q > 0$ be given. In view of (4.257) there exists a natural number T_0 such that

$$\sum_{t=0}^{T_0-1} w(y_t) - T_0\mu < -Q - 2M. \tag{4.258}$$

It follows from (4.258), (4.255), (4.256), (4.251), Theorem 4.24, and the choice of M that for each integer $T > T_0$, we have

$$\sum_{t=0}^{T-1} w(y_t) - T\mu = \sum_{t=0}^{T_0-1} w(y_t) - T_0\mu + \sum_{t=T_0}^{T-1} w(y_t) - \hat{U}(0, M_0, T_0, T)$$

$$+ \hat{U}(m_0, M_0, T_0, T) - (T - T_0)\mu < -Q - 2M + 2M < -Q.$$

Theorem 4.26 is proved. □

It is not difficult to see that (4.245) and (4.246) hold for an increasing continuous functions $f_1 : [0, \infty) \to [0, \infty)$ satisfying (4.240) if the function $f(x)/x$ is strictly decreasing on $(0, \infty)$:

$$\lim_{x\to\infty} f(x)/x = 0$$

and

$$\liminf_{x\to 0^+} f(x)/x > 1.$$

4.9 Auxiliary Results for Theorems 4.23–4.25

Assume that real numbers m_0, M_0 satisfy

$$2m_0 < x^* < M_0. \tag{4.259}$$

In view of (4.259) and (4.245),

$$f_1(2^{-1}x^*) > (1 - v_1)x^*/2. \tag{4.260}$$

Put

$$\Lambda = f_1(x^*/2) - (1 - v_1)x^*/2. \tag{4.261}$$

Relations (4.260) and (4.261) imply that $\Lambda > 0$. Choose a real number $\delta_0 > 0$ such that

$$(f_1(\delta_0) + f_2(\delta_0))(1 - v_2)^{-1} + \delta_0 < \min\{m_0/4, \Lambda/4\}. \tag{4.262}$$

Choose a real number $\gamma_* > 0$ such that

$$\gamma_* < x^*/8, \ 2(1-v_2)^{-1}(f_1(x^*+\gamma_*) - (1-v_1)x^* + 4\gamma_*) < \Lambda/2. \qquad (4.263)$$

Assumption (A) implies that there exists a number $c_0 > 1$ such that

$$|f_2(z_1) - f_2(z_2)| \le c_0|z_1 - z_2| \text{ for all } z_1, z_2 \in [0, M_0(1-v_2)^{-1}], \qquad (4.264)$$

$$|w(z_1) - w(z_2)| \le c_0|z_1 - z_2| \text{ for all } z_1, z_2 \in [0, f_2(M_0(1-v_2)^{-1})]. \qquad (4.265)$$

Equations (4.242)–(4.244) and (4.246) imply the following result.

Lemma 4.27. *Let $\tilde{M}_0 > x^*$, $T \ge 1$ be an integer and $(\{x_{1,t}\}_{t=0}^T, \{x_{2,t}\}_{t=0}^T, \{y_t\}_{t=0}^{T-1})$ be a program such that $x_{1,0}, x_{2,0} \le \tilde{M}_0$. Then*

$$x_{1,t} \le \tilde{M}_0, \ t = 0, \dots, T, \ x_{2,t} \le \tilde{M}_0(1-v_2)^{-1}, \ t = 0, \dots, T,$$

$$y_t \le f_2(\tilde{M}_0(1-v_2)^{-1}), \ t = 0, \dots, T-1.$$

Lemma 4.28. *Let M_1 be a positive number. Then there exists an integer $L \ge 1$ such that for each integer $T \ge 1$ and each program*

$$\left(\{x_{1,t}\}_{t=0}^T, \{x_{2,t}\}_{t=0}^T, \{y_t\}_{t=0}^{T-1}\right)$$

which satisfies

$$m_0 \le x_{1,0} \le M_0, \ 0 \le x_{2,0} \le M_0, \qquad (4.266)$$

$$\sum_{t=0}^{T-1} w(y_t) \ge U(x_{1,0}, x_{2,0}, 0, T) - M_1,$$

the following property holds:
  For each integer $\tau \in \{0, \dots, T\}$ there exists a nonnegative integer $\tau_0 \in [\tau - L, \tau]$ such that $x_{1,\tau_0} \ge \delta_0$.

Proof. In view of (4.249) there exists an integer $L_0 > 4$ such that

$$g^i(\delta_0) > x^*/2 \text{ for all integers } i \ge L_0, \qquad (4.267)$$

$$v_2^{L_0-1}M_0(1-v_2)^{-1} < \delta_0. \qquad (4.268)$$

Choose an integer

$$L > 4L_0 + 4 \qquad (4.269)$$

such that

$$(L - L_0 - 2)(w(f_2(\Lambda)) - w(f_2(\Lambda/4))) > M_1 + (L_0 + 2)w(f_2(M_0(1-v_2)^{-1})). \qquad (4.270)$$

Assume that $T \geq 1$ is an integer, a program

$$(\{x_{1,t}\}_{t=0}^{T}, \{x_{2,t}\}_{t=0}^{T}, \{y_t\}_{t=0}^{T-1})$$

satisfies (4.266), and an integer $\tau \in \{0, \ldots, T\}$. It follows from (4.262) and (4.266) that there exists an integer τ_1 such that

$$\tau_1 \in \{0, \ldots, \tau\}, \tag{4.271}$$

$$x_{1,\tau_1} \geq \delta_0 \text{ and } x_{1,t} < \delta_0 \text{ for all integers } t \text{ satisfying } \tau_1 < t \leq \tau \tag{4.272}$$

(it may happen that $\tau_1 = \tau$). In order to complete the proof of the lemma it is sufficient to show that $\tau_1 \geq \tau - L$.

Assume the contrary. Then

$$\tau_1 < \tau - L. \tag{4.273}$$

Define a program $(\{\bar{x}_{1,t}\}_{t=0}^{T}, \{\bar{x}_{2,t}\}_{t=0}^{T}, \{\bar{y}_t\}_{t=0}^{T-1})$. Set

$$\bar{x}_{1,t} = x_{1,t}, \ \bar{x}_{2,t} = x_{2,t}, \ t = 0, \ldots, \tau_1, \tag{4.274}$$

$$\text{if } \tau_1 \geq 1 \text{ put } \bar{y}_t = y_t, \ t = 0, \ldots, \tau_1 - 1 \tag{4.275}$$

and for $t = \tau_1, \ldots, \tau_1 + L_0 - 1$ put

$$\bar{x}_{1,t+1} = v_1 \bar{x}_{1,t} + f_1(\bar{x}_{1,t}), \ \bar{x}_{2,t+1} = v_2 \bar{x}_{2,t}, \ \bar{y}_t = 0. \tag{4.276}$$

It follows from (4.266), (4.274), (4.248), (4.272), (4.277) and the monotonicity of the functions f_1 and g that

$$\bar{x}_{1,\tau_1+L_0} = g^{L_0}(\bar{x}_{1,\tau_1}) \geq g^{L_0}(\delta_0) > x^*/2. \tag{4.277}$$

For all integers $t = \tau_1 + L_0, \ldots, \tau - 1$ set

$$\bar{x}_{1,t+1} = v_1 \bar{x}_{1,t} + f_1(\bar{x}_{1,t}) - \Lambda, \ \bar{x}_{2,t+1} = v_2 \bar{x}_{2,t} + \Lambda, \ \bar{y}_t = f(\bar{x}_{2,t}). \tag{4.278}$$

In view of (4.277), (4.278), (4.261), (4.274)–(4.276), and the monotonicity of the function f_1,

$$\bar{x}_{1,t} \geq x^*/2, \ t = \tau_1 + L_0, \ldots, \tau \tag{4.279}$$

and $(\{\bar{x}_{1,t}\}_{t=0}^{\tau}, \{\bar{x}_{2,t}\}_{t=0}^{\tau}, \{\bar{y}_t\}_{t=0}^{\tau-1}))$ is a program. It follows from (4.266), (4.259), and Lemma 4.27 that

$$x_{1,\tau_1+1} \leq M_0, \ x_{2,\tau_1+1} \leq M_0(1 - v_2)^{-1}. \tag{4.280}$$

Relations (4.272), (4.273), and (4.242)–(4.244) imply that for all integers t satisfying $\tau_1 < t \le \tau - 1$, we have

$$x_{2,t+1} \le v_2 x_{2,t} + f_1(x_{1,t}) \le v_2 x_{2,t} + f_1(\delta_0).$$

When combined with (4.280) this inequality implies that for all integers $t \in [\tau_1 + 1, \tau]$,

$$x_{2,t} \le v_2^{t-\tau_1-1} M_0 (1 - v_2)^{-1} + f_1(\delta_0) \sum_{j=0}^{\infty} v_2^j$$

$$\le v_2^{t-\tau_1-1} M_0 (1 - v_2)^{-1} + f_1(\delta_0)(1 - v_2)^{-1}. \tag{4.281}$$

It follows from (4.281), (4.268), and (4.262) that for all integers $t \in [\tau_1 + L_0, \tau]$,

$$x_{2,t} \le v_2^{L_0-1} M_0 (1 - v_2)^{-1} + f_1(\delta_0)(1 - v_2)^{-1} < \delta_0 + f_1(\delta_0)(1 - v_2)^{-1} < \Lambda/4. \tag{4.282}$$

In view of (4.282) and (4.278), we have

$$x_{2,\tau} < \bar{x}_{2,\tau}. \tag{4.283}$$

Relations (4.272), (4.273) (4.262), (4.259), and (4.279) imply that

$$x_{1,\tau} < \delta_0 < m_0/4 < x^*/8 < \bar{x}_{1,\tau}. \tag{4.284}$$

For all integers t satisfying $\tau \le t < T$ set

$$\bar{x}_{1,t+1} = v_1 \bar{x}_{1,t} + x_{1,t+1} - v_1 x_{1,t}, \quad \bar{x}_{2,t+1} = v_2 \bar{x}_{2,t} + x_{2,t+1} - v_2 x_{2,t}, \quad \bar{y}_t = y_t. \tag{4.285}$$

In view of (4.284), (4.283), (4.285), for all integers $t = \tau, \ldots, T$,

$$\bar{x}_{1,t} \ge x_{1,t}, \quad \bar{x}_{2,t} \ge x_{2,t}$$

and $(\{\bar{x}_{1,t}\}_{t=0}^{T}, \{\bar{x}_{2,t}\}_{t=0}^{T}, \{\bar{y}_t\}_{t=0}^{T-1})$ is a program. Relations (4.274) and (4.266) imply that

$$\bar{x}_{1,0} = x_{1,0} \in [m_0, M_0], \quad \bar{x}_{2,0} = x_{2,0} \in [0, M_0]. \tag{4.286}$$

It follows from (4.286), (4.266), (4.275), (4.285), (4.242)–(4.246), (4.273), (4.269), (4.282), (4.278) and Lemma 4.27 that

$$M_1 \ge \sum_{t=0}^{T-1} w(\bar{y}_t) - \sum_{t=0}^{T-1} w(y_t) = \sum_{t=\tau_1}^{\tau-1} w(\bar{y}_t) - \sum_{t=\tau_1}^{\tau-1} w(y_t)$$

$$\ge \sum_{t=\tau_1+L_0+1}^{\tau-1} w(f_2(\bar{x}_{2,t})) - \sum_{t=\tau_1}^{\tau_1+L_0} w(f_2(x_{2,t})) - \sum_{t=\tau_1+L_0+1}^{\tau-1} w(f_2(x_{2,t}))$$

$$\geq (\tau - \tau_1 - L_0 - 2)w(f_2(\Lambda)) - (L_0 + 1)w(f_2(M_0(1 - v_2)^{-1}))$$
$$- (\tau - \tau_1 - L_0 - 1)w(f_2(\Lambda/4))$$
$$= (\tau - \tau_1 - L_0 - 2)[w(f_2(\Lambda)) - w(f_2(\Lambda/4))] - (L_0 + 2)w(f_2(M_0(1 - v_2)^{-1}))$$
$$> (L - L_0 - 1)[w(f_2(\Lambda)) - w(f_2(\Lambda/4))] - (L_0 + 2)w(f_2(M_0(1 - v_2)^{-1}))$$

and

$$(L - L_0 - 1)[w(f_2(\Lambda)) - w(f_2(\Lambda/4))] < M_1 + (L_0 + 2)w(f_2(M_0(1 - v_2)^{-1})).$$

This contradicts (4.270). The contradiction we have reached proves that $\tau_1 \geq \tau - L$ and Lemma 4.28 itself. $\qquad\square$

Lemma 4.29. *Let* $M_1 > 0$ *and let a natural number* L *be as guaranteed by Lemma 4.28. Then there exists a real number* $\delta \in (0, 1)$ *such that for each integer* $T \geq L$ *and each program* $(\{x_{1,t}\}_{t=0}^{T}, \{x_{2,t}\}_{t=0}^{T}, \{y_t\}_{t=0}^{T-1})$ *which satisfies*

$$m_0 \leq x_{1,0} \leq M_0, \ 0 \leq x_{2,0} \leq M_0, \ \sum_{t=0}^{T-1} w(y_t) \geq U(x_{1,0}, x_{2,0}, 0, T) - M_1$$

$$(4.287)$$

the following inequality holds:

$$x_{1,t} \geq \delta, \ t = 0, \ldots, T - L. \tag{4.288}$$

Proof. Choose a real $\delta \in (0, \delta_0)$ such that $\delta < 1$ and

$$g^i(\delta) < \delta_0, \ i = 1, \ldots, L. \tag{4.289}$$

Assume that a natural number $T \geq L$ and that a program $(\{x_{1,t}\}_{t=0}^{T}, \{x_{2,t}\}_{t=0}^{T}, \{y_t\}_{t=0}^{T-1})$ satisfies (4.287). We claim that (4.288) holds. Assume the contrary. Then there exists an integer t_1 such that

$$t_1 \in \{0, \ldots, T - L\}, \tag{4.290}$$

$$x_{1,t_1} < \delta. \tag{4.291}$$

In view of the choice of L, Lemma 4.28, (4.290), and (4.287), there exists an integer τ_0 such that

$$\tau_0 \leq t_1 + L, \ \tau_0 \geq t_1, \ x_{1,\tau_0} \geq \delta_0. \tag{4.292}$$

Relations (4.292) and (4.291) imply that

$$\tau_0 > t_1. \tag{4.293}$$

It follows from (4.242) and (4.244), (4.248), the monotonicity of the function g, (4.293), (4.292), (4.289), and (4.291) that

$$x_{1,\tau_0} \leq g^{\tau_0 - t_1}(\delta) < \delta_0.$$

This inequality contradicts (4.292). The contradiction we have reached proves (4.288). Lemma 4.29 is proved.   □

Lemma 4.30. *Let M_1 be a positive number. Then there exists a pair of natural numbers L_1, L_2 such that for each integer $T \geq L_1 + L_2$, each program $(\{x_{1,t}\}_{t=0}^T, \{x_{2,t}\}_{t=0}^T, \{y_t\}_{t=0}^{T-1})$ which satisfies*

$$x_{1,0} \in [m_0, M_0], \quad x_{2,0} \in [0, M_0], \tag{4.294}$$

$$\sum_{t=0}^{T-1} w(y_t) \geq U(x_{1,0}, x_{2,0}, 0, T) - M_1,$$

and each integer $\tau \in \{L_1, \ldots, T - L_2\}$, the inequality

$$\min\{x_{1,t} : t = \tau, \ldots, \tau + L_2\} \leq x^* - \gamma_* \tag{4.295}$$

holds.

Proof. In view (4.249) there exists a natural number $L_1 \geq 4$ such that for each integer $i \geq L_1$, we have

$$g^i(M_0) < x^* + \gamma_*/8, \tag{4.296}$$

$$g^i(x^*/4) \geq x^* - \gamma_*/8. \tag{4.297}$$

It follows from (4.247), monotonicity of the function f_1, (4.239) and (4.249) that there exists a natural number $\tilde{L} \geq 4$ such that

$$v_2^{\tilde{L}} M_0 < f_1(x^* + \gamma_*/8) - (1 - v_1)x^* + 2\gamma_*, \tag{4.298}$$

$$g^{\tilde{L}}(x^*/4) > x^* - \gamma_*/4. \tag{4.299}$$

Choose an integer $L_2 \geq 1$ such that

$$L_2 > 4 + 8L_1 + 8\tilde{L}, \tag{4.300}$$

$$2^{-1} L_2(w(f_2(\Lambda)) - w(f_2(\Lambda/2))) > (\tilde{L} + 1)w(f_2(M_0(1 - v_2)^{-1}))$$
$$+ M_1 + 4 + c_0^2 M_0(1 - v_2)^{-2}. \tag{4.301}$$

Assume that an integer $T \geq L_1 + L_2$, a program $(\{x_{1,t}\}_{t=0}^T, \{x_{2,t}\}_{t=0}^T, \{y_t\}_{t=0}^{T-1})$ satisfies (4.294), and an integer $\tau \in \{L_1, \ldots, T - L_2\}$. In order to complete the proof of the lemma it is sufficient to show that inequality (4.295) holds. Assume the contrary. Then

$$x_{1,t} > x^* - \gamma_*, \quad t = \tau, \ldots, \tau + L_2. \tag{4.302}$$

In view of (4.294), (4.248), the choice of L_1 (see (4.296), (4.297)), (4.242)–(4.244), and monotonicity of the functions f_1, g,

$$x_{1,t} \leq x^* + \gamma_*/8, \ t = L_1, \ldots, T. \tag{4.303}$$

There are two cases:

(1)

$$x_{1,t} > x^* - \gamma_*, \ t = \tau, \ldots, T; \tag{4.304}$$

(2) there is an integer $\tilde{\tau}$ such that

$$T \geq \tilde{\tau} > \tau + L_2, \ x_{1,\tilde{\tau}} \leq x^* - \gamma_*, \tag{4.305}$$

$$x_{1,t} > x^* - \gamma_* \text{ for all integers } t \text{ satisfying } \tau \leq t < \tilde{\tau}. \tag{4.306}$$

In the case (1) set

$$\tilde{\tau} = T + 1. \tag{4.307}$$

It follows from (4.242)–(4.244), (4.303)–(4.307) and monotonicity of the function f_1 that for all integers t satisfying $\tau \leq t < \tilde{\tau} - 1$, we have

$$x_{2,t+1} - v_2 x_{2,t} \leq f_1(x_{1,t}) - [x_{1,t+1} - v_1 x_{1,t}] \leq f_1(x^* + \gamma_*/8)$$
$$+ v_1(x^* + \gamma_*/8) - (x^* - \gamma_*) \leq f_1(x^* + \gamma_*/8) - (1 - v_1)x^* + 2\gamma_*. \tag{4.308}$$

Lemma 4.27 and (4.294) imply that

$$x_{2,\tau} \leq M_0(1 - v_2)^{-1}. \tag{4.309}$$

In view of (4.309) and (4.308), for all integers t satisfying $\tau \leq t \leq \tilde{\tau} - 1$,

$$x_{2,t} \leq v_2^{t-\tau} M_0(1-v_2)^{-1} + (f_1(x^* + \gamma_*/8) - (1-v_1)x^* + 2\gamma_*)(1-v_2)^{-1}. \tag{4.310}$$

It follows from (4.310) and (4.298) that for all integers t satisfying $\tau + \tilde{L} \leq t \leq \tilde{\tau} - 1$,

$$x_{2,t} \leq v_2^{\tilde{L}} M_0(1 - v_2)^{-1} + (1 - v_2)^{-1}(f_1(x^* + \gamma_*/8) - (1 - v_1)x^* + 2\gamma_*)$$
$$< 2(1 - v_2)^{-1}(f_1(x^* + \gamma_*/8) - (1 - v_2)x^* + 2\gamma_*). \tag{4.311}$$

For all integers $t = 0, \ldots, \tau + \tilde{L}$ put

$$\tilde{x}_{1,t} = x_{1,t}, \ \tilde{x}_{2,t} = x_{2,t}, \tag{4.312}$$

$$\tilde{y}_t = y_t, \ t = 0, \ldots, \tau + \tilde{L} - 1. \tag{4.313}$$

For all integers $t = \tau + \tilde{L}, \ldots, \tilde{\tau} - \tilde{L} - 1$ put

$$\tilde{x}_{1,t+1} = v_1 \tilde{x}_{1,t} + f_1(\tilde{x}_{1,t}) - \Lambda, \ \tilde{x}_{2,t+1} = v_2 \tilde{x}_{2,t} + \Lambda, \ \tilde{y}_t = f_2(\tilde{x}_{2,t}). \quad (4.314)$$

It follows from (4.312), (3.314), (4.302), (4.300), (4.263), and (4.261) that

$$\tilde{x}_{1,\tau+\tilde{L}} \geq x^*/2, \ \tilde{x}_{1,t} \geq x^*/2, \ t = \tau + \tilde{L}, \ldots, \tilde{\tau} - \tilde{L}, \quad (4.315)$$

$$\tilde{y}_t \geq f_2(\Lambda), \ t = \tau + \tilde{L}, \ldots, \tilde{\tau} - \tilde{L} - 1 \quad (4.316)$$

and $(\{\tilde{x}_{1,t}\}_{t=0}^{\tilde{\tau}-\tilde{L}}, \{\tilde{x}_{2,t}\}_{t=0}^{\tilde{\tau}-\tilde{L}}, \{\tilde{y}_t\}_{t=0}^{\tilde{\tau}-\tilde{L}-1})$ is a program. For all $t = \tilde{\tau} - \tilde{L}, \ldots, \tilde{\tau} - 1$, put

$$\tilde{x}_{1,t+1} = v_1 \tilde{x}_{1,t} + f_1(\tilde{x}_{1,t}), \ \tilde{x}_{2,t+1} = v_2 \tilde{x}_{2,t}, \ \tilde{y}_t = f_2(\tilde{x}_{2,t}). \quad (4.317)$$

It is clear that

$$(\{\tilde{x}_{1,t}\}_{t=0}^{\tilde{\tau}}, \{\tilde{x}_{2,t}\}_{t=0}^{\tilde{\tau}}, \{\tilde{y}_t\}_{t=0}^{\tilde{\tau}-1})$$

is a program. In view of (4.317), (4.315), (4.299), (4.248) and monotonicity of the function g, we have

$$\tilde{x}_{1,\tilde{\tau}} = g^{\tilde{L}}(\tilde{x}_{1,\tilde{\tau}-\tilde{L}}) \geq g^{\tilde{L}}(x^*/2) > x^* - \gamma_*/4. \quad (4.318)$$

Put

$$\tau_* = \min\{\tilde{\tau}, T\}. \quad (4.319)$$

Relations (4.319), (4.307), (4.305), (4.312), and (4.316) imply that

$$\sum_{t=0}^{\tau_*-1} w(\tilde{y}_t) - \sum_{t=0}^{\tau_*-1} w(y_t) = \sum_{t=\tau+\tilde{L}}^{\tau_*-1} w(\tilde{y}_t) - \sum_{t=\tau+\tilde{L}}^{\tau_*-1} w(y_t)$$

$$\geq \sum_{t=\tau+\tilde{L}}^{\tilde{\tau}-\tilde{L}} w(f_2(\Lambda)) - \sum_{t=\tau+\tilde{L}}^{\tau_*-1} w(y_t)$$

$$\geq (\tilde{\tau} - \tau - 2\tilde{L})w(f_2(\Lambda)) - \sum_{t=\tau+\tilde{L}}^{\tau_*-1} w(y_t). \quad (4.320)$$

It follows from (4.319), (4.307), (4.305), (4.300), Lemma 4.27, (4.294), (4.311), (4.242)–(4.244), (4.263), and monotonicity of the functions f_2, w that

$$\sum_{t=\tau+\tilde{L}}^{\tau_*-1} w(y_t) \leq \sum_{t=\tau+\tilde{L}}^{\tau-2} w(y_t) + w(f_2(M_0(1-v_2)^{-1}))$$

$$\leq (\tilde{\tau} - \tau - \tilde{L})w(f_2(2(1-v_2)^{-1}(f_1(x^* + \gamma_*/8)$$

$$- (1-v_1)x^* + 2\gamma_*))) + w(f_2(M_0(1-v_2)^{-1}))$$

$$\leq (\tilde{\tau} - \tau - \tilde{L})w(f_2(\Lambda/2)) + w(f_2(M_0(1-v_2)^{-1})). \quad (4.321)$$

In view of (4.320) and (4.321), we have

$$\sum_{t=0}^{\tau_*-1} w(\tilde{y}_t) - \sum_{t=0}^{\tau_*-1} w(y_t) \geq (\tilde{\tau} - \tau - 2\tilde{L})(w(f_2(\Lambda)) - w(f_2(\Lambda/2)))$$

$$- \tilde{L}w(f_2(\Lambda/2)) - w(f_2(M_0(1 - v_2)^{-1})). \qquad (4.322)$$

Assume that the case (1) holds. Then [see (4.307), (4.317)]

$$(\{\tilde{x}_{1,t}\}_{t=0}^{T}, \{\tilde{x}_{2,t}\}_{t=0}^{T}, \{\tilde{y}_t\}_{t=0}^{T-1})$$

is a program and by (4.294), (4.319), (4.307), and (4.322), we have

$$M_1 + \sum_{t=0}^{T-1} w(\tilde{y}_t) - U(x_{1,0}, x_{2,0}, 0, T) \geq \sum_{t=0}^{T-1} w(\tilde{y}_t) - \sum_{t=0}^{T-1} w(y_t)$$

$$\geq (\tilde{\tau} - \tau - 2\tilde{L})(w(f_2(\Lambda)) - w(f_2(\Lambda/2)))$$

$$- \tilde{L}w(f_2(\Lambda/2)) - w(f_2(M_0(1 - v_2)^{-1}))$$

$$\geq (L_2 - 2\tilde{L})(w(f_2(\Lambda)) - w(f_2(\Lambda/2)))$$

$$- \tilde{L}w(f_2(\Lambda/2)) - w(f_2(M_0(1 - v_2)^{-1}))$$

and

$$\sum_{t=0}^{T-1} w(\tilde{y}_t) - U(x_{1,0}, x_{2,0}, 0, T)$$

$$\geq (L_2 - 2\tilde{L})(w(f_2(\Lambda)) - w(f_2(\Lambda/2)))$$

$$- \tilde{L}w(f_2(\Lambda/2)) - w(f_2(M_0(1 - v_2)^{-1})) - M_1 > 4.$$

This relation contradicts (4.312). The contradiction we have reached proves that the case (1) does not holds. Hence the case (2) holds. It follows from (4.320), (4.319), (4.305) and (4.322) that $\tau_* = \tilde{\tau}$ and

$$\sum_{t=0}^{\tilde{\tau}-1} w(\tilde{y}_t) - \sum_{t=0}^{\tilde{\tau}-1} w(y_t) \geq (L_2 - 2\tilde{L})w(f_2(\Lambda))$$

$$- w(f_2(\Lambda/2)) - \tilde{L}w(f_2(\Lambda/2)) - w(f_2(M_0(1 - v_2)^{-1})). \qquad (4.323)$$

Relations (4.318) and (4.305) imply that

$$\tilde{x}_{1,\tilde{\tau}} > x_{1,\tilde{\tau}}. \qquad (4.324)$$

For all integers t satisfying $\tilde{\tau} \le t \le T - 1$ put

$$\tilde{x}_{1,t+1} = v_1\tilde{x}_{1,t} + x_{1,t+1} - v_1 x_{1,t}, \tag{4.325}$$

$$\tilde{x}_{2,t+1} = v_2\tilde{x}_{2,t} + x_{2,t+1} - v_2 x_{2,t}, \quad \tilde{y}_t = f_2(\tilde{x}_{2,t}).$$

In view of (4.324) and (4.325),

$$\tilde{x}_{1,t} \ge x_{1,t}, \quad t = \tilde{\tau}, \dots, T \tag{4.326}$$

and $(\{\tilde{x}_{1,t}\}_{t=0}^{T}, \{\tilde{x}_{2,t}\}_{t=0}^{T}, \{\tilde{y}_t\}_{t=0}^{T-1})$ is a program. In view of (4.325), for all integers t satisfying $\tilde{\tau} \le t \le T - 1$, we have

$$x_{2,t} - \tilde{x}_{2,t} \le v_2^{t-\tilde{\tau}}(x_{2,\tilde{\tau}} - \tilde{x}_{2,\tilde{\tau}}). \tag{4.327}$$

Lemma 4.27, (4.294) and (4.312) imply that

$$x_{2,t}, \tilde{x}_{2,t}, \le M_0(1 - v_2)^{-1}, \quad t = 0, \dots, T, \tag{4.328}$$

$$y_t, \tilde{y}_t \le f_2(M_0(1 - v_2)^{-1}), \quad t = 0, \dots, T - 1. \tag{4.329}$$

Assume that an integer t satisfies

$$\tilde{\tau} \le t \le T - 1, \quad y_t > \tilde{y}_t. \tag{4.330}$$

Relations (4.242)–(4.244) and (4.325) imply that

$$f_2(x_{2,t}) \ge y_t > \tilde{y}_t = f_2(\tilde{x}_{2,t}) \tag{4.331}$$

and combined with the strict monotonicity of the function f_2 this implies that

$$x_{2,t} > \tilde{x}_{2,t}. \tag{4.332}$$

It follows from (4.330), (4.242)–(4.244), (4.332), (4.328), (4.327) and the choice of c_0 [see (4.264)] that

$$0 < y_t - \tilde{y}_t \le f_2(x_{2,t}) - f_2(\tilde{x}_{2,t}) \le c_0(x_{2,t} - \tilde{x}_{2,t})$$

$$\le c_0 v_2^{t-\tilde{\tau}}(x_{2,\tilde{\tau}} - \tilde{x}_{2,\tilde{\tau}}) \le c_0 v_2^{t-\tilde{\tau}} M_0(1 - v_2)^{-1}. \tag{4.333}$$

In view of (4.330), monotonicity of the function w, (4.265), (4.329), and (4.333), we have

$$0 < w(y_t) - w(\tilde{y}_t) \le c_0(y_t - \tilde{y}_t) \le c_0^2 v_2^{t-\tilde{\tau}} M_0(1 - v_2)^{-1}. \tag{4.334}$$

Relations (4.334) and (4.330) imply that

$$\sum \{w(y_t) - w(\tilde{y}_t) : \text{ an integer } t \text{ satisfies } \tilde{\tau} \leq t \leq T - 1\} \leq \sum \{w(y_t) - w(\tilde{y}_t) :$$

$$\text{an integer } t \text{ satisfies } \tilde{\tau} \leq t \leq T - 1, \ y_t > \tilde{y}_t\} \leq \sum \{c_0^2 v_2^{t-\tilde{\tau}} M_0 (1 - v_2)^{-1} :$$

$$\text{an integer } t \text{ satisfies } \tilde{\tau} \leq t \leq T - 1, \ y_t > \tilde{y}_t\} \leq c_0^2 M_0 (1 - v_2)^{-2}. \tag{4.335}$$

It follows from (4.294), (4.323), (4.335), (4.300), and (4.261) that

$$M_1 + \sum_{t=0}^{T-1} w(\tilde{y}_t) - U(x_{1,0}, x_{2,0}, 0, T) \geq \sum_{t=0}^{T-1} w(\tilde{y}_t) - \sum_{t=0}^{T-1} w(y_t)$$

$$= \sum_{t=0}^{\tilde{\tau}-1} w(\tilde{y}_t) - \sum_{t=0}^{\tilde{\tau}-1} w(y_t)$$

$$- \sum \{w(y_t) - w(\tilde{y}_t) : \text{ an integer } t \text{ satisfies } \tilde{\tau} \leq t \leq T - 1\}$$

$$\geq (L_2 - 2\tilde{L})(w(f_2(\Lambda)) - w(f_2(\Lambda/2)))$$

$$- \tilde{L}w(f_2(\Lambda/2)) - w(f_2(M_0(1 - v_2)^{-1})) - c_0^2 M_0 (1 - v_2)^{-2}$$

and

$$\sum_{t=0}^{T-1} w(\tilde{y}_t) - U(x_{1,0}, x_{2,0}, 0, T)$$

$$\geq 2^{-1} L_2 (w(f_2(\Lambda)) - w(f_2(\Lambda/2)))$$

$$- (\tilde{L} + 1)(f_2(M_0(1 - v_2)^{-1})) - M_1 - c_0^2 M_0 (1 - v_2)^{-2} > 4.$$

This relation contradicts (4.312). The contradiction we have reached proves (4.295) and Lemma 4.30 itself. □

Lemma 4.31. *Let M_1 be a positive number. Then there exists a pair of natural numbers $\tilde{L}_1, \tilde{L}_2$ and a positive number M_2 such that for each integer $T \geq \tilde{L}_1 + \tilde{L}_2$, each program*

$$(\{x_{1,t}\}_{t=0}^{T}, \{x_{2,t}\}_{t=0}^{T}, \{y_t\}_{t=0}^{T-1})$$

which satisfies

$$x_{1,0} \in [m_0, M_0], \ x_{2,0} \in [0, M_0], \ \sum_{t=0}^{T-1} w(y_t) \geq U(x_{1,0}, x_{2,0}, 0, T) - M_1 \tag{4.336}$$

and each pair of integers T_1, T_2 satisfying

$$0 \leq T_1 < T_2 \leq T - \tilde{L}_2, \ T_2 - T_1 \geq \tilde{L}_1, \tag{4.337}$$

the inequality

$$\sum_{t=T_1}^{T_2-1} w(y_t) \geq U(x_{1,T_1}, x_{2,T_1}, T_1, T_2) - M_2 \qquad (4.338)$$

holds.

Proof. Lemma 4.29 implies that there exist an integer $L_1 \geq 1$ and a real number $\delta_1 \in (0, 1)$ such that for each integer $T \geq L_1$ and each program $(\{x_{1,t}\}_{t=0}^T, \{x_{2,t}\}_{t=0}^T, \{y_t\}_{t=0}^{T-1})$ which satisfies (4.336), we have

$$x_{1,t} \geq \delta_1, \ t = 0, \dots, T - L_1. \qquad (4.339)$$

We may assume without loss of generality that

$$\delta_1 < m_0, \ \delta_1 < \delta_0. \qquad (4.340)$$

It follows from Lemma 4.29 (applied with $m_0 = \delta_1$) that there exist an integer $L_2 \geq 1$ and a real number $\delta_2 \in (0, \delta_1)$ such that for each integer $T \geq L_2$ and each program $(\{x_{1,t}\}_{t=0}^T, \{x_{2,t}\}_{t=0}^T, \{y_t\}_{t=0}^{T-1})$ which satisfies

$$\delta_1 \leq x_{1,0} \leq M_0, \ 0 \leq x_{2,0} \leq M_0(1 - v_2)^{-1}, \qquad (4.341)$$

$$\sum_{t=0}^{T-1} w(y_t) \geq U(x_{1,0}, x_{2,0}, 0, T) - M_1 - 1, \qquad (4.342)$$

the inequality

$$x_{1,t} \geq \delta_2, \ t = 0, \dots, T - L_2 \qquad (4.343)$$

holds.

   In view of Lemma 4.30 there exists a pair of natural numbers L_3, L_4 such that for each integer $T \geq L_3 + L_4$, each program $(\{x_{1,t}\}_{t=0}^T, \{x_{2,t}\}_{t=0}^T, \{y_t\}_{t=0}^{T-1})$ which satisfies

$$m_0 \leq x_{1,0} \leq M_0, \ 0 \leq x_{2,0} \leq M_0, \ \sum_{t=0}^{T-1} w(y_t) \geq U(x_{1,0}, x_{2,0}, 0, T) - M_1,$$

and each integer $\tau \in \{L_3, \dots, T - L_4\}$, we have

$$\min\{x_{1,t} : t \in \{\tau, \dots, \tau + L_4\}\} \leq x^* - \gamma_*. \qquad (4.344)$$

Relations (4.263) and (4.249) imply that there exists an integer $L_5 \geq 1$ such that

$$g^t(\delta_2) > x^* - \gamma_*/8 \text{ for all integers } t \geq L_5. \qquad (4.345)$$

Choose a pair of natural numbers $\tilde{L}_1, \tilde{L}_2$ and a real number M_2 such that

$$\tilde{L}_1 > \sum_{i=1}^{5} L_i + 4, \ \tilde{L}_2 > 1 + L_1 + L_2 + 2L_4 + 4, \quad (4.346)$$

$$M_2 \geq M_1 + w(f_2(M_0(1 - v_2)^{-2}))(L_2 + L_5 + 2L_4) + c_0 M_0(1 - v_2)^{-2}. \quad (4.347)$$

Assume that an integer $T \geq \tilde{L}_1 + \tilde{L}_2$, a program

$$(\{x_{1,t}\}_{t=0}^{T}, \{x_{2,t}\}_{t=0}^{T}, \{y_t\}_{t=0}^{T-1})$$

satisfies (4.336), and a pair of integers T_1, T_2 satisfies (4.337). In order to prove the lemma it is sufficient to show that (4.338) holds.

Proposition 4.22 implies that there exists a program

$$(\{\tilde{x}_{1,t}\}_{t=T_1}^{T_2}, \{\tilde{x}_{2,t}\}_{t=T_1}^{T_2}, \{\tilde{y}_t\}_{t=T_1}^{T_2-1})$$

such that

$$\tilde{x}_{1,T_1} = x_{1,T_1}, \ \tilde{x}_{2,T_1} = x_{2,T_1}, \ \sum_{t=T_1}^{T_2-1} w(\tilde{y}_t) = U(x_{1,T_1}, x_{2,T_1}, T_1, T_2). \quad (4.348)$$

In view of the choice of L_1, δ_1 [see (4.339)], (4.336), and (4.346), we have

$$x_{1,t} \geq \delta_1, \ t = 0, \ldots, T - L_1. \quad (4.349)$$

It follows from the choice of L_2 and δ_2 [see (4.341)–(4.343)], (4.337), (4.346), (4.336), (4.348), Lemma 4.27, and (4.349) that

$$\tilde{x}_{1,t} \geq \delta_2, \ t = T_1, \ldots, T_2 - L_2. \quad (4.350)$$

Relations (4.337) and (4.346) imply that

$$T_2 + 2L_4 \leq T - \tilde{L}_2 + 2L_4 \leq T, \ L_3 < \tilde{L}_1 \leq T_2 < T_2 + L_4 \leq T - L_4. \quad (4.351)$$

In view of the choice of L_3, L_4 (see (4.344) with $\tau = T_2 + L_4$), (4.346), (4.336), and (4.351), there exists an integer t_0 such that

$$t_0 \in \{T_2 + L_4, \ldots, T_2 + 2L_4\}, \ x_{1,t_0} \leq x^* - \gamma_*. \quad (4.352)$$

Define a program $(\{\bar{x}_{1,t}\}_{t=0}^{T}, \{\bar{x}_{2,t}\}_{t=0}^{T}, \{\bar{y}_t\}_{t=0}^{T-1})$ as follows. Put

$$\bar{x}_{1,t} = x_{1,t}, \ \bar{x}_{2,t} = x_{2,t}, \ t = 0, \ldots, T_1, \ \bar{y}_t = y_t, \ t = 0, \ldots, T_1 - 1 \text{ if } T_1 > 0, \tag{4.353}$$

$$\bar{x}_{1,t} = \tilde{x}_{1,t}, \ \bar{x}_{2,t} = \tilde{x}_{2,t}, \ t = T_1 + 1, \ldots, T_2 - L_2 - L_5,$$
$$\bar{y}_t = \tilde{y}_t, \ t = T_1, \ldots, T_2 - L_2 - L_5 - 1.$$

In view of (4.353), (4.348), (4.337), and (4.346), the sequence

$$\left(\{\bar{x}_{1,t}\}_{t=0}^{T_2-L_2-L_5}, \{\bar{x}_{2,t}\}_{t=0}^{T_2-L_2-L_5}, \{\bar{y}_t\}_{t=0}^{T_2-L_2-L_5-1} \right)$$

is a program. For all integers $t = T_2 - L_2 - L_5, \ldots, t_0 - 1$ put

$$\bar{x}_{1,t+1} = v_1 \bar{x}_{1,t} + f_1(\bar{x}_{1,t}), \ \bar{x}_{2,t+1} = v_2 \bar{x}_{2,t}, \ \bar{y}_t = f_2(\bar{x}_{2,t}). \tag{4.354}$$

It is clear that $(\{\bar{x}_{1,t}\}_{t=0}^{t_0}, \{\bar{x}_{2,t}\}_{t=0}^{t_0}, \{\bar{y}_t\}_{t=0}^{t_0-1})$ is a program. Relations (4.353), (4.350), (4.337) and (4.346) imply that

$$\bar{x}_{1,T_2-L_2-L_5} = \tilde{x}_{1,T_2-L_2-L_5} \geq \delta_2. \tag{4.355}$$

It follows from (4.354), (4.248), (4.355), monotonicity of the function g, (4.352) and (4.345) that

$$\bar{x}_{1,t_0} = g^{t_0-(T_2-L_2-L_5)}(\bar{x}_{1,T_2-L_2-L_5})$$
$$\geq g^{t_0-(T_2-L_2-L_5)}(\delta_2) \geq x^* - \gamma_*/8 > x_{1,t_0}. \tag{4.356}$$

For all integers $t = t_0, \ldots, T - 1$ put

$$\bar{x}_{1,t+1} = v_1 \bar{x}_{1,t} + x_{1,t+1} - v_1 x_{1,t},$$
$$\bar{x}_{2,t+1} = v_2 \bar{x}_{2,t} + x_{2,t+1} - v_2 x_{1,t}, \ \bar{y}_t = f_2(\bar{x}_{2,t}). \tag{4.357}$$

By (4.356) and (4.357),

$$\bar{x}_{1,t} \geq x_{1,t} \text{ for all } t = t_0, \ldots, T \tag{4.358}$$

and $(\{\bar{x}_{1,t}\}_{t=0}^{T}, \{\bar{x}_{2,t}\}_{t=0}^{T}, \{\bar{y}_t\}_{t=0}^{T-1})$ is a program. Relations (4.353), (4.336), (4.352), (4.337), (4.346) imply that

$$M_1 \geq \sum_{t=0}^{T-1} w(\bar{y}_t) - \sum_{t=0}^{T-1} w(y_t) = \sum_{t=T_1}^{T-1} w(\bar{y}_t) - \sum_{t=T_1}^{T-1} w(y_t)$$

$$= \sum_{t=T_1}^{t_0-1} w(\bar{y}_t) - \sum_{t=T_1}^{t_0-1} w(y_t) + \sum_{t=t_0}^{T-1} w(\bar{y}_t) - \sum_{t=t_0}^{T-1} w(y_t)$$

$$\geq \sum_{t=T_1}^{T_2-L_2-L_5-1} w(\bar{y}_t) - \sum_{t=T_1}^{t_0-1} w(y_t) + \sum_{t=t_0}^{T-1} w(\bar{y}_t) - \sum_{t=t_0}^{T-1} w(y_t). \tag{4.359}$$

In view of (4.348), (4.352), (4.336), and Lemma 4.27, we have

$$\sum_{t=T_1}^{T_2-L_2-L_5-1} w(\tilde{y}_t) - \sum_{t=T_1}^{t_0-1} w(y_t) = U(x_{1,T_1}, x_{2,T_1}, T_1, T_2) - \sum_{t=T_2-L_2-L_5}^{T_2-1} w(\tilde{y}_t)$$

$$- \sum_{t=T_1}^{T_2-1} w(y_t) - \sum_{t=T_2}^{t_0-1} w(y_t)$$

$$\geq U(x_{1,T_1}, x_{2,T_1}, T_1, T_2) - \sum_{t=T_1}^{T_2-1} w(y_t)$$

$$- \sum_{t=T_2-L_2-L_5}^{T_2-1} w(\tilde{y}_t) - 2L_4 w(f_2(M_0(1-v_2)^{-1})).$$

$$(4.360)$$

In view of (4.336), Lemma 4.27, and (4.348),

$$\tilde{x}_{1,T_1} \leq M_0, \ \tilde{x}_{2,T_1} \leq M_0(1-v_2)^{-1}, \ \tilde{x}_{2,t} \leq M_0(1-v_2)^{-2}, \ t = T_1 \ldots, T_2.$$
$$(4.361)$$

Relations (4.360), (4.361), and (4.242)–(4.244) imply that

$$\sum_{t=T_1}^{T_2-L_2-L_5-1} w(\tilde{y}_t) - \sum_{t=T_1}^{t_0-1} w(y_t) \geq U(x_{1,T_1}, x_{2,T_1}, T_1, T_2)$$

$$- \sum_{t=T_1}^{T_2-1} w(y_t) - (L_2 + L_5) w(f_2(M_0(1-v_2)^{-2})) - 2L_4 w(f_2(M_0(1-v_2)^{-1})).$$

$$(4.362)$$

It follows from (4.336), (4.353), and Lemma 4.27 that

$$x_{2,t}, \ \bar{x}_{2,t} \leq M_0(1-v_2)^{-1}, \ t = 0, \ldots, T,$$

$$y_t, \ \bar{y}_t \leq f_2(M_0(1-v_2)^{-1}), \ t = 0, \ldots, T-1. \qquad (4.363)$$

In view of (4.357), for all integers t satisfying $t_0 \leq t < T$,

$$x_{2,t} - \bar{x}_{2,t} \leq v_2^{t-t_0}(x_{2,t_0} - \bar{x}_{2,t_0}). \qquad (4.364)$$

Assume that an integer t satisfies

$$t_0 \leq t < T, \ y_t > \bar{y}_t. \qquad (4.365)$$

Relations (4.242)–(4.244), (4.365), and (3.357) imply that

$$f_2(x_{2,t}) \geq y_t > \bar{y}_t = f_2(\bar{x}_{2,t}) \qquad (4.366)$$

and together with the monotonicity of the function f_2 this inequality implies that

$$x_{2,t} > \bar{x}_{2,t}, \; x_{2,t_0} > \bar{x}_{2,t_0}. \tag{4.367}$$

In view of (4.366), (4.363), the choice of c_0 (see (4.264)), (4.361), (4.365), and (4.364),

$$0 < y_t - \bar{y}_t \leq f_2(x_{2,t}) - f_2(\bar{x}_{2,t}) \leq c_0(x_{2,t} - \bar{x}_{2,t})$$
$$\leq c_0 v_2^{t-t_0}(x_{2,t_0} - \bar{x}_{2,t_0}) \leq c_0 v_2^{t-t_0} M_0(1 - v_2)^{-1}. \tag{4.368}$$

In view of monotonicity of the function w, (4.368), (4.363), and (4.265), we have

$$0 < w(y_t) - w(\bar{y}_t) \leq c_0(y_t - \bar{y}_t) \leq c_0^2 v_2^{t-t_0} M_0(1 - v_2)^{-1}. \tag{4.369}$$

Relations (4.365) and (4.369) imply that

$$\sum_{t=t_0}^{T-1} w(y_t) - \sum_{t=0}^{T-1} w(\bar{y}_t) \leq \sum_{t=t_0}^{T-1} c_0^2 v_2^{t-t_0} M_0(1 - v_2)^{-1} \leq c_0^2 M_0(1 - v_2)^{-2}. \tag{4.370}$$

It follows from (4.359), (4.362), (4.370), and (4.347) that

$$M_1 \geq U(x_{1,T_1}, x_{2,T_1}, T_1, T_2) - \sum_{t=T_1}^{T_2-1} w(y_t)$$
$$- (L_2 + L_5 + 2L_4)w(f_2(M_0(1 - v_2)^{-2})) - c_0^2 M_0(1 - v_2)^{-2},$$

$$\sum_{t=T_1}^{T_2-1} w(y_t) \geq U(x_{1,T_1}, x_{2,T_1}, T_1, T_2) - M_1 - (L_2 + L_5 + 2L_4)w(f_2(M_0(1 - v_2)^{-2}))$$
$$- c_0^2 M_0(1 - v_2)^{-2} \geq U(x_{1,T_1}, x_{2,T_1}, T_1, T_2) - M_2.$$

Lemma 4.31 is proved.   □

Lemma 4.32. *There exist an integer $L \geq 1$ and a positive number $\tilde{M}$ such that for each $x_{1,0}, \tilde{x}_{1,0} \in [m_0, M_0]$, $x_{2,0}, \tilde{x}_{2,0} \in [0, M_0]$ and each integer $T \geq L$,*

$$|U(x_{1,0}, x_{2,0}, 0, T) - U(\tilde{x}_{1,0}, \tilde{x}_{2,0}, 0, T)| \leq \tilde{M}.$$

Proof. Let natural numbers L_1, L_2 be as guaranteed by Lemma 4.30 with $M_1 = 1$. In view of (4.259) and (4.249), there exists an integer $L_3 \geq 1$ such that

$$g^t(m_0) > x^* - \gamma_* \text{ for all integers } t \geq L_3. \tag{4.371}$$

Fix a natural number

$$L > L_1 + L_2 + L_3 \tag{4.372}$$

and choose a number

$$\tilde{M} > Lw(f_2(M_0(1 - v_2)^{-1})) + c_0^2 M_0(1 - v_2)^{-2}. \tag{4.373}$$

Assume that an integer

$$T \geq L, \; x_{1,0}, \; \tilde{x}_{1,0} \in [m_0, M_0], \; x_{2,0}, \; \tilde{x}_{2,0} \in [0, M_0]. \tag{4.374}$$

Proposition 4.22 implies that there exists a program

$$(\{x_{1,t}\}_{t=0}^{T}, \{x_{2,t}\}_{t=0}^{T}, \{y_t\}_{t=0}^{T-1})$$

such that

$$\sum_{t=0}^{T-1} w(y_t) = U(x_{1,0}, x_{2,0}, 0, T). \tag{4.375}$$

In view of (3.372),

$$L_1 + L_3 > L_1, \; L_1 + L_3 < T - L_2. \tag{4.376}$$

Relations (4.374), (4.375), (4.372), (4.376) and Lemma 4.30 (applied with $\tau = L_1 + L_3$) imply that there exists an integer

$$t_0 \in [L_1 + L_3, L_1 + L_2 + L_3] \tag{4.377}$$

such that

$$x_{1,t_0} < x^* - \gamma_*. \tag{4.378}$$

For all integers $t = 0, \ldots, t_0 - 1$ put

$$\tilde{x}_{1,t+1} = v_1 \tilde{x}_{1,t} + f_1(\tilde{x}_{1,t}), \; \tilde{x}_{2,t+1} = v_2 \tilde{x}_{2,t}, \; \tilde{y}_t = f_2(\tilde{x}_{2,t}). \tag{4.379}$$

It is clear that $(\{\tilde{x}_{1,t}\}_{t=0}^{t_0}, \{\tilde{x}_{2,t}\}_{t=0}^{t_0}, \{\tilde{y}_t\}_{t=0}^{t_0-1})$ is a program. In view of (4.248), (4.379), monotonicity of the function g, (4.374), (4.377), (4.371), and (4.378),

$$\tilde{x}_{1,t_0} = g^{t_0}(\tilde{x}_{1,0}) \geq g^{t_0}(m_0) > x^* - \gamma_* > x_{1,t_0}. \tag{4.380}$$

For all integers $t = t_0, \ldots, T - 1$ put

$$\tilde{x}_{1,t+1} = v_1 \tilde{x}_{1,t} + x_{1,t+1} - v_1 x_{1,t}, \; \tilde{x}_{2,t+1} = v_2 \tilde{x}_{2,t} + x_{2,t+1} - v_2 x_{2,t}, \; \tilde{y}_t = f_2(\tilde{x}_{2,t}). \tag{4.381}$$

Relations (4.381) and (4.380) imply that

$$\tilde{x}_{1,t} \geq x_{1,t} \text{ for all } t = t_0, \ldots, T \tag{4.382}$$

and that $(\{\tilde{x}_{1,t}\}_{t=0}^T, \{\tilde{x}_{2,t}\}_{t=0}^T, \{\tilde{y}_t\}_{t=0}^{T-1})$ is a program. In view of (4.374) and Lemma 4.27,

$$x_{1,t}, \ \tilde{x}_{1,t} \le M_0, \ x_{2,t}, \ \tilde{x}_{2,t} \le M_0(1 - v_2)^{-1}, \ t = 0, 1, \dots, T. \qquad (4.383)$$

In view of (4.381), for all integers $t = t_0, \dots, T$, we have

$$x_{2,t} - \tilde{x}_{2,t} \le v_2^{t-t_0}(x_{2,t_0} - \tilde{x}_{2,t_0}). \qquad (4.384)$$

Assume that an integer t satisfies

$$t_0 \le t < T, \ y_t > \tilde{y}_t. \qquad (4.385)$$

It follows from (4.385), (4.242)–(4.244), (4.381), and strong monotonicity of the function f_2 that

$$f_2(x_{2,t}) \ge y_t > \tilde{y}_t = f_2(\tilde{x}_{2,t}), \ x_{2,t} > \tilde{x}_{2,t}. \qquad (4.386)$$

Relations (4.385), (4.383), (4.264), and (4.384) imply that

$$0 < y_t - \tilde{y}_t \le f_2(x_{2,t}) - f_2(\tilde{x}_{2,t}) \le c_0(x_{2,t} - \tilde{x}_{2,t}) \le c_0 v_2^{t-t_0} M_0(1 - v_2)^{-1}. \qquad (4.387)$$

In view of monotonicity of the function w, (4.387), (4.383), (4.242)–(4.244), (4.265), and (4.385),

$$0 < w(y_t) - w(\tilde{y}_t) \le c_0(y_t - \tilde{y}_t) \le v_2^{t-t_0} c_0^2 M_0(1 - v_2)^{-1}. \qquad (4.388)$$

It follows from (4.377), (4.383), (4.242)–(4.244), and (4.362) that

$$\sum_{t=0}^{T-1} w(y_t) - \sum_{t=0}^{T-1} w(\tilde{y}_t) \le (L_1 + L_2 + L_3)w(f_2(M_0(1 - v_2)^{-1}))$$

$$+ \sum_{t=t_0}^{T-1} w(y_t) - \sum_{t=t_0}^{T-1} w(\tilde{y}_t)$$

$$\le (L_1 + L_2 + L_3)w(f_2(M_0(1 - v_2)^{-1}))$$

$$+ \sum_{t=t_0}^{\infty} v_2^{t-t_0} c_0^2 M_0(1 - v_2)^{-1} < \tilde{M}. \qquad (4.389)$$

In view of (4.389) and (4.375),

$$U(\tilde{x}_{1,0}, \tilde{x}_{2,0}, 0, T) \ge \sum_{t=0}^{T-1} w(\tilde{y}_t) \ge \sum_{t=0}^{T-1} w(y_t) - \tilde{M}$$

$$= U(x_{1,0}, x_{2,0}, 0, T) - \tilde{M}.$$

This completes the proof of Lemma 4.32.   □

Lemma 4.32 implies the following result.

Lemma 4.33. *Let a natural number L and $\tilde{M} > 0$ be as guaranteed by Lemma 4.32. Then for each $x_{1,0} \in [m_0, M_0]$, $x_{2,0} \in [0, M_0]$, and each integer $T \geq L$,*

$$|U(x_{1,0}, x_{2,0}, 0, T) - \hat{U}(m_0, M_0, 0, T)| \leq \tilde{M}.$$

Lemma 4.34. *Let M_1 be a positive number. Then there exists a pair of natural numbers $\tilde{L}_1, \tilde{L}_2$ and a positive number $\tilde{M}_2$ such that for each integer $T \geq \tilde{L}_1 + \tilde{L}_2$, each program*

$$(\{x_{1,t}\}_{t=0}^T, \{x_{2,t}\}_{t=0}^T, \{y_t\}_{t=0}^{T-1})$$

which satisfies

$$x_{1,0} \in [m_0, M_0], \ x_{2,0} \in [0, M_0], \ \sum_{t=0}^{T-1} w(y_t) \geq U(x_{1,0}, x_{2,0}, 0, T) - M_1 \qquad (4.390)$$

and each pair of integers T_1, T_2 satisfying $0 \leq T_1 < T_2 \leq T - \tilde{L}_2$, $T_2 - T_1 \geq \tilde{L}_1$ the inequality

$$\sum_{t=T_1}^{T_2-1} w(y_t) \geq \hat{U}(m_0, M_0, T_1, T_2) - \tilde{M}_2$$

holds.

Proof. Lemma 4.29 implies that there exist an integer $L_1 \geq 1$ and a real number $\delta \in (0, 1)$ such that for each integer $T \geq L_1$ and each program $(\{x_{1,t}\}_{t=0}^T, \{x_{2,t}\}_{t=0}^T, \{y_t\}_{t=0}^{T-1})$ which satisfies (4.390),

$$x_{1,t} \geq \delta, \ t = 0, \ldots, T - L_1. \qquad (4.391)$$

We may assume that

$$\delta < m_0/4. \qquad (4.392)$$

Lemma 4.33 implies that there exist an integer $L_2 \geq 1$ and a positive number $\tilde{M}$ such that for each pair of real numbers $x_{1,0} \in [\delta, M_0]$,

$$x_{2,0} \in [0, M_0(1 - v_2)^{-1}]$$

and each integer $T \geq L_2$,

$$|U(x_{1,0}, x_{2,0}, 0, T) - \hat{U}(\delta, M_0, 0, T)| \leq \tilde{M}. \qquad (4.393)$$

Let natural numbers $\tilde{L}_1$, $\tilde{L}_2$ and a number $M_2 > 0$ be as guaranteed by Lemma 4.31. We may assume without loss of generality that

$$\tilde{L}_1, \tilde{L}_2 > L_1 + L_2. \tag{4.394}$$

Choose

$$\tilde{M}_2 \geq \tilde{M} + M_2. \tag{4.395}$$

Assume that an integer $T \geq \tilde{L}_1 + \tilde{L}_2$, a program

$$(\{x_{1,t}\}_{t=0}^{T}, \{x_{2,t}\}_{t=0}^{T}, \{y_t\}_{t=0}^{T-1})$$

satisfies (4.390), and a pair of integers T_1, T_2 satisfies

$$0 \leq T_1 < T_2 \leq T - \tilde{L}_2, \ T_2 - T_1 \geq \tilde{L}_1. \tag{4.396}$$

Then in view of the choice of $\tilde{L}_1$, $\tilde{L}_2$, M_2, Lemma 4.31, (4.390), and (4.396), we have

$$\sum_{t=T_1}^{T_2-1} w(y_t) \geq U(x_{1,T_1}, x_{2,T_2}, T_1, T_2) - M_2. \tag{4.397}$$

In view of the choice of δ [see (4.391)] and (4.394), (4.391) holds. By (4.391), (4.394), and (4.396),

$$x_{1,T_1} \geq \delta. \tag{4.398}$$

In view of (4.390) and Lemma 4.27,

$$x_{1,T_1} \leq M_0, \ x_{2,T_1} \leq M_0(1 - v_2)^{-1}. \tag{4.399}$$

It follows from the choice of L_2 and $\tilde{M}$, (4.398), (4.399), (4.396), and (4.394) that

$$|U(x_{1,T_1}, x_{2,T_2}, T_1, T_2) - \hat{U}(\delta, M_0, T_1, T_2)| \leq \tilde{M}. \tag{4.400}$$

In view of (4.392) and monotonicity of the functions w, f_1, f_2,

$$\hat{U}(\delta, M_0, T_1, T_2) = \hat{U}(m_0, M_0, T_1, T_2).$$

By (4.397), (4.400), (4.395) and the equality above,

$$\sum_{t=T_1}^{T_2-1} w(y_t) \geq \hat{U}(m_0, M_0, T_1, T_2) - \tilde{M} - M_2 \geq \hat{U}(m_0, M_0, T_1, T_2) - \tilde{M}_2.$$

Lemma 4.34 is proved.   □

Lemma 4.35. *For each natural number T,*

$$\hat{U}(m_0, M_0, 0, T) \geq (T-1)w(f_2(f_1(2^{-1}x^*) - (1-v_1)x^*/2)).$$

Proof. Put

$$x_{1,t} = x^*/2 \text{ for all integers } t \geq 0, \ x_{2,0} = x^*/2,$$

$$x_{2,t+1} = v_2 x_{2,t} + f_1(x^*/2) - (1-v_1)x^*/2, \ y_t = f_2(x_{2,t}) \text{ for all integers } t \geq 0.$$

It is not difficult to see that $\{x_{1,t}, x_{2,t}, y_t\}_{t=0}^{\infty}$ is a program and that for all natural numbers T, we have

$$\hat{U}(m_0, M_0, 0, T) \geq \sum_{t=0}^{T-1} w(y_t) = (T-1)w(f_2(f_1(2^{-1}x^*) - (1-v_1)x^*/2)).$$

Lemma 4.35 is proved. $\square$

4.10   Proof of Theorem 4.23

We may assume without loss of generality that

$$2m_0 < x^*. \tag{4.401}$$

Let $M_1 = 1$ and let natural numbers $\tilde{L}_1, \tilde{L}_2$ and $\tilde{M}_2 > 0$ be as guaranteed by Lemma 4.34. Let

$$x_{1,0} \in [m_0, M_0], \ x_{2,0} \in [0, M_0] \tag{4.402}$$

be given. Proposition 4.22 implies that for each natural number k there exists a program

$$(\{x_{1,t}^{(k)}\}_{t=0}^{k}, \{x_{2,t}^{(k)}\}_{t=0}^{k}, \{y_t^{(k)}\}_{t=0}^{k-1})$$

such that

$$x_{1,0}^{(k)} = x_{1,0}, \ x_{2,0}^{(k)} = x_{2,0}, \ \sum_{t=0}^{k-1} w(y_t^{(k)}) = U(x_{1,0}, x_{2,0}, 0, k). \tag{4.403}$$

By (4.403), (4.402), the choice of $\tilde{L}_1, \tilde{L}_2$ and $\tilde{M}_2$, and Lemma 4.34, the following property holds:

(P3) for each integer $k \geq \tilde{L}_1 + \tilde{L}_2$ and each pair of integers $T_1, T_2 \in [0, k - \tilde{L}_2]$ satisfying $T_2 - T_1 \geq \tilde{L}_1$, we have $\sum_{t=T_1}^{T_2-1} w(y_t^{(k)}) \geq \hat{U}(m_0, M_0, T_1, T_2) - \tilde{M}_2$.

Evidently there exists a strictly increasing sequence of natural numbers $\{k_j\}_{j=1}^{\infty}$ such that for each nonnegative integer t there exists

$$x_{1,t} = \lim_{j\to\infty} x_{1,t}^{(k_j)}, \quad x_{2,t} = \lim_{j\to\infty} x_{2,t}^{(k_j)}, \quad y_t = \lim_{j\to\infty} y_t^{(k_j)}. \tag{4.404}$$

It is clear that $\{x_{1,t}, x_{2,t}, y_t\}_{t=0}^{\infty}$ is a program. By (4.404), (4.402) and property (P3), for each pair of integers $T_1, T_2 \geq 0$ satisfying $T_2 - T_1 \geq \tilde{L}_1$,

$$\left| \sum_{t=T_1}^{T_2-1} w(y_t) - \hat{U}(m_0, M_0, T_1, T_2) \right| \leq \tilde{M}_2. \tag{4.405}$$

In order to complete the proof of Theorem 4.23 it is sufficient to show that for each natural number T,

$$\sum_{t=0}^{T-1} w(y_t) = U(x_{1,0}, x_{2,0}, x_{1,T}, x_{2,T}, 0, T). \tag{4.406}$$

Assume the contrary. Then there exists an integer $T \geq 1$ such that

$$\Delta_0 := U(x_{1,0}, x_{2,0}, x_{1,T}, x_{2,T}, 0, T) - \sum_{t=0}^{T-1} w(y_t) > 0. \tag{4.407}$$

It follows from (4.405), (4.242)–(4.244), Lemma 4.35, and strict monotonicity of the function w, f_2 that the following property holds:

(P4) for each natural number i there exists a natural number $j > i$ such that

$$x_{2,t} \geq 2^{-1}(f_1(x^*/2) - (1 - v_1)x^*/2).$$

In view of (4.402) and Lemma 4.27, we have

$$x_{1,t} \leq M_0, \quad x_{2,t} \leq M_0(1 - v_2)^{-1}, \quad t = 0, 1, \ldots. \tag{4.408}$$

Let an integer i be given. Assume that for all integers $t > i$,

$$x_{2,t+1} - v_2 x_{2,t} \leq 4^{-1}(f_1(x^*/2) - (1 - v_1)x^*/2)(1 - v_2).$$

When combined with (4.408) this inequality implies that for all integers $t \geq i$, we have

$$x_{2,t} \leq v_2^{t-i} M_0 + (1 - v_2)^{-1} 4^{-1}(f_1(x^*/2) - (1 - v_1)x^*/2)(1 - v_2).$$

This inequality implies that

$$\limsup_{t \to \infty} x_{2,t} \leq 4^{-1}(f_1(x^*/2) - (1 - v_1)x^*/2)(1 - v_2).$$

The inequality above contradicts property (P4). This implies that the following property holds:

(P5) For each nonnegative integer i there is an integer $j > i$ such that

$$x_{2,j+1} - v_2 x_{2,j} > 4^{-1}(f_1(x^*/2) - (1 - v_1)x^*/2)(1 - v_2).$$

Property (P5) implies that there exists a natural number $S > T + 8$ such that

$$x_{2,S} - v_2 x_{2,S-1} > 4^{-1}(1 - v_2)(f_1(x^*/2) - (1 - v_1)x^*/2). \tag{4.409}$$

Choose a real number $\Delta_1 > 0$ such that

$$\Delta_1 < 8^{-1}(1 - v_2)(f_1(x^*/2) - (1 - v_1)x^*/2),$$

$$(S + 8)(1 + c_0)\Delta_1 < \Delta_0/8, \ 2\Delta_1 c_0 (1 - v_2)^{-1} < \Delta_0/8. \tag{4.410}$$

In view of (4.404) there exists a natural number $k > S + 4$ such that for all $t = 0, \ldots, S + 3$, we have

$$|x_{1,t} - x_{1,t}^{(k)}| \leq \Delta_1/2, \ |x_{2,t} - x_{2,t}^{(k)}| \leq \Delta_1/2, \ |y_t - y_t^{(k)}| \leq \Delta_1/2. \tag{4.411}$$

In view of (4.252) there exists a program $(\{\tilde{x}_{1,t}\}_{t=0}^T, \{\tilde{x}_{2,t}\}_{t=0}^T, \{\tilde{y}_t\}_{t=0}^{T-1})$ such that

$$\tilde{x}_{1,T} \geq x_{1,T}, \ \tilde{x}_{2,T} \geq x_{2,T}, \ \tilde{x}_{1,0} = x_{1,0}, \ \tilde{x}_{2,0} = x_{2,0}$$

$$\sum_{t=0}^{T-1} w(\tilde{y}_t) = U(x_{1,0}, x_{2,0}, x_{1,T}, x_{2,T}, 0, T). \tag{4.412}$$

For all integers t satisfying $T \leq t < S - 1$ set

$$\tilde{x}_{1,t+1} = v_1 \tilde{x}_{1,t} + x_{1,t+1} - v_1 x_{1,t}, \ \tilde{x}_{2,t+1} = v_2 \tilde{x}_{2,t} + x_{2,t+1} - v_2 x_{2,t}, \ \tilde{y}_t = f_2(\tilde{x}_{2,t}).$$

$$\tag{4.413}$$

Relations (4.412) and (4.413) imply that

$$\tilde{x}_{1,t} \geq x_{1,t}, \ \tilde{x}_{2,t} \geq x_{2,t}, \ t = T, \ldots, S - 1 \tag{4.414}$$

and that $(\{\tilde{x}_{1,t}\}_{t=0}^{S-1}, \{\tilde{x}_{2,t}\}_{t=0}^{S-1}, \{\tilde{y}_t\}_{t=0}^{S-2})$ is a program. Put

$$\tilde{x}_{1,S} = v_1 \tilde{x}_{1,S-1} + x_{1,S} - v_1 x_{1,S-1} + \Delta_1, \tag{4.415}$$

$$\tilde{x}_{2,S} = v_2 \tilde{x}_{2,S-1} + f_1(\tilde{x}_{1,S-1}) - (\tilde{x}_{1,S} - v_1 \tilde{x}_{1,S-1}), \ \tilde{y}_{S-1} = f_2(\tilde{x}_{2,S-1}).$$

By (4.415), (4.414), and (4.242)–(4.244), we have

$$\tilde{x}_{1,S} - v_1\tilde{x}_{1,S-1} \geq \Delta_1, \ \tilde{x}_{1,S} \geq x_{1,S} + \Delta_1. \tag{4.416}$$

It follows from (4.415), (4.414), (4.242)–(4.244), (4.409), and (4.10) that

$$\tilde{x}_{2,S} - v_2\tilde{x}_{2,S-1} \geq f_1(x_{2,S-1}) - (x_{1,S} - v_1 x_{1,S-1} + \Delta_1)$$

$$\geq x_{2,S} - v_2 x_{2,S-1} - \Delta_1 > 0. \tag{4.417}$$

In view of (4.415),

$$\tilde{x}_{1,S} - v_1\tilde{x}_{1,S-1} + \tilde{x}_{2,S} - v_2\tilde{x}_{2,S-1} = f_1(\tilde{x}_{1,S-1}). \tag{4.418}$$

Relations (4.415)–(4.418) imply that $(\{\tilde{x}_{1,t}\}_{t=0}^{S}, \{\tilde{x}_{2,t}\}_{t=0}^{S}, \{\tilde{y}_t\}_{t=0}^{S-1})$ is a program. It follows from (4.415), (4.414), and (4.242)–(4.244) that

$$\tilde{x}_{2,S} - x_{2,S} \geq v_2\tilde{x}_{2,S-1} + f_1(\tilde{x}_{1,S-1}) - (\tilde{x}_{1,S} - v_1\tilde{x}_{1,S-1}) - x_{2,S}$$

$$\geq v_2 x_{2,S-1} + f_1(x_{1,S-1}) - (x_{1,S} - v_1 x_{1,S-1} + \Delta_1) - x_{1,S} \geq -\Delta_1. \tag{4.419}$$

In view of (4.416), (4.419), and (4.411), we have

$$\tilde{x}_{1,S} \geq x_{1,S}^{(k)}, \ \tilde{x}_{2,S} \geq x_{2,S}^{(k)} - 2\Delta_1. \tag{4.420}$$

For all integers t satisfying $S \leq t < k$ put

$$\tilde{x}_{1,t+1} = v_1\tilde{x}_{1,t} + x_{1,t+1}^{(k)} - v_1 x_{1,t}^{(k)},$$

$$\tilde{x}_{2,t+1} = v_2\tilde{x}_{2,t} + x_{2,t+1}^{(k)} - v_2 x_{2,t}^{(k)}, \ \tilde{y}_t = f_2(\tilde{x}_{2,t}). \tag{4.421}$$

Relations (4.420), (4.421) imply that for all integers $t = S, \ldots, k$,

$$\tilde{x}_{1,t} \geq x_{1,t}^{(k)}$$

and

$$(\{\tilde{x}_{1,t}\}_{t=0}^{k}, \{\tilde{x}_{2,t}\}_{t=0}^{k}, \{\tilde{y}_t\}_{t=0}^{k-1})$$

is a program. In view of (4.421) and (4.420) for all integers t satisfying $S \leq t \leq k$, we have

$$\tilde{x}_{2,t} - x_{2,t}^{(k)} = v_2^{t-S}(\tilde{x}_{2,S} - x_{2,S}^{(k)}) \geq -2\Delta_1 v_2^{t-S}. \tag{4.422}$$

It follows from (4.402), (4.403), Lemma 4.27, and (4.412) that

$$\tilde{x}_{1,t} \leq M_0, \ t = 0, 1, \ldots, \ \tilde{x}_{2,t} \leq M_0(1 - v_2)^{-1}, \ t = 0, 1, \ldots,$$

$$x_{1,t}^{(k)} \leq M_0, \ t = 0, 1, \ldots, k, \ x_{2,t}^{(k)} \leq M_0(1 - v_2)^{-1}, \ t = 0, 1, \ldots, k. \tag{4.423}$$

By (4.403), (4.412), and (4.407),

$$0 \geq \sum_{t=0}^{k-1} w(\tilde{y}_t) - \sum_{t=0}^{k-1} w(y_t^{(k)})$$

$$= \sum_{t=0}^{T-1} w(\tilde{y}_t) - \sum_{t=0}^{T-1} w(y_t) + \sum_{t=0}^{T-1} w(y_t) - \sum_{t=0}^{T-1} w(y_t^{(k)}) + \sum_{t=T}^{k-1} w(\tilde{y}_t) - \sum_{t=T}^{k-1} w(y_t^{(k)})$$

$$\geq \Delta_0 + \sum_{t=0}^{T-1} w(y_t) - \sum_{t=0}^{T-1} w(y_t^{(k)}) + \sum_{t=T}^{k-1} w(\tilde{y}_t) - \sum_{t=T}^{k-1} w(y_t^{(k)}). \tag{4.424}$$

In view of (4.411), (4.423), (4.408), and (4.464), for all integers $t = 0, \ldots, S - 1$,

$$|w(y_t) - w(y_t^{(k)})| \leq c_0 |y_t - y_t^{(k)}| \leq c_0 \Delta_1 / 2. \tag{4.425}$$

It follows from (4.413), (4.414), and (4.242)–(4.244) that for all integers $t = T, \ldots, S - 1$,

$$\tilde{y}_t = f_2(\tilde{x}_{2,t}) \geq f_2(x_{2,t}) \geq y_t, \ w(\tilde{y}_t) \geq w(y_t) \geq w(y_t^{(k)}) - 2^{-1}\Delta_1 c_0. \tag{4.426}$$

Relations (4.424)–(4.426) imply that

$$0 \geq \Delta_0 - T c_0 2^{-1} \Delta_1 + \sum_{t=S}^{k-1} w(\tilde{y}_t) - \sum_{t=S}^{k-1} w(y_t^{(k)}) - (S - T) 2^{-1} c_0 \Delta_1. \tag{4.427}$$

Let an integer t satisfy

$$S \leq t < k, \ \tilde{y}_t < y_t^{(k)}. \tag{4.428}$$

In view of (4.428), (4.423), (4.264), (4.265), monotonicity of the function w, (4.242)–(4.244), and (4.421),

$$0 < w(y_t^{(k)}) - w(\tilde{y}_t) \leq c_0(y_t^{(k)} - \tilde{y}_t) \leq c_0(f_2(x_{2,t}^{(k)}) - f_2(\tilde{x}_{2,t}))$$

$$\leq c_0^2(x_{2,t}^{(k)} - \tilde{x}_{2,t}) \leq c_0^2 2 \Delta_1 v_2^{t-S}. \tag{4.429}$$

It follows from (4.427), (4.428), (4.429), and (4.10) that

$$0 \geq \Delta_0 - S c_0 \Delta_1 - 2 \Delta_1 c_0 (1 - v_2)^{-1} \geq \Delta_0 - \Delta_0/8 - \Delta_0/8 > \Delta_0/2,$$

a contradiction. The contradiction we have reached proves (4.406) and Theorem 4.23 itself.

4.11   Proof of Theorem 4.24

Let

$$x_{1,0} \in [m_0, M_0], \ x_{2,0} \in [m_0, M_0] \qquad (4.430)$$

and let M_* be as guaranteed by Theorem 4.23. Theorem 4.23 and (4.430) imply that there exists a program $\{x_{1,t}, x_{2,t}, y_t\}_{t=0}^{\infty}$ such that for each pair of integers $T_1, T_2 \geq 0$ satisfying $T_1 < T_2$, we have

$$\left| \sum_{t=T_1}^{T_2-1} w(y_t) - \hat{U}(m_0, M_0, T_1, T_2) \right| \leq M_*. \qquad (4.431)$$

Let $p \geq 1$ be an integer. We claim that for all sufficiently large natural numbers T,

$$\left| p^{-1} \hat{U}(m_0, M_0, 0, p) - T^{-1} \sum_{t=0}^{T-1} w(y_t) \right| \leq 2M_* p^{-1}. \qquad (4.432)$$

Assume that $T \geq p$ is an integer. Then there exists a pair of integers q, s such that

$$q \geq 1, \ 0 \leq s < p, \ T = pq + s. \qquad (4.433)$$

By (4.433),

$$T^{-1} \sum_{t=0}^{T-1} w(y_t) - p^{-1} \hat{U}(m_0, M_0, 0, p)$$

$$= T^{-1} \left(\sum_{t=0}^{pq-1} w(y_t) + \sum \{w(y_t) : \ t \text{ is an integer such that} \right.$$

$$\left. pq \leq t \leq T-1 \} \right) - p^{-1} \hat{U}(m_0, M_0, 0, p)$$

$$= T^{-1} \sum \{w(y_t) : \ t \text{ is an integer such that } pq \leq t \leq T-1\}$$

$$+ (T^{-1} pq)(pq)^{-1} \sum_{i=0}^{q-1} \sum_{t=ip}^{(i+1)p-1} w(y_t) - p^{-1} \hat{U}(m_0, M_0, 0, p)$$

$$= T^{-1} \sum \{w(y_t) : \ t \text{ is an integer such that } pq \leq t \leq T-1\}$$

$$+ (T^{-1} pq)(pq)^{-1} \left[\sum_{i=0}^{q-1} \left(\sum_{t=ip}^{(i+1)p-1} w(y_t) \right. \right.$$

$$\left. \left. - \hat{U}(m_0, M_0, 0, p) \right) + q\hat{U}(m_0, M_0, 0, p) \right] - p^{-1} \hat{U}(m_0, M_0, 0, p).$$

$$(4.434)$$

In view of (4.434), the inclusion $x_{1,0}, x_{2,0} \in [m_0, M_0]$, (4.431), (4.433), and Lemma 4.27, we have

$$\left| T^{-1} \sum_{t=0}^{T-1} w(y_t) - p^{-1} \hat{U}(m_0, M_0, 0, p) \right|$$

$$\leq T^{-1} p w(f_2(M_0(1-v_2)^{-1})) + (pq)^{-1} q M_* + \hat{U}(m_0, M_0, 0, p)|q/T - 1/p|$$

$$\leq T^{-1} p w(f_2(M_0(1-v_2)^{-1})) + M_*/p$$

$$+ \hat{U}(m_0, M_0, 0, p) s (pT)^{-1} \to M_*/p \text{ as } T \to \infty.$$

Thus (4.432) is valid for all sufficiently large natural numbers T, as claimed.

Since p is any natural number we conclude that $\{T^{-1} \sum_{t=0}^{T-1} w(y_t)\}_{T=1}^{\infty}$ is a Cauchy sequence. It is clear that there exists $\lim_{T \to \infty} T^{-1} \sum_{t=0}^{T-1} w(y_t)$ and for each natural number $p \geq 1$, we have

$$\left| p^{-1} \hat{U}(m_0, M_0, 0, p) - \lim_{T \to \infty} T^{-1} \sum_{t=0}^{T-1} w(y_t) \right| \leq p^{-1}(2M_*). \tag{4.435}$$

Since inequality (4.435) is true for each natural number p we conclude that

$$\lim_{T \to \infty} T^{-1} \sum_{t=0}^{T-1} w(y_t) = \lim_{p \to \infty} \hat{U}(m_0, M_0, 0, p)/p. \tag{4.436}$$

Set

$$\mu = \lim_{T \to \infty} T^{-1} \sum_{t=0}^{T-1} w(y_t). \tag{4.437}$$

In view of (4.435) and (4.437), for all integers $p \geq 1$,

$$|\hat{U}(m_0, M_0, 0, p)/p - \mu| \leq 2M_*/p.$$

Theorem 4.24 is proved.

4.12 Two-Dimensional Nonautonomous Problems

Let

$$v_1, v_2 \in [0, \infty) \tag{4.438}$$

and let $f_1, f_2 : [0, \infty) \to [0, \infty)$ be monotone increasing functions such that for $i = 1, 2$,

$$f_i(0) = 0 \text{ and } f_i(z) > 0 \text{ for all numbers } z > 0. \tag{4.439}$$

Assume that $w_t : [0, \infty) \to [0, \infty)$, $t = 0, 1, \ldots$ are monotone increasing continuous functions and that for all nonnegative integers t,

$$w_t(0) = 0 \text{ and } w_t(x) > 0 \text{ for all numbers } x > 0. \tag{4.440}$$

We suppose that the following assumption holds:

(A4) for each positive number M, $\lim_{t \to \infty} w_t(M) = 0$.

Recall that a sequence $\{x_{1,t}, x_{2,t}, y_t\}_{t=0}^{\infty}$ is called a program if for all integers $t \geq 0$,

$$x_{1,t}, x_{2,t}, y_t \in R_+^1, \tag{4.441}$$

$$x_{1,t+1} \geq v_1 x_{1,t}, \ x_{2,t+1} \geq v_2 x_{2,t}, \tag{4.442}$$

$$x_{1,t+1} - v_1 x_{1,t} + x_{2,t+1} - v_2 x_{2,t} \leq f_1(x_{1,t}), \tag{4.443}$$

$$0 \leq y_t \leq f_2(x_{2,t}). \tag{4.444}$$

Let integers T_1, T_2 satisfy $T_2 > T_1 \geq 0$. A sequence

$$(\{x_{1,t}\}_{t=T_1}^{T_2}, \{x_{2,t}\}_{t=T_1}^{T_2}, \{y_t\}_{t=T_1}^{T_2-1})$$

is called a program if

$$x_{1,t}, \ x_{2,t} \in R_+^1, \ t = T_1, \ldots, T_2, \ y_t \in R_+^1, \ t = T_1, \ldots, T_2 - 1$$

and if for all integers $t = T_1, \ldots, T_2 - 1$ inequalities (4.442)–(4.444) hold.

We study an infinite horizon optimal control problem which corresponds to a finite horizon problem:

$$\sum_{t=0}^{T-1} w_t(y_t) \to \max, \ (\{x_{1,t}\}_{t=T_1}^{T_2}, \{x_{2,t}\}_{t=T_1}^{T_2}, \{y_t\}_{t=T_1}^{T_2-1})$$

is a program such that $x_{1,0} = z_1$, $x_{2,0} = z_2$,

where T is a natural number and $z_1, z_2 \in R_+^1$.

Recall that these optimal control systems describe a two-sector model of economic dynamics where the first sector produces funds, the second sector produces consumption, $x_{1,t}$ is funds of the first sector at moment t, $x_{2,t}$ is funds of the second sector at moment t, y_t is consumption at moment t and $w_t(y_t)$ evaluates consumption at moment t.

The dynamics of the model is described by (4.442)–(4.444).

Assume that there exists a real number $x^* > 0$ such that

$$f_1(x) > (1 - v_1)x \text{ for all numbers } x \in (0, x^*), \tag{4.445}$$

$$f_1(x) < (1 - v_1)x \text{ for all numbers } x \in (x^*, \infty). \tag{4.446}$$

In view of (4.445) and (4.446),

$$f_1(x^*) = (1 - v_1)x^*. \tag{4.447}$$

Define

$$g(z) = f_1(z) + v_1 z, \ z \in R_+^1, \ g^0 = g, \ g^{i+1} = g \circ g^i \tag{4.448}$$

for all nonnegative integers i. By (4.445) and (4.446),

$$\lim_{i \to \infty} g^i(z) = x^* \text{ for all } z > 0. \tag{4.449}$$

A program $\{x_{1,t}, x_{2,t}, y_t\}_{t=0}^{\infty}$ is called overtaking optimal if for each program $\{x'_{1,t}, x'_{2,t}, y'_t\}_{t=0}^{\infty}$ satisfying $x'_{1,0} = x_{1,0}, x'_{2,0} = x_{2,0}$ the inequality

$$\limsup_{T \to \infty} \sum_{t=0}^{T-1} [w_t(y'_t) - w_t(y_t)] \leq 0$$

holds.

In the section which is based on [72] we use the following assumptions:

(A5) For each positive number M, $\sum_{t=0}^{\infty} w_t(M) < \infty$.
(A6) For each pair of real numbers M_1, M_2 which satisfy $M_2 > M_1 > 0$, the equation

$$\sum_{t=0}^{\infty} (w_t(M_2) - w_t(M_1)) = \infty$$

holds.

We prove the results obtained in [72] which show the existence of an overtaking optimal program for any initial state if at least one of the Assumptions (A5) and (A6) holds. In the case of Assumption (A5) the proof of the existence result is not difficult and standard while in the case of assumption (A6) the situation is more difficult and less understood, and in order to prove our existence result we need an additional Assumption (A7) stated below. This assumption means that the functions f_2 and $w_t, t = 0, 1, \ldots$ are Lipschitz on bounded sets.

Remark 4.36. Let $w : [0, \infty) \to [0, \infty)$ be a continuous increasing function which satisfies $w(0) = 0$ and $w(x) > 0$ for all positive numbers x and $\{\lambda_t\}_{t=0}^{\infty} \subset (0, \infty)$ satisfy $\lim_{t \to \infty} \lambda_t = 0$ and $w_t = \lambda_t w, t = 0, 1, \ldots$. Then Assumption (A4) holds. It is clear that Assumption (A5) holds if $\sum_{t=0}^{\infty} \lambda_t < \infty$ and Assumption (A6) holds if $\sum_{t=0}^{\infty} \lambda_t = \infty$ and the function w is strictly increasing. The case with $\sum_{t=0}^{\infty} \lambda_t < \infty$ is studied in the economic literature. Here our main interest is in the case when $\sum_{t=0}^{\infty} \lambda_t = \infty$.

For each pair of real numbers $x_{1,0}, x_{2,0} \in R^1_+$ and each natural number T define

$$U(x_{1,0}, x_{2,0}, T) = \sup \left\{ \sum_{t=0}^{T-1} w_t(y_t) : (\{x_{1,t}\}_{t=0}^{T}, \{x_{2,t}\}_{t=0}^{T}, \{y_t\}_{t=0}^{T-1}) \text{ is a program} \right\}.$$

(4.450)

The next proposition follows immediately from the continuity of the functions $f_1, f_2, w_t, t = 0, 1, \ldots$.

Proposition 4.37. *Let* $x_{1,0}, x_{2,0} \in R^1_+$ *and let* $T \geq 1$ *be an integer. Then there exists a program* $(\{x_{1,t}\}_{t=0}^{T}, \{x_{2,t}\}_{t=0}^{T}, \{y_t\}_{t=0}^{T-1})$ *such that*

$$\sum_{t=0}^{T-1} w_t(y_t) = U(x_{1,0}, x_{2,0}, T).$$

The following two theorems were obtained in [72].

Theorem 4.38. *Assume that Assumption (A5) holds and let* $M_0 > x^*$. *Then for each point* $z = (z_1, z_2) \in [0, M_0] \times [0, M_0]$ *there exists a program* $\{x_{1,t}^{(z)}, x_{2,t}^{(z)}, y_t^{(z)}\}_{t=0}^{\infty}$ *such that*

$$(x_{1,0}^{(z)}, x_{2,0}^{(z)}) = z$$

and that the following assertion holds.

Let δ *be a positive number. Then there exists an integer* $L^{(\delta)} \geq 1$ *such that for each integer* $S \geq L^{(\delta)}$ *and each* $z = (z_1, z_2) \in [0, M_0] \times [0, M_0]$ *the inequality*

$$\sum_{t=0}^{S-1} w_t(y_t^{(z)}) \geq U(z, S) - \delta$$

holds.

We also use the following assumption:

(A7) for each positive number M there exists a positive number c_M such that for all pairs of real numbers $x_1, x_2 \in [0, M]$ and all nonnegative integers t,

$$|f_2(x_1) - f_2(x_2)| \leq c_M |x_1 - x_2|,$$

$$|w_t(x_1) - w_t(x_2)| \leq c_M |x_1 - x_2|.$$

Theorem 4.39. *Suppose that Assumptions (A6) and (A7) hold and let* $0 < m_0 < x^* < M_0$. *Then for each point* $z = (z_1, z_2) \in [m_0, M_0] \times [0, M_0]$ *there exists a program* $\{x_{1,t}^{(z)}, x_{2,t}^{(z)}, y_t^{(z)}\}_{t=0}^{\infty}$ *such that* $(x_{1,0}^{(z)}, x_{2,0}^{(z)}) = z$ *and that the following assertion holds.*

Let δ be a positive number. Then there exists an integer $L^{(\delta)} \geq 1$ such that for each integer $S \geq L^{(\delta)}$ and each point $z = (z_1, z_2) \in [m_0, M_0] \times [0, M_0]$ the inequality

$$\sum_{t=0}^{S-1} w_t(y_t^{(z)}) \geq U(z, S) - \delta$$

holds.

It is clear that the program $\{x_{1,t}^{(z)}, x_{2,t}^{(z)}, y_t^{(z)}\}_{t=0}^{\infty}$ in the statement of Theorem 4.38 (Theorem 4.39 respectively) is overtaking optimal.

4.13   Proof of Theorem 4.38

Let

$$z = (z_1, z_2) \in [0, M_0] \times [0, M_0] \tag{4.451}$$

be given. Proposition 4.37 implies that for each natural number k there exists a program

$$(\{x_{1,t}^{(z,k)}\}_{t=0}^{k}, \{x_{2,t}^{(z,k)}\}_{t=0}^{k}, \{y_t^{(z,k)}\}_{t=0}^{k-1})$$

such that

$$(x_{1,0}^{(z,k)}, x_{2,0}^{(z,k)}) = z, \quad \sum_{t=0}^{k-1} w_t(y_t^{(z,k)}) = U(z, k). \tag{4.452}$$

It follows from (4.451), (4.452), (4.441)–(4.443), (4.445) and the inequality $M_0 > x^*$ that for each natural number k,

$$x_{1,t}^{(z,k)} \leq M_0, \ t = 0, \ldots, k, \tag{4.453}$$

$$x_{2,t}^{(z,k)} \leq M_0 \sum_{j=0}^{\infty} v_2^j, \ t = 0, \ldots, k, \tag{4.454}$$

$$y_t^{(z,k)} \leq f_2 \left(M_0 \sum_{j=0}^{\infty} v_2^j \right), \ t = 0, \ldots, k - 1. \tag{4.455}$$

Relations (4.453)–(4.455) imply that there exists a strictly increasing sequence of natural numbers $\{k_i\}_{i=1}^{\infty}$ such that for each nonnegative integer t there exist

$$x_{1,t}^{(z)} = \lim_{i \to \infty} x_{1,t}^{(z,k_i)}, \ x_{2,t}^{(z)} = \lim_{i \to \infty} x_{2,t}^{(z,k_i)}, \ y_t^{(z)} = \lim_{i \to \infty} y_t^{(z,k_i)}. \tag{4.456}$$

It is not difficult to see that $\{x_{1,t}^{(z)}, x_{2,t}^{(z)}, y_t^{(z)}\}_{t=0}^{\infty}$ is a program and that

$$(x_{1,0}^{(z)}, x_{2,0}^{(z)}) = z. \tag{4.457}$$

Let $\delta > 0$ be given. Assumption (A5) implies that there exists an integer $L^{(\delta)} \geq 1$ such that

$$\sum_{t=L^{(\delta)}}^{\infty} w_t(f_2(M_0(1-v_2)^{-1})) < \delta. \tag{4.458}$$

Assume that an integer $S \geq L^{(\delta)}$ and that (4.451) holds. Proposition 4.37 implies that there exists a program $(\{x_{1,t}\}_{t=0}^{S}, \{x_{2,t}\}_{t=0}^{S}, \{y_t\}_{t=0}^{S-1})$ such that

$$(x_{1,0}, x_{2,0}) = z, \ \sum_{t=0}^{S-1} w_t(y_t) = U(z, S). \tag{4.459}$$

For all integers $t \geq S$ put

$$y_t = 0, \ x_{2,t+1} = v_2 x_{2,t}, \ x_{1,t+1} = v_1 x_{1,t} + f_1(x_{1,t}). \tag{4.460}$$

It is clear that $\{x_{1,t}, x_{2,t}, y_t\}_{t=0}^{\infty}$ is a program. In view of (4.455), (4.452), (4.459), (4.460), and (4.458), for each natural number i satisfying $k_i > S$, we have

$$\sum_{t=0}^{S-1} w_t(y_t^{(z,k_i)}) \geq \sum_{t=0}^{k_i-1} w_t(y_t^{(z,k_i)}) - \sum_{t=S}^{k_i-1} w_t(f_2(M_0(1-v_2)^{-1}))$$

$$\geq \sum_{t=0}^{k_i-1} w_t(y_t) - \sum_{t=S}^{\infty} w_t(f_2(M_0(1-v_2)^{-1}))$$

$$\geq \sum_{t=0}^{S-1} w_t(y_t) - \sum_{t=L^{\delta}}^{\infty} w_t(f_2(M_0(1-v_2)^{-1})) \geq U(z, S) - \delta.$$

When combined with (4.456) this implies that

$$\sum_{t=0}^{S-1} w_t(y_t^{(z)}) \geq U(z, S) - \delta.$$

Theorem 4.38 is proved.

4.14   Auxiliary Results for Theorem 4.39

Equations (4.441)–(4.446) easily imply the following result.

Lemma 4.40. *Let $M_0 > x^*$, $T \geq 1$ be an integer and let*

$$(\{x_{1,t}\}_{t=0}^{T}, \{x_{2,t}\}_{t=0}^{T}, \{y_t\}_{t=0}^{T-1})$$

be a program such that $x_{1,0}, x_{2,0} \leq M_0$. Then for all integers $t = 0, \ldots, T$,

$$x_{1,t} \leq M_0, \ x_{2,t} \leq M_0(1 - v_2)^{-1}$$

and for all integers $t = 0, \ldots, T - 1$,

$$y_t \leq f_2(M_0(1 - v_2)^{-1}).$$

Let

$$0 < m_0 < x^* < M_0. \tag{4.461}$$

Relations (4.445) and (4.461) imply that

$$f_1(m_0) > (1 - v_1)m_0. \tag{4.462}$$

Choose a real number $\gamma_0 > 0$ such that

$$f_1(\gamma_0) + \gamma_0 < m_0/4, \tag{4.463}$$

$$f_1(\gamma_0) < 4^{-1}(1 - v_2)[f_1(m_0) - (1 - v_1)m_0], \tag{4.464}$$

$$f_2(2f_1(\gamma_0)(1 - v_2)^{-1}) < 4^{-1}f_2(f_1(m_0) - (1 - v_1)m_0). \tag{4.465}$$

In view of (4.445), (4.461), and (4.463), we have

$$f_1(\gamma_0) > (1 - v_1)\gamma_0. \tag{4.466}$$

Assumption (A7) implies that there exists a positive number c_0 such that

$$|f_2(x_1) - f_2(x_2)| \leq c_0|x_1 - x_2| \text{ for all } x_1, x_2 \in [0, M_0(1 - v_2)^{-1}] \tag{4.467}$$

and that for all nonnegative integers t,

$$|w_t(y_1) - w_t(y_2)| \leq c_0|y_1 - y_2| \text{ for all } y_1, y_2 \in [0, f_2(M_0(1 - v_2)^{-1})]. \tag{4.468}$$

Lemma 4.41. *Let $p \geq 1$ be an integer. Then there exists an integer $L_0 \geq 1$ such that for each integer $T \geq L_0 + p$, each program*

$$(\{x_{1,t}\}_{t=0}^{T}, \{x_{2,t}\}_{t=0}^{T}, \{y_t\}_{t=0}^{T-1})$$

satisfying

$$x_{1,0} \in [m_0, M_0], \; x_{2,0} \in [0, M_0], \; \sum_{t=0}^{T-1} w_t(y_t) = U(x_{1,0}, x_{2,0}, T), \qquad (4.469)$$

and each integer $S \in [0, p]$ the inequality

$$\max\{x_{1,t} : t \in \{S, \ldots, S + L_0 - 1\}\} \geq \gamma_0 \qquad (4.470)$$

holds.

Proof. In view of (4.461) and (4.449) there exists a natural number $L_1 \geq 4$ such that

$$g^{L_1}(\gamma_0) > m_0, \; v_2^{L_1} M_0 < f_1(\gamma_0). \qquad (4.471)$$

Assumption (A6) and (4.465) imply that there exists an integer $L_2 \geq 4$ such that for all integers $S \in [0, p]$, we have

$$\sum_{t=S+L_1}^{S+L_1+L_2} [w_t(f_2(f_1(m_0) - (1 - v_1)m_0)) - w_t(f_2(2f_1(\gamma_0)(1 - v_2)^{-1}))]$$

$$> \sum_{t=0}^{p+L_1} w_t(f_2(M_0(1 - v_2)^{-1})) + 4 \qquad (4.472)$$

and a natural number

$$L_0 > 4(L_1 + L_2 + 4). \qquad (4.473)$$

Assume that an integer $T \geq L_0 + p$, a program $(\{x_{1,t}\}_{t=0}^T, \{x_{2,t}\}_{t=0}^T, \{y_t\}_{t=0}^{T-1})$ satisfies (4.469), and an integer $S \in [0, p]$. We claim that inequality (4.470) holds.

Assume the contrary. Then

$$x_{1,t} < \gamma_0 \text{ for all integers } t = S, \ldots, S + L_0 - 1. \qquad (4.474)$$

In view of (4.463), (4.469), and (4.474), there exists a nonnegative integer τ_0 such that

$$\tau_0 < S, \; x_{1,\tau_0} \geq \gamma_0, \; x_{1,t} < \gamma_0, \; t = \tau_0 + 1, \ldots, S + L_0 - 1. \qquad (4.475)$$

Set

$$\bar{x}_{1,t} = x_{1,t}, \; \bar{x}_{2,t} = x_{2,t} \text{ for all } t = 0, \ldots, \tau_0,$$

$$\bar{y}_t = y_t \text{ for all integers } t \text{ satisfying } 0 \leq t < \tau_0. \qquad (4.476)$$

For all integers t satisfying $\tau_0 \le t < \tau_0 + L_1$ set

$$\bar{x}_{1,t+1} = v_1\bar{x}_{1,t} + f_1(\bar{x}_{1,t}),$$
$$\bar{x}_{2,t+1} = v_2\bar{x}_{2,t}, \ \bar{y}_t = 0. \tag{4.477}$$

It is clear that

$$(\{\bar{x}_{1,t}\}_{t=0}^{\tau_0+L_1}, \{\bar{x}_{2,t}\}_{t=0}^{\tau_0+L_1}, \{\bar{y}_t\}_{t=0}^{\tau_0+L_1-1})$$

is a program. It follows from (4.448), (4.477), (4.476), (4.475), and (4.471) that

$$\bar{x}_{1,\tau_0+L_1} = g^{L_1}(\bar{x}_{1,\tau_0}) = g^{L_1}(x_{1,\tau_0}) \ge g^{L_1}(\gamma_0) > m_0. \tag{4.478}$$

There are two cases:

$$\max\{x_{1,t} : t = S, \ldots, T-1\} < \gamma_0, \tag{4.479}$$
$$\max\{x_{1,t} : t = S, \ldots, T-1\} \ge \gamma_0. \tag{4.480}$$

Assume that (4.479) holds. For all integers t satisfying $\tau_0 + L_1 \le t \le T - 1$, set

$$\bar{y}_t = f_2(\bar{x}_{2,t}), \ \bar{x}_{2,t+1} = v_2\bar{x}_{2,t} + f_1(m_0) - (1-v_1)m_0,$$
$$\bar{x}_{1,t+1} = v_1\bar{x}_{1,t} + f_1(\bar{x}_{1,t}) - f_1(m_0) + (1-v_1)m_0. \tag{4.481}$$

By (4.481) and (4.478),

$$\bar{x}_{1,t} \ge m_0, \ t = \tau_0 + L_1, \ldots, T \tag{4.482}$$

and that $(\{\bar{x}_{1,t}\}_{t=0}^{T}, \{\bar{x}_{2,t}\}_{t=0}^{T}, \{\bar{y}_t\}_{t=0}^{T-1})$ is a program. By (4.476), (4.469), (4.481), (4.475), and (4.473), we have

$$U(x_{1,0}, x_{2,0}, T) \ge \sum_{t=0}^{T-1} w_t(\bar{y}_t) = \sum_{t=0}^{T-1} w_t(y_t) + \sum_{t=0}^{T-1} w_t(\bar{y}_t) - \sum_{t=0}^{T-1} w_t(y_t)$$

$$= U(x_{1,0}, x_{2,0}, T) + \sum_{t=\tau_0}^{T-1} w_t(\bar{y}_t) - \sum_{t=\tau_0}^{T-1} w_t(y_t)$$

$$\ge U(x_{1,0}, x_{2,0}, T) + \sum_{t=\tau_0+L_1+1}^{T-1} w_t(f_2(f_1(m_0) - (1-v_1)m_0))$$

$$- \sum_{t=\tau_0}^{T-1} w_t(y_t). \tag{4.483}$$

In view of Lemma 4.40 and (4.469),

$$x_{2,t} \le M_0(1 - v_2)^{-1}, \ t = 0, \ldots .T, \tag{4.484}$$

$$y_t \le f_2(M_0(1 - v_2)^{-1}), \ t = 0, \ldots, T - 1.$$

It follows from (4.479), (4.475), and (4.441)–(4.444) that for all integers t satisfying $\tau_0 + 1 \le t < T - 1$,

$$x_{2,t+1} \le v_2 x_{2,t} + f_1(x_{1,t}) \le v_2 x_{2,t} + f_1(\gamma_0). \tag{4.485}$$

Relations (4.484) and (4.485) imply that for all integers t satisfying $\tau_0 + 2 \le t \le T - 1$,

$$x_{2,t} \le M_0(1 - v_2)^{-1} v_2^{t-1-\tau_0} + f_1(\gamma_0)(1 - v_2)^{-1},$$

$$y_{2,t} \le f_2(x_{2,t}) \le f_2(M_0(1 - v_2)^{-1} v_2^{t-1-\tau_0} + f_1(\gamma_0)(1 - v_2)^{-1}).$$

$$\tag{4.486}$$

In view of (4.14) and (4.471), for all integers t satisfying $\tau_0 + 1 + L_1 \le t \le T - 1$, we have

$$y_{2,t} \le f_2(2 f_1(\gamma_0)(1 - v_2)^{-1}). \tag{4.487}$$

It follows from (4.483), (4.484), (4.487), (4.475), (4.465), and (4.472) that

$U(x_{1,0}, x_{2,0}, T)$

$$\ge U(x_{1,0}, x_{2,0}, T) + \sum_{t=\tau_0+L_1+1}^{T-1} w_t(f_2(f_1(m_0) - (1 - v_1)m_0))$$

$$- \sum_{t=\tau_0}^{\tau_0+L_1} w_t(f_2(M_0(1 - v_2)^{-1})) - \sum_{t=\tau_0+1+L_1}^{T-1} w_t(f_2(2 f_1(\gamma_0)(1 - v_2)^{-1}))$$

$$\ge U(x_{1,0}, x_{2,0}, T)$$

$$+ \sum_{t=\tau_0+1+L_1}^{T-1} [w_t(f_2(f_1(m_0) - (1 - v_1)m_0)) - w_t(f_2(2 f_1(\gamma_0)(1 - v_2)^{-1}))]$$

$$- \sum_{t=\tau_0}^{\tau_0+L_1} w_t(f_2(M_0(1 - v_2)^{-1}))$$

$$\ge U(x_{1,0}, x_{2,0}, T) + \sum_{t=p+L_1}^{T-1} [w_t(f_2(f_1(m_0) - (1 - v_1)m_0))$$

$$- w_t(f_2(2f_1(\gamma_0)(1 - v_2)^{-1}))]$$

$$- \sum_{t=0}^{p+L_1} w_t(f_2(M_0(1 - v_2)^{-1})) \geq U(x_{1,0}, x_{2,0}, T) + 4,$$

a contradiction. The contradiction we have reached shows that (4.479) does not hold. Therefore (4.480) is true.

In view of (4.480) and (4.464), there exists an integer τ_1 such that

$$S + L_0 \leq \tau_1 < T, \; x_{1,\tau_1} \geq \gamma_0, \; x_{1,t} < \gamma_0, \; t = S, \dots, \tau_1 - 1. \tag{4.488}$$

For all integers t satisfying $\tau_0 + L_1 \leq t \leq \tau_1 - 1$ put

$$\bar{y}_t = f_2(\bar{x}_{2,t}), \; \bar{x}_{2,t+1} = v_2 \bar{x}_{2,t} + f_1(m_0) - (1 - v_1)m_0,$$

$$\bar{x}_{1,t+1} = v_1 \bar{x}_{1,t} + f_1(\bar{x}_{1,t}) - f_1(m_0) + (1 - v_1)m_0. \tag{4.489}$$

By (4.478), (4.462) and (4.441)–(4.444),

$$\bar{x}_{1,t} \geq m_0, \; t = \tau_0 + L_1, \dots, \tau_1 \tag{4.490}$$

and $(\{\bar{x}_{1,t}\}_{t=0}^{\tau_1}, \{\bar{x}_{2,t}\}_{t=0}^{\tau_1}, \{\bar{y}_t\}_{t=0}^{\tau_1-1})$ is a program. Relations (4.490) and (4.489) imply that

$$\bar{x}_{2,t} \geq f_1(m_0) - (1 - v_1)m_0, \; t = \tau_0 + L_1 + 1, \dots, \tau_1, \tag{4.491}$$

$$\bar{y}_t \geq f_2(f_1(m_0) - (1 - v_1)m_0), \; t = \tau_0 + L_1 + 1, \dots, \tau_1 - 1. \tag{4.492}$$

In view of Lemma 4.40 and (4.469), (4.484) holds. It follows from (4.475), (4.488), and (4.14) that for all integers t satisfying $\tau_0 + 2 \leq t \leq \tau_1$, we have

$$x_{2,t} \leq M_0(1 - v_2)^{-1} v_2^{t-1-\tau_0} + f_1(\gamma_0)(1 - v_2)^{-1}. \tag{4.493}$$

In view of (4.493), (4.488), and (4.471), for all integers t satisfying $\tau_0 + 1 + L_1 \leq t \leq \tau_1$,

$$x_{2,t} \leq 2 f_1(\gamma_0)(1 - v_2)^{-1}. \tag{4.494}$$

Relations (4.494), (4.489), (4.488), (4.475), and (4.473) imply that for all integers t satisfying $\bar{L}_0 + 1 + L_1 \leq t \leq \tau_1 - 1$,

$$y_t \leq f_2(2f_1(\gamma_0)(1 - v_2)^{-1}). \tag{4.495}$$

By (4.488), (4.441)–(4.444), (4.490) and (4.463),

$$x_{1,\tau_1} \leq v_1 \gamma_0 + f_1(\gamma_0) < m_0 \leq \bar{x}_{1,\tau_1}. \tag{4.496}$$

It follows from (4.494), (4.491), and (4.464) that

$$x_{2,\tau_1} \leq 2f_1(\gamma_0)(1 - v_2)^{-1} < f_1(m_0) - (1 - v_1)m_0 \leq \bar{x}_{2,\tau_1}. \tag{4.497}$$

For all integers t satisfying $\tau_1 \leq t < T$ put

$$\bar{y}_t = y_t, \quad \bar{x}_{1,t+1} = v_1\bar{x}_{1,t} + x_{1,t+1} - v_1x_{1,t},$$

$$\bar{x}_{2,t+1} = v_2\bar{x}_{2,t} + x_{2,t+1} - v_2x_{2,t}. \tag{4.498}$$

In view of (4.496), (4.497), and (4.498)

$$\bar{x}_{1,t} \geq x_{1,t}, \quad \bar{x}_{2,t} \geq x_{2,t}$$

for all integers $t = \tau_1, \ldots, T$ and $(\{\bar{x}_{1,t}\}_{t=0}^T, \{\bar{x}_{2,t}\}_{t=0}^T, \{\bar{y}_t\}_{t=0}^{T-1})$ is a program.

By (4.476), (4.469), (4.498), (4.484), (4.482), (4.495), (4.475), (4.473), and (4.472),

$$U(x_{1,0}, x_{2,0}, T) \geq \sum_{t=0}^{T-1} w_t(\bar{y}_t) = \sum_{t=0}^{T-1} w_t(y_t) + \sum_{t=0}^{T-1} w_t(\bar{y}_t) - \sum_{t=0}^{T-1} w_t(y_t)$$

$$= U(x_{1,0}, x_{2,0}, T) + \sum_{t=0}^{T-1} w_t(\bar{y}_t) - \sum_{t=0}^{T-1} w_t(y_t)$$

$$= U(x_{1,0}, x_{2,0}, T) + \sum_{t=\tau_0}^{\tau_1-1} w_t(\bar{y}_t) - \sum_{t=\tau_0}^{\tau_1-1} w_t(y_t)$$

$$\geq U(x_{1,0}, x_{2,0}, T) + \sum_{t=\tau_0+L_1+1}^{\tau_1-1} w_t(f_2(f_1(m_0) - (1 - v_1)m_0))$$

$$- \sum_{t=\tau_0}^{\tau_0+L_1} w_t(f_2(M_0(1 - v_2)^{-1}))$$

$$- \sum_{t=\tau_0+L_1+1}^{\tau_1-1} w_t(f_2(2f_1(\gamma_0)(1 - v_2)^{-1}))$$

$$\geq U(x_{1,0}, x_{2,0}, T) - \sum_{t=0}^{p+L_1} w_t(f_2(M_0(1 - v_2)^{-1}))$$

$$+ \sum_{t=S+L_1}^{S+L_0-1} [w_t(f_2(f_1(m_0) - (1 - v_1)m_0))$$

$$- w_t(f_2(2f_1(\gamma_0)(1 - v_2)^{-1}))]$$
$$\geq U(x_{1,0}, x_{2,0}, T) + 4,$$

a contradiction. The contradiction we have reached proves (4.470) and Lemma 4.41 itself. ∎

We suppose that the sum over an empty set is zero.

Lemma 4.42. *Let δ be a positive number. Then there exists an integer $\bar{L} \geq 1$ such that for each integer $L \geq \bar{L}$ there exists an integer $\tau \geq L$ such that the following assertion holds:*

For each integer $T \geq \tau$ and each program $(\{x_{1,t}\}_{t=0}^{T}, \{x_{2,t}\}_{t=0}^{T}, \{y_t\}_{t=0}^{T-1})$ satisfying

$$x_{1,0} \in [m_0, M_0], \quad x_{2,0} \in [0, M_0], \tag{4.499}$$

$$\sum_{t=0}^{T-1} w_t(y_t) = U(x_{1,0}, x_{2,0}, T),$$

the inequality

$$\sum_{t=0}^{L-1} w_t(y_t) \geq U(x_{1,0}, x_{2,0}, L) - \delta \tag{4.500}$$

holds.

Proof. Choose a real number $\gamma_1 > 0$ such that

$$\gamma_1 < \gamma_0/4, \quad \gamma_1 < [f_1(\gamma_0) - (1 - v_1)\gamma_0]/4 \tag{4.501}$$

and a real number $\gamma_2 > 0$ such that

$$(1 - v_2)^{-1}\gamma_2 < \gamma_1/4. \tag{4.502}$$

In view of (4.449) and (4.461) there exists an integer $L_1^* \geq 4$ such that for all integers $i \geq L_1^*$, we have

$$g^i(\gamma_0) > x^* - \gamma_1/4, \quad g^i(\gamma_0) > m_0. \tag{4.503}$$

It follows from (4.464), (4.448), and (4.449) that there exists an integer $L_2^* \geq 4$ such that

$$v_2^{L_2^*} M_0(1 - v_2)^{-1} + f_1(\gamma_0)(1 - v_2)^{-1} < [f_1(m_0) - (1 - v_1)m_0]4^{-1},$$

$$v_2^{L_2^*} M_0(1 - v_2)^{-1} \leq \gamma_1/4, \tag{4.504}$$

$$g^i(M_0) \le x^* + \gamma_2/4 \text{ for all integers } i \ge L_2^*, \qquad (4.505)$$

$$g^i(\gamma_0) > x^* - \gamma_2/4 \text{ for all integers } i \ge L_2^*, \qquad (4.506)$$

$$c_0^2 v_2^{L_2^*}(M_0(1 - v_2)^{-2}) < \delta/8. \qquad (4.507)$$

Assumption (A4) implies that there exists an integer $p \ge 4$ such that for all integers $j \ge p$,

$$w_j(f_2(M_0(1 - v_2)^{-1})) \le (\delta/4)(L_1^* + L_2^* + 2)^{-1}. \qquad (4.508)$$

Lemma 4.41 implies that there exists an integer $L_0 \ge 1$ such that the following property holds:

(P6) For each integer $T \ge L_0 + p$, each program

$$(\{x_{1,t}\}_{t=0}^T, \{x_{2,t}\}_{t=0}^T, \{y_t\}_{t=0}^{T-1})$$

satisfying (4.499), and each integer $S \in [0, p]$, the inequality

$$\max\{x_{1,t} : t = S, \dots, S + L_0 - 1\} \ge \gamma_0$$

holds.

Choose an integer

$$\bar{L} > L_0 + p + 4 + 4L_1^* + 4L_2^*. \qquad (4.509)$$

Assume that a natural number $L \ge \bar{L}$ is given and choose an integer

$$\tau > L + 4 + 4L_1^* + 4L_2^*. \qquad (4.510)$$

Assume that an integer $T \ge \tau$ is given and that a program

$$(\{x_{1,t}\}_{t=0}^T, \{x_{2,t}\}_{t=0}^T, \{y_t\}_{t=0}^{T-1})$$

satisfies (4.499). We claim that (4.500) holds. Assume the contrary. Then

$$\sum_{t=0}^{L-1} w_t(y_t) < U(x_{1,0}, x_{2,0}, L) - \delta. \qquad (4.511)$$

Proposition 4.37 implies that there exists a program

$$(\{\tilde{x}_{1,t}\}_{t=0}^L, \{\tilde{x}_{2,t}\}_{t=0}^L, \{\tilde{y}_t\}_{t=0}^{L-1})$$

such that

$$\tilde{x}_{1,0} = x_{1,0}, \ \tilde{x}_{2,0} = x_{2,0}, \tag{4.512}$$

$$\sum_{t=0}^{L-1} w_t(\tilde{y}_t) = U(x_{1,0}, x_{2,0}, L). \tag{4.513}$$

In view of (4.512), (4.499), and (4.463), there exists a nonnegative integer τ_0 such that

$$\tau_0 \leq L, \ \tilde{x}_{1,\tau_0} \geq \gamma_0, \ \text{and if an integer } t \text{ satisfies } \tau_0 < t \leq L, \text{ then } \tilde{x}_{1,t} < \gamma_0. \tag{4.514}$$

It follows from (4.509), (4.512), (4.513) and (4.499), (4.514) and property (P6) that

$$\tau_0 \geq p. \tag{4.515}$$

Define a program $(\{\bar{x}_{1,t}\}_{t=0}^{L}, \{\bar{x}_{2,t}\}_{t=0}^{L}, \{\bar{y}_t\}_{t=0}^{L-1})$. There are two cases:

$$L - \tau_0 \leq L_1^* + L_2^*; \tag{4.516}$$

$$L - \tau_0 > L_1^* + L_2^*. \tag{4.517}$$

If inequality (4.516) is valid, put

$$\bar{x}_{1,t} = \tilde{x}_{1,t}, \ \bar{x}_{2,t} = \tilde{x}_{2,t}, \ t = 0,\dots,L, \ \bar{y}_t = \tilde{y}_t, \ t = 0,\dots,L-1. \tag{4.518}$$

Assume that inequality (4.517) is true. Put

$$\bar{x}_{1,t} = \tilde{x}_{1,t}, \ \bar{x}_{2,t} = \tilde{x}_{2,t}, \ t = 0,\dots,\tau_0, \ \bar{y}_t = \tilde{y}_t, \ t = 0,\dots,\tau_0-1, \tag{4.519}$$

and for all integers $t = \tau_0,\dots,\tau_0 + L_1^* - 1$ set

$$\bar{y}_t = 0, \ \bar{x}_{1,t+1} = v_1\bar{x}_{1,t} + f_1(\bar{x}_{1,t}), \ \bar{x}_{2,t+1} = v_2\bar{x}_{2,t}. \tag{4.520}$$

It is clear that

$$(\{\bar{x}_{1,t}\}_{t=0}^{\tau_0+L_1^*}, \{\bar{x}_{2,t}\}_{t=0}^{\tau_0+L_1^*}, \{\bar{y}_t\}_{t=0}^{\tau_0+L_1^*-1})$$

is a program. In view of (4.520), (4.514), (4.503), and (4.448),

$$\bar{x}_{1,\tau_0+L_1^*} = g^{L_1^*}(\bar{x}_{1,\tau_0}) \geq g^{L_1^*}(\gamma_0) > m_0. \tag{4.521}$$

For all integers t satisfying $\tau_0 + L_1^* \leq t < L$ set

$$\bar{y}_t = f_2(\bar{x}_{2,t}), \ \bar{x}_{1,t+1} = v_1\bar{x}_{1,t} + (1-v_1)m_0,$$
$$\bar{x}_{2,t+1} = v_2\bar{x}_{2,t} + f_1(\bar{x}_{1,t}) - (1-v_1)m_0. \tag{4.522}$$

Relations (4.521) and (4.522) imply that

$$\bar{x}_{1,t} \geq m_0, \ t = \tau_0 + L_1^*,\dots,L. \tag{4.523}$$

It follows from (4.522), (4.523), and (4.462) that for all integers $t = \tau_0 + L_1^*, \ldots, L-1$,

$$\bar{x}_{2,t+1} - v_2\bar{x}_{2,t} \geq f_1(m_0) - (1-v_1)m_0 > 0. \tag{4.524}$$

In view of (4.522)–(4.524), $(\{\bar{x}_{1,t}\}_{t=0}^L, \{\bar{x}_{2,t}\}_{t=0}^L, \{\bar{y}_t\}_{t=0}^{L-1})$ is a program. By (4.522) and (4.523), for all integers $t = \tau_0 + L_1^* + 1, \ldots, L-1$, we have

$$\bar{x}_{2,t} \geq f_1(m_0) - (1-v_1)m_0, \quad \bar{y}_{2,t} \geq f_2(f_1(m_0) - (1-v_1)m_0). \tag{4.525}$$

Lemma 4.40, (4.499) and (4.512) imply that

$$\tilde{x}_{2,\tau_0+L_1^*-1}, \ x_{2,\tau_0+L_1^*-1} \leq M_0(1-v_2)^{-1}. \tag{4.526}$$

In view of (4.514), for all integers t satisfying $\tau_0 < t < L$,

$$\tilde{x}_{2,t+1} \leq v_2\tilde{x}_{2,t} + f_1(\gamma_0). \tag{4.527}$$

By (4.526) and (4.527), for all integers t satisfying $L_1^* + \tau_0 \leq t \leq L$,

$$\tilde{x}_{2,t} \leq f_1(\gamma_0)(1-v_2)^{-1} + v_2^{t-L_1^*-\tau_0} M_0(1-v_2)^{-1}. \tag{4.528}$$

In view of (4.528), (4.504), and (4.525), for all integers t satisfying $L_1^* + L_2^* + \tau_0 \leq t \leq L-1$,

$$\tilde{x}_{2,t} \leq f_1(\gamma_0)(1-v_2)^{-1} + v_2^{L_2^*} M_0(1-v_2)^{-1}$$
$$\leq 4^{-1}(f_1(m_0) - (1-v_1)m_0) \leq \bar{x}_{2,t}. \tag{4.529}$$

By (4.529) and (4.522), for all integers t satisfying $L_1^* + L_2^* + \tau_0 \leq t \leq L-1$, we have

$$\tilde{y}_t \leq f_2(\tilde{x}_{2,t}) \leq f_2(\bar{x}_{2,t}) = \bar{y}_t. \tag{4.530}$$

It follows from (4.513), (4.519), (4.530), (4.512), Lemma 4.40, (4.499), (4.508), and (4.515) that

$$U(x_{1,0}, x_{2,0}, L) - \sum_{t=0}^{L-1} w_t(\bar{y}_t) = \sum_{t=0}^{L-1} w_t(\tilde{y}_t) - \sum_{t=0}^{L-1} w_t(\bar{y}_t)$$

$$\leq \sum_{t=\tau_0}^{L_1^*+L_2^*+\tau_0-1} w_t(\tilde{y}_t) - \sum_{t=\tau_0}^{L_1^*+L_2^*+\tau_0-1} w_t(\bar{y}_t)$$

$$\leq \sum_{t=\tau_0}^{L_1^*+L_2^*+\tau_0-1} w_t(f_2(M_0(1-v_2)^{-1})) \leq \delta/8.$$

Thus in both cases we have constructed a program

$$(\{\bar{x}_{1,t}\}_{t=0}^{L}, \{\bar{x}_{2,t}\}_{t=0}^{L}, \{\bar{y}_t\}_{t=0}^{L-1})$$

such that

$$\bar{x}_{1,0} = x_{1,0}, \ \bar{x}_{2,0} = x_{2,0}, \ U(x_{1,0}, x_{2,0}, L) - \delta/8 < \sum_{t=0}^{L-1} w_t(\bar{y}_t) \tag{4.531}$$

and for which there is an integer

$$S_0 \in [L - L_1^* - L_2^*, L] \text{ such that } \bar{x}_{1,S_0} \geq \gamma_0 \tag{4.532}$$

[see (4.512)–(4.514), and (4.518) in the case of (4.516) and see (4.519) and (4.523) in the case of (4.517)].

There are two cases:

(1) there is an integer

$$S_1 \in [S_0 + L_1^* + 2L_2^*, T - 2] \text{ such that } x_{2,S_1+1} \geq \gamma_1; \tag{4.533}$$

(2)

$$x_{2,t} < \gamma_1 \text{ for all integers } t \in [S_0 + L_1^* + 2L_2^* + 1, T - 1]. \tag{4.534}$$

Assume that the case (2) holds. Set

$$x'_{1,t} = \bar{x}_{1,t}, \ x'_{2,t} = \bar{x}_{2,t}, \ t = 0, \ldots, S_0, \ y'_t = \bar{y}_t, \ t = 0, \ldots, S_0 - 1. \tag{4.535}$$

For all integers t satisfying $S_0 \leq t < T$ we put

$$y'_t = f_2(x'_{2,t}), \ x'_{1,t+1} = v_1 x'_{1,t} + (1 - v_1)\gamma_0,$$
$$x'_{2,t+1} = v_2 x'_{2,t} + f_1(\gamma_0)(1 - v_1)\gamma_0. \tag{4.536}$$

In view of (4.532), (4.535), and (4.536),

$$x'_{1,t} \geq \gamma_0, \ t = S_0, \ldots, T. \tag{4.537}$$

It is not difficult to see that $(\{x'_{1,t}\}_{t=0}^{T}, \{x'_{2,t}\}_{t=0}^{T}, \{y'_t\}_{t=0}^{T-1})$ is a program. Relations (4.535) and (4.531) imply that

$$x'_{1,0} = x_{1,0}, \ x'_{2,0} = x_{2,0}. \tag{4.538}$$

It follows from (4.538), (4.499), (4.534), (4.536), (4.501), Lemma 4.40, (4.509), (4.532), (4.508), (4.535), (4.531), and (4.512) that

$$0 \leq U(x_{1,0}, x_{2,0}, T) - \sum_{t=0}^{T-1} w_t(y_t') = \sum_{t=0}^{T-1} w_t(y_t) - \sum_{t=0}^{T-1} w_t(y_t')$$

$$\leq \sum_{t=0}^{S_0+L_1^*+2L_2^*} w_t(y_t) + \sum_{t=S_0+L_1^*+2L_2^*+1}^{T-1} w_t(f_2(\gamma_1)) - \sum_{t=0}^{S_0} w_t(y_t')$$

$$- \sum_{t=S_0+1}^{T-1} w_t(f_2(f_1(\gamma_0) - (1-v_1)\gamma_0))$$

$$\leq \sum_{t=0}^{S_0+L_1^*+2L_2^*} w_t(y_t) - \sum_{t=0}^{S_0-1} w_t(y_t')$$

$$\leq \sum_{t=0}^{S_0-1} w_t(y_t) + (L_1^* + 2L_2^* + 1) \sup\{w_j(f_2(M_0(1-v_2)^{-1})) : \text{an integer } j \geq p\}$$

$$- \sum_{t=0}^{S_0-1} w_t(y_t') \leq \sum_{t=0}^{L-1} w_t(y_t) - \sum_{t=0}^{S_0-1} w_t(\bar{y}_t) + \delta/8$$

$$\leq \sum_{t=0}^{L-1} w_t(y_t) - \sum_{t=0}^{L-1} w_t(\bar{y}_t)$$

$$+ (L_1^* + L_2^* + 1) \sup\{w_j(f_2(M_0(1-v_2)^{-1})) : \text{an integer } j \geq p\} + \delta/8$$

$$\leq \sum_{t=0}^{L-1} w_t(y_t) - U(x_{1,0}, x_{2,0}, L) + \delta/8 + \delta/8 + \delta/8 < -\delta/2,$$

a contradiction. Therefore the case (1) holds. Set

$$x_{1,t}' = \bar{x}_{1,t}, \ x_{2,t}' = \bar{x}_{2,t}, \ t = 0, \ldots, S_0, \ y_t' = \bar{y}_t, \ t = 0, \ldots, S_0 - 1. \tag{4.539}$$

We may assume without loss of generality that

$$x_{2,t} < \gamma_1 \text{ for all integers } t \text{ satisfying } S_0 + L_1^* + 2L_2^* + 1 \leq t \leq S_1. \tag{4.540}$$

We claim that there is an integer $t \in [S_1 - L_2^*, S_1]$ such that

$$x_{2,t+1} - v_2 x_{2,t} \geq \gamma_2.$$

Assume the contrary. Then

$$x_{2,t+1} - v_2 x_{2,t} \leq \gamma_2, \ t = S_1 - L_2^*, \ldots, S_1. \tag{4.541}$$

Lemma 4.40 and (4.499) imply that

$$x_{2,S_1-L_2^*} \le M_0(1-v_2)^{-1}.$$

When combined with (4.541), (4.504), and (4.502) this inequality implies that

$$x_{2,S_1+1} \le v_2^{L_2^*} M_0(1-v_2)^{-1} + (1-v_2)^{-1}\gamma_2 \le \gamma_1/2.$$

This contradicts (4.533). The contradiction we have reached proves that there exists an integer

$$S_2 \in [S_1 - L_2^*, S_1] \text{ such that } x_{2,S_2+1} - v_2 x_{2,S_2} \ge \gamma_2. \tag{4.542}$$

In view of (4.441)–(4.443), (4.448) (4.505), and (4.499), we have

$$x_{1,S_2+1} + x_{2,S_2+1} - v_2 x_{2,S_2} \le g(x_{1,S_2}) \le g^{L_2^*+1}(x_{1,S_2-L_2^*}) \le x^* + \gamma_2/4.$$

When combined with (4.542) this inequality implies that

$$x_{1,S_2+1} \le x^* - 3\gamma_2/4. \tag{4.543}$$

For all integers t satisfying $S_0 \le t \le S_2 - L_2^*$ we set

$$y_t' = f_2(x_{2t}'), \quad x_{1,t+1}' = v_1 x_{1,t}' + (1-v_1)\gamma_0,$$
$$x_{2,t+1}' = v_2 x_{2,t}' + f_1(\gamma_0) - (1-v_1)\gamma_0. \tag{4.544}$$

Relations (4.544), (4.539), and (4.532) imply that

$$x_{1,t}' \ge \gamma_0, \quad t = S_0 \dots, S_2 - L_2^* + 1, \tag{4.545}$$
$$x_{2,t}' \ge f_1(\gamma_0) - (1-v_1)\gamma_0, \quad t = S_0 + 1, \dots, S_2 - L_2^* + 1. \tag{4.546}$$

It is now easy to see that $(\{x_{1,t}'\}_{t=0}^{S_2-L_2^*+1}, \{x_{2,t}'\}_{t=0}^{S_2-L_2^*+1}, \{y_t'\}_{t=0}^{S_2-L_2^*})$ is a program.
For all integers t satisfying $S_2 - L_2^* + 1 \le t \le S_2$ we put

$$y_t' = f_2(x_{2t}'), \quad x_{1,t+1}' = v_1 x_{1,t}' + f_1(x_{1,t}'), \quad x_{2,t+1}' = v_2 x_{2,t}'. \tag{4.547}$$

In view of (4.547), (4.448), (4.545), (4.506), and (4.543), we have

$$x_{1,S_2+1}' = g^{L_2^*}(x_{1,S_2-L_2^*+1}') \ge g^{L_2^*}(\gamma_0) > x^* - \gamma_2/4 > x_{1,S_2+1}. \tag{4.548}$$

It is clear that $(\{x_{1,t}'\}_{t=0}^{S_2+1}, \{x_{2,t}'\}_{t=0}^{S_2+1}, \{y_t'\}_{t=0}^{S_2})$ is a program. For all integers t satisfying $S_2 + 1 \le t < T$ we set

$$y_t' = f_2(x_{2,t}'), \quad x_{1,t+1}' = v_1 x_{1,t} + x_{1,t+1} - v_1 x_{1,t}, \tag{4.549}$$
$$x_{2,t+1}' = v_2 x_{2,t}' + x_{2,t+1}' - v_2 x_{2,t}'.$$

It follows from (4.548), (4.549), (4.499) and Lemma 4.40 that

$$x'_{1,t} \geq x_{1,t}, \ t = S_2 + 1, \ldots, T \tag{4.550}$$

and for all integers t satisfying $S_2 + 1 < t \leq T$,

$$x_{2,t} - x'_{2,t} = v_2^{t-S_2-1}(x_{2,S_2+1} - x'_{2,S_2+1}) \leq v_2^{t-S_2-1}(M_0(1-v_2)^{-1}). \tag{4.551}$$

It is not difficult to see that $(\{x'_{1,t}\}_{t=0}^T, \{x'_{2,t}\}_{t=0}^T, \{y'_t\}_{t=0}^{T-1})$ is a program. Relations (4.539), (4.538) and (4.499) imply that

$$0 \leq U(x_{1,0}, x_{2,0}, T) - \sum_{t=0}^{T-1} w_t(y'_t) = \sum_{t=0}^{T-1} w_t(y_t) - \sum_{t=0}^{T-1} w_t(y'_t)$$

$$= \sum_{t=0}^{T-1} w_t(y_t) - \sum_{t=0}^{S_0-1} w_t(\bar{y}_t) - \sum_{t=S_0}^{T-1} w_t(y'_t). \tag{4.552}$$

In view of (4.532), Lemma 4.40, (4.499), (4.509), (4.511), (4.531), and (4.508), we have

$$\sum_{t=0}^{S_0-1} w_t(y_t) - \sum_{t=0}^{S_0-1} w_t(\bar{y}_t) \leq \sum_{t=0}^{L-1} w_t(y_t) - \sum_{t=0}^{L-1} w_t(\bar{y}_t)$$

$$+ (L_1^* + L_2^*) \sup\{w_i(f_2(M_0(1-v_2)^{-1})) : \text{an integer } i \geq p\}$$

$$\leq U(x_{1,0}, x_{2,0}, L) - \delta - U(x_{1,0}, x_{2,0}, L) + \delta/8 + \delta/8 = -(3/4)\delta. \tag{4.553}$$

It follows from (4.499), Lemma 4.40, (4.544), (4.542), (4.533), (4.534), (4.549), (4.546), (4.509), (4.508), (4.501), and (4.532) that

$$\sum_{t=S_0}^{S_2-L_2^*} w_t(y_t) - \sum_{t=S_0}^{S_2-L_2^*} w_t(y'_t) \leq \sum_{t=S_0}^{S_0+L_1^*+2L_2^*} w_t(f_2(M_0(1-v_2)^{-1}))$$

$$+ \sum \{w_t(f_2(\gamma_1)) : t \text{ is an integer such that } S_0 + L_1^* + 2L_2^* < t \leq S_2 - L_2^*\}$$

$$- \sum_{t=S_0+1}^{S_2-L_2^*} w_t(f_2(f_1(\gamma_0) - (1-v_1)\gamma_0)) \leq \sum_{t=S_0}^{S_0+L_1^*+2L_2^*} w_t(f_2(M_0(1-v_2)^{-1}))$$

$$\leq (L_1^*+2L_2^*+1) \sup\{w_i(f_2(M_0(1-v_2)^{-1})) : i \text{ is an integer and } i \geq p\} \leq \delta/8. \tag{4.554}$$

In view of (4.499), Lemma 4.40, (4.468), (4.533), (4.532), (4.509) and (4.508), we have

$$\sum_{t=S_2-L_2^*}^{S_2} w_t(y_t) - \sum_{t=S_2-L_2^*}^{S_2} w_t(y_t')$$

$$\leq (L_2^* + 1)\sup\{w_i(f_2(M_0(1-v_2)^{-1})) : i \text{ is an integer and } i \geq p\} \leq \delta/8. \tag{4.555}$$

It follows from (4.499), Lemma 4.40, (4.542), (4.533), (4.532), (4.509), (4.508), (4.539), (4.531) and (4.468) that

$$\sum_{t=S_2+1}^{T-1} w_t(y_t) - \sum_{t=S_2+1}^{T-1} w_t(y_t')$$

$$= \sum\{w_t(y_t) - w_t(y_t') : \text{ an integer } t \in [S_2+1, T-1] \text{ and } t \leq 1 + S_2 + L_2^*\}$$

$$+ \sum\{w_t(y_t) - w_t(y_t') : \text{ an integer } t \in [S_2+1, T-1] \text{ and } t > S_2 + L_2^* + 1\}$$

$$\leq (L_2^* + 1)\sup\{w_i(f_2(M_0(1-v_2)^{-1})) : \text{ an integer } i \geq p\}$$

$$+ \sum\{w_t(y_t) - w_t(y_t') : \text{ an integer } t \text{ satisfies } S_2 + L_2^* + 1 < t \leq T-1\}$$

$$\leq \delta/8 + \sum\{w_t(y_t) - w_t(y_t') : \text{ an integer } t \text{ satisfies}$$

$$1 + S_2 + L_2^* < t \leq T-1 \text{ and } y_t > y_t'\}$$

$$\leq \delta/8 + \sum\{c_0(y_t - y_t') : \text{ an integer } t$$

$$\text{satisfies } 1 + S_2 + L_2^* < t \leq T-1 \text{ and } y_t > y_t'\}.$$

In view of (4.549), (4.467), (4.499), Lemma 4.40, (4.551), (4.507), and the relation above, we have

$$\sum_{t=S_2+1}^{T-1} w_t(y_t) - \sum_{t=S_2+1}^{T-1} w_t(y_t')$$

$$\leq \delta/8 + c_0 \sum\{y_t - y_t' : \text{ an integer } t$$

$$\text{satisfies } 1 + S_2 + L_2^* < t \leq T-1 \text{ and } y_t > y_t'\}$$

$$\leq \delta/8 + c_0 \sum\{y_t - f_2(x_{2,t}') :$$

$$\text{an integer } t \text{ satisfies } 1 + S_2 + L_2^* < t \leq T-1 \text{ and } y_t > f_2(x_{2,t}')\}$$

$$\leq \delta/8 + c_0 \sum\{f(x_{2,t}) - f_2(x_{2,t}') :$$

$$\text{an integer } t \text{ satisfies } 1 + S_2 + L_2^* < t \leq T-1 \text{ and } x_{2,t} > x_{2,t}'\}$$

$$\le \delta/8 + c_0^2 \sum \{x_{2,t} - x_{2,t}' :$$

an integer t satisfies $1 + S_2 + L_2^* < t \le T - 1$ and $x_{2,t} > x_{2,t}'\}$

$$\le \delta/8 + c_0^2 \sum \{v_2^{t-S_2-1}(M_0(1-v_2)^{-1}) :$$

an integer t satisfies $1 + S_2 + L_2^* < t \le T - 1\}$

$$\le \delta/8 + c_0^2 v_2^{2+L_2^*}(M_0(1-v_2)^{-1})(1-v_2)^{-1} < \delta/4.$$

It follows from (4.552), (4.553), (4.554), (4.555), and the relation above that

$$0 \le U(x_{1,0}, x_{2,0}, T) - \sum_{t=0}^{T-1} w_t(y_t') \le -(3/4)\delta + \delta/8 + \delta/8 + \delta/4 < -\delta/4,$$

a contradiction. The contradiction we have reached proves (4.500) and Lemma 4.42 itself. □

4.15　Proof of Theorem 4.39

Proposition 4.37 implies that for each point $z = (z_1, z_2) \in R_+^1 \times R_+^1$ and each natural number T there exists a program

$$(\{x_{1,t}^{(z,T)}\}_{t=0}^T, \{x_{2,t}^{(z,T)}\}_{t=0}^T, \{y_t^{(z,T)}\}_{t=0}^{T-1})$$

such that

$$x_{1,0}^{(z,T)} = z_1, \quad x_{2,0}^{(z,T)} = z_2, \quad \sum_{t=0}^{T-1} w_t(y_t^{(z,T)}) = U(z_0, z_1, T). \tag{4.556}$$

Let

$$z_1 \in [m_0, M_0], \quad z_2 \in [0, M_0] \tag{4.557}$$

be given. In view of (4.556), (4.557) and Lemma 4.40, there exist a strictly increasing sequence of natural numbers $\{T_k\}_{k=1}^\infty$ and a program

$$\{x_{1,t}^{(z)}, x_{2,t}^{(z)}, y_t^{(z)}\}_{t=0}^\infty$$

such that for each nonnegative integer t, we have

$$x_{1,t}^{(z)} = \lim_{k\to\infty} x_{1,t}^{(z,T_k)}, \quad x_{2,t}^{(z)} = \lim_{k\to\infty} x_{2,t}^{(z,T_k)}, \quad y_t^{(z)} = \lim_{k\to\infty} y_t^{(z,T_k)}. \tag{4.558}$$

It is clear that

$$x_{1,0}^{(z)} = z_1, \ x_{2,0}^{(z)} = z_2. \tag{4.559}$$

Let $\delta > 0$ be given. Lemma 4.42 and (4.556) imply that there exists an integer $L_\delta \geq 1$ such that the following property holds:

(P7) For each integer $L \geq L_\delta$ there exists a natural number $\tau_L \geq L$ such that for each integer $T \geq \tau_L$, each real number $z_1 \in [m_0, M_0]$, and each real number $z_2 \in [0, M_0]$, we have

$$\sum_{t=0}^{T-1} w_t(y_t^{(z,T)}) \geq U(z_1, z_2, L) - \delta/4.$$

Let (4.557) hold, an integer $L \geq L_\delta$, and let an integer $\tau_L \geq L$ be as guaranteed by property (P7). By (4.558) there exists an integer $k \geq 1$ such that

$$T_k > \tau_L,$$

$$\left| \sum_{t=0}^{L-1} w_t(y_t^{(z)}) - \sum_{t=0}^{L-1} w_t(y_t^{(z,T_k)}) \right| \leq \delta/4. \tag{4.560}$$

In view of (4.556), (4.557), (4.560), and (P7),

$$\sum_{t=0}^{L-1} w_t(y_t^{(z,T_k)}) \geq U(z_1, z_2, L) - \delta/4.$$

When combined with inequality (4.560) this relation implies that

$$\sum_{t=0}^{L-1} w_t(y_t^{(z)}) \geq U(z_1, z_2, L) - \delta.$$

Theorem 4.39 is proved.

4.16 Autonomous Discrete-Time Periodic Problems

Let

$$|x| = \max\{|x_i| : \ i = 1, \ldots, n\} \text{ for all } x = (x_1, \ldots, x_n) \in R^n$$

and let $\mathbf{Z}$ be the set of all integers. Assume that $v : R^n \times R^n \to R^1$ is a lower semicontinuous function (i.e., $v(\lim_{k \to \infty}(x_k, y_k)) \leq \liminf_{k \to \infty} v(x_k, y_k)$) which satisfies the following assumptions:

$$\sup\{v(x, y) : \ x, y \in R^n, \ 0 \leq x_i \leq 1$$

$$\text{and } 0 \leq y_i - x_i \leq 1 \text{ for } i = 1, \ldots, n\} = a < \infty, \tag{4.561}$$

$$\inf\{v(x, y) : \ x, y \in R^n\} = b > -\infty, \tag{4.562}$$

$$v(x + m, y + m) = v(x, y) \text{ for each } x, y \in R^n \text{ and each } m \in \mathbf{Z}^n; \tag{4.563}$$

there exists a positive number Γ such that

$$\inf\{v(x, y) : \ x, y \in R^n \text{ and } |x - y| \geq \Gamma\} \geq a. \tag{4.564}$$

We prove the following result obtained in [31].

Theorem 4.43. *There exists a constant μ such that:*

(1) For every sequence $\{z_i\}_{i=0}^{\infty} \subset R^n$ and every nonnegative integer N the inequality

$$\sum_{i=0}^{N} [v(z_i, z_{i+1}) - \mu] \geq b - a$$

holds.

(2) For every sequence $\{z_i\}_{i=0}^{\infty} \subset R^n$ the sequence

$$\left\{ \sum_{i=0}^{N} [v(z_i, z_{i+1}) - \mu] \right\}_{N=0}^{\infty}$$

is either bounded or it diverges to infinity.

(3) For every initial point $z_0 \in R^n$ there exists a sequence $\{z_i^*\}_{i=0}^{\infty}$ with $z_0^* = z_0$ which satisfies

$$\left| \sum_{i=0}^{N} [v(z_i^*, x_{i+1}^*) - \mu] \right| \leq 4(a - b)$$

for all nonnegative integers N.

We precede the proof of the theorem by auxiliary lemmas.
Set

$$\mu = \inf \left\{ \liminf_{N \to \infty} N^{-1} \sum_{i=0}^{N-1} v(z_i, z_{i+1}) : \ \{z_i\}_{i=0}^{\infty} \subset R^n \right\}. \tag{4.565}$$

For any integer number $N \geq 1$ put

$$\lambda(N) = \inf \left\{ N^{-1} \sum_{i=0}^{N-1} v(z_i, z_{i+1}) : \ \{z_i\}_{i=0}^{N} \subset R^n \text{ and } z_N - z_0 \in \mathbf{Z}^n \right\},$$

$$\tag{4.566}$$

$$\rho(N) = \inf \left\{ N^{-1} \sum_{i=0}^{N-1} v(z_i, z_{i+1}) : \{z_i\}_{i=0}^N \subset R^n \right\}. \tag{4.567}$$

Remark 4.44. Let $N \geq 1$ be an integer and let $\{z_i\}_{i=0}^N \subset R^n$ satisfy $z_N - z_0 \in \mathbf{Z}^n$. We can associate with $\{z_i\}_{i=0}^N$ a sequence $\{y_i\}_{i=0}^\infty \subset R^n$ such that

$$y_i = z_i, \ i = 0, \dots, N,$$

$$y_{i+jN} = y_i + j(z_N - z_0) \text{ for all integers } i, j \geq 0.$$

It follows from Remark 4.44 and relations (4.562), (4.563), (4.565), (4.566), and (4.567) that

$$\rho(N) \leq \mu \leq \lambda(N), \ N = 1, 2, \dots \tag{4.568}$$

Put

$$A = \{(x + m, y + m) : \ x = (x_1, \dots, x_n), \ y = (y_1, \dots, y_n) \in R^n \text{ satisfy}$$

$$0 \leq x_i \leq 1, \ 0 \leq y_i - x_i \leq 1 \text{ for } i = 1, \dots, n \text{ and } m \in \mathbf{Z}^n\}.$$

Lemma 4.45. $N(\lambda(N) - \rho(N)) \leq a - b$ for all integers $N \geq 1$.

Proof. Let $N \geq 1$ be an integer and $\{z_i\}_{i=0}^N \subset R^n$. It is clear that there exists a sequence $\{y_i\}_{i=0}^N \subset R^n$ such that

$$y_i = z_i, \ i = 0, \dots, N - 1, \ y_N - y_0 \in \mathbf{Z}^n \text{ and } (y_{N-1}, y_N) \in A.$$

Relations (4.561)–(4.563) and (4.566) imply that

$$N\lambda(N) \leq \sum_{i=0}^{N-1} v(y_i, y_{i+1}) \leq \sum_{i=0}^{N-1} v(z_i, z_{i+1}) - b + a.$$

Since this inequality holds for an arbitrary sequence $\{z_i\}_{i=0}^N \subset R^n$, this completes the proof of the lemma. □

Lemma 4.46. *Let* $\{z_i\}_{i=0}^\infty \subset R^n$ *and let* $q \geq 1$ *be an integer such that* $|z_q - z_{q-1}| \geq \Gamma$. *Assume that a sequence* $\{y_i\}_{i=0}^\infty \subset R^n$ *satisfies*

$$y_i = z_i, \ i = 0, \dots, q - 1, \ (y_{q-1}, y_q) \in A,$$

$$y_i - z_i = y_q - z_q \in \mathbf{Z}^n \text{ for all integers } i \geq q.$$

Then $v(z_i, z_{i+1}) \geq v(y_i, y_{i+1})$ *for all nonnegative integers* i.

The validity of Lemma 4.46 follows from relations (4.561), (4.563), and (4.564).

Proof of Theorem 4.43. Let $\{z_i\}_{i=0}^\infty \subset R^n$ and $N \geq 1$ be an integer. There exists a sequence of points $\{y_i\}_{i=0}^N \subset R^n$ such that

$$y_i = z_i, \; i = 0, \ldots, N-1, \; (y_{N-1}, y_N) \in A \text{ and } y_N - y_0 \in \mathbf{Z}^n.$$

By (4.561)–(4.563), (4.566), and (4.568), we have

$$\sum_{i=0}^{N-1} v(z_i, z_{i+1}) \geq \sum_{i=0}^{N-1} v(y_i, y_{i+1}) + b - a \geq N\lambda(N) + b - a \geq N\mu + b - a.$$

Assertion 1 of Theorem 4.43 is established.

Assertion 2 follows from Assertion 1. Let us prove Assertion 3. It is sufficient to establish the existence of a sequence $\{z_i\}_{i=0}^\infty \subset R^n$ such that

$$\left| \sum_{i=0}^N [v(z_i, z_{i+1}) - \mu] \right| \leq 2(a - b) \text{ for all integers } N \geq 0.$$

We can assume without loss of generality that $\Gamma > 2$. Let $N \geq 1$ be an integer. By Lemma 4.46, there exists a sequence $\{z_i^N\}_{i=0}^N \subset R^n$ such that

$$|z_i^N - z_{i+1}^N| \leq \Gamma, \; i = 0 \ldots, N-1, \; z_0^N - z_N^N \in \mathbf{Z}^n, \; |z_0^N| \leq 1$$

and

$$\sum_{i=0}^{N-1} v(z_i^N, z_{i+1}^N) = N\lambda(N).$$

Lemma 4.45 and (4.568) imply that

$$\sum_{i=0}^{N-1} [v(z_i^N, z_{i+1}^N) - \mu] \leq a - b, \; N = 1, 2, \ldots. \tag{4.569}$$

It is clear that there exists a strictly increasing sequence of natural numbers $\{N_j\}_{j=1}^\infty$ such that for every nonnegative integer i, we have

$$z_i^{N_j} \to y_i \in R^n \text{ as } j \to \infty.$$

Let a natural number N be given. For all large integers $j \geq 1$ assertion 1 and (4.569) imply that

$$\sum_{i=0}^{N_j-1} [v(z_i^{N_j}, z_{i+1}^{N_j}) - \mu] \leq a - b,$$

$$\sum_{i=N}^{N_j-1} [v(z_i^{N_j}, z_{i+1}^{N_j}) - \mu] \geq -a + b,$$

$$\sum_{i=0}^{N-1} [v(z_i^{N_j}, z_{i+1}^{N_j}) - \mu] \leq 2(a - b).$$

This relation implies that

$$\sum_{i=0}^{N-1} [v(y_i, y_{i+1}) - \mu] \leq 2(a - b),$$

which completes the proof of the theorem. $\square$

The next result was also obtained in [31].

Theorem 4.47. *Let v be a continuous function. We define*

$$\pi(x) = \inf \left\{ \liminf_{N \to \infty} \sum_{i=0}^{N-1} [v(z_i, z_{i+1}) - \mu] : \{z_i\}_{i=0}^{\infty} \subset R^n, \ z_0 = x \right\},$$

$$\theta(x, y) = v(x, y) - \mu + \pi(y) - \pi(x) \qquad (4.570)$$

for each pair of points $x, y \in R^n$. Then $\pi : R^n \to R^1$ and $\theta : R^n \times R^n \to R^1$ are continuous functions,

$$\pi(x + m) = \pi(x), \ \theta(x + m, y + m) = \theta(x, y)$$

for each $x, y \in R^n$ and each $m \in \mathbf{Z}^n$,

the function θ is nonnegative, and

$$E(x) = \{y \in R^n : \ \theta(x, y) = 0\}$$

is nonempty for any point $x \in R^n$.

Proof. We can assume without loss of generality that $\Gamma > 2$. For every point $x \in R^n$ put

$$\Lambda(x) = \{\{z_i\}_{i=0}^{\infty} \subset R^n : \ z_0 = x \text{ and } |z_1 - z_0| \leq \Gamma\}.$$

It is not difficult to see that

$$\pi(x + m) = \pi(x), \ \theta(x + m, y + m) = \theta(x, y)$$

for each pair of points $x, y \in R^n$ and each $m \in \mathbf{Z}^n$ and

$$\pi(x) \le v(x, y) - \mu + \pi(y) \text{ for all } x, y \in R^n.$$

Thus the function θ is nonnegative. By Lemma 4.46, we have

$$\pi(x) = \inf \left\{ \liminf_{N \to \infty} \sum_{i=0}^{N-1} [v(z_i, z_{i+1}) - \mu] : \{z_i\}_{i=0}^{\infty} \in \Lambda(x) \right\}, \ x \in R^n.$$

This equality and the uniform continuity of the function v on bounded subsets of $R^n \times R^n$ imply the continuity of the function π.

It only remains to prove that the set $E(x) \ne \emptyset$ for every point $x \in R^n$. Suppose to the contrary that for some point $x \in R^n$ the set $E(x) = \emptyset$. There exists a sequence $\{x_i\}_{i=1}^{\infty} \subset R^n$ such that $\theta(x, x_i) \to \inf\{\theta(x, y) : y \in R^n\}$ as $i \to \infty$.

Let $i \ge 1$ be an integer. If $|x_i - x| > \Gamma$ we choose a point $y_i \in R^n$ such that $(x, y_i) \in A$ and $y_i - x_i \in \mathbf{Z}^n$. If $|x_i - x| \le \Gamma$, we set $y_i = x_i$. By (4.561), (4.563), and (4.564),

$$\theta(x, y_i) \le \theta(x, x_i), \ i = 1, 2, \dots .$$

Now it is not difficult to see that there exists a point $\bar{x} \in R^n$ such that

$$\theta(x, \bar{x}) = \inf\{\theta(x, y) : y \in R^n\} = \delta > 0.$$

There exists a sequence $\{z_i\}_{i=1}^{\infty} \subset R^n$ such that $z_0 = x$ and that

$$\liminf_{N \to \infty} \sum_{i=0}^{N-1} [v(z_i, z_{i+1}) - \mu] \le \pi(x) + 2^{-1}\delta.$$

It is clear that

$$\pi(x) + 2^{-1}\delta \ge [\theta(x, z_1) + \pi(x) - \pi(z_1)]$$

$$+ \liminf_{N \to \infty} \sum_{i=1}^{N} [v(z_i, z_{i+1}) - \mu] \ge [\delta + \pi(x) - \pi(z_1)] + \pi(z_1).$$

We obtained a contradiction; therefore the set $E(x) \ne \emptyset$ for all points $x \in R^n$. The theorem is proved. $\square$

4.17   Variational Problems with Periodic Integrands

Let $L : R^n \times R^n \to R^1$ be a bounded below Borel function which is bounded on any compact subset of R^{2n}. We assume that

$$L(x + m, v) = L(x, v) \text{ for all } x, v \in R^n \text{ and all } m \in \mathbf{Z}^n \qquad (4.571)$$

and that there exists a pair of positive numbers c_1, c_2 such that

$$L(z, y) \geq c_1 |y| \text{ for all } z, y \in R^n \text{ such that } |y| \geq c_2. \tag{4.572}$$

A trajectory is an absolutely continuous function $z : \Delta \to R^n$ where Δ is either $[a, b] \subset R^1$ or $[a, \infty)$.

We prove the following result which was obtained in [31].

Theorem 4.48. *There exist a pair of real numbers $M(L) > 0$ and $\mu(L)$ such that:*

(1) For every trajectory $z : [0, \infty) \to R^n$ and any positive number T,

$$\int_0^T [L(z(t), z'(t)) - \mu(L)]dt \geq -M(L).$$

(2) For every trajectory $z : [0, \infty) \to R^n$ the function

$$T \to \int_0^T [L(z(t), z'(t)) - \mu(L)]dt, \ T \in (0, \infty)$$

is either bounded or diverges to infinity as $T \to \infty$.

(3) For every point $z_0 \in R^n$ there exists a trajectory $z : [0, \infty) \to R^n$ such that $z(0) = z_0$ and for any positive number T,

$$\left| \int_0^T [L(z(t), z'(t)) - \mu(L)]dt \right| \leq M(L).$$

We precede the proof of Theorem 4.48 by several auxiliary results. Put

$$d_L = \inf\{L(x, y) : x, y \in R^n\}. \tag{4.573}$$

For each pair of points $x, y \in K$ we set

$$u(x, y) = \inf \left\{ \int_0^1 L(z(t), z'(t))dt : \right.$$

$$\left. z : [0, 1] \to R^n \text{ is a trajectory, } z(0) = x, \ z(1) = y \right\}.$$

It is not difficult to see that

$$\inf\{u(x, y) : x, y \in R^n\} \geq d_L, \tag{4.574}$$

the function $u : R^n \times R^n \to R^1$ is bounded on any compact subset of R^{2n} and

$$u(x + m, y + m) = u(x, y) \text{ for all } x, y \in R^n \text{ and all } m \in \mathbf{Z}^n. \tag{4.575}$$

Lemma 4.49. *For every real number $K > 0$ there exists a number $\Gamma \geq 0$ such that*

$$u(x, y) \geq K \text{ for all } x, y \in R^n \text{ satisfying } |x - y| \geq \Gamma.$$

Proof. Let $K > 0$ be a real number. Choose a real number $\Gamma > 0$ such that

$$\Gamma \geq c_2 + c_1^{-1}(K + \sup\{|L(x, y)| : x, y \in R^n, |y| \leq c_2\}). \tag{4.576}$$

Let $x, y \in R^n$ satisfy $|x - y| \geq \Gamma$ and let $z : [0, 1] \to R^n$ be a trajectory which satisfy

$$z(0) = x, \; z(1) = y.$$

Put

$$F_1 = \{t \in [0, 1] : |z'(t)| < c_2\}, \; F_2 = [0, 1] \setminus F_1.$$

In view of (4.572) and (4.576), we have

$$\int_0^1 L(z(t), z'(t))dt \geq \int_{F_2} L(z(t), z'(t))dt - \sup\{|L(\xi, \eta)| : \xi, \eta \in R^n, |\eta| \leq c_2\},$$

$$\Gamma \leq |x - y| \leq \int_0^1 |z'(t)|dt \leq c_2 + \int_{F_2} |z'(t)|dt \leq c_2 + \int_{F_2} c_1^{-1}L(z(t), z'(t))dt$$

$$\leq c_2 + c_1^{-1}\left[\int_0^1 L(z(t), z'(t))dt + \sup\{|L(\xi, \eta)| : \xi, \eta \in R^n, |\eta| \leq c_2\}\right],$$

$$\int_0^1 L(z(t), z'(t))dt \geq K.$$

Thus $u(x, y) \geq K$ and the lemma is proved.   □

For any pair of points $x, y \in R^n$ define

$$v(x, y) = \liminf_{(\xi, \eta) \to (x, y)} u(\xi, \eta),$$

where $\xi, \eta \in R^n$. It is clear that the function $v : R^n \times R^n \to R^1$ is bounded from below, lower semicontinuous function which is bounded on any compact subset of R^{2n}. By (4.575),

$$v(x + m, y + m) = v(x, y) \text{ for all } x, y \in R^n \text{ and all } m \in \mathbf{Z}^n.$$

Put

$$b = \inf\{v(x, y) : x, y \in R^n\},$$

$$a = \sup\{v(x, y) : x = (x_1, \ldots, x_n), \ y = (y_1, \ldots, y_n) \in R^n,$$

$$0 \le x_i \le 1, \ 0 \le y_i - x_i \le 1 \text{ for } i = 1, \ldots, n\},$$

$$\mu = \inf\{\liminf_{N \to \infty} N^{-1} \sum_{i=0}^{N-1} v(x_i, x_{i+1}) : \{x_i\}_{i=0}^{\infty} \subset R^n\}.$$

Lemma 4.49 implies that there exists a real number $\Gamma > 0$ such that

$$\inf\{v(x, y) : x, y \in R^n, \ |x - y| \ge \Gamma\} \ge a + 1.$$

It is not difficult to see that Theorem 4.43 is valid with v, μ, a, b.

Lemma 4.50. *Let $x, y \in R^n$ and let $\epsilon \in (0, 1/2)$. Then there exists a real number $\gamma \in (0, \epsilon)$ and a trajectory $z : [0, 1 + \gamma] \to R^n$ such that*

$$z(0) = x, \ z(1 + \gamma) = y, \ \int_0^{1+\gamma} L(z(t), z'(t))dt \le v(x, y) + \epsilon.$$

Proof. Put

$$K = \sup\{|L(z, v)| : z, v \in R^n, \ |v| \le 16\}. \tag{4.577}$$

Fix a real number $\gamma \in (0, \epsilon)$ such that

$$\gamma K < \epsilon/8. \tag{4.578}$$

Clearly, there exists a pair of points $x_1, y_1 \in R^n$ such that

$$|x - x_1| \le 8^{-1}\gamma, \ |y - y_1| \le \gamma/8, \tag{4.579}$$

$$u(x_1, y_1) < v(x, y) + \gamma/8.$$

There exists a trajectory $z_0 : [0, 1] \to R^n$ such that

$$z_0(0) = x_1, \ z_0(1) = y_1, \ \int_0^1 L(z_0(t), z_0'(t))dt < v(x, y) + \gamma/8.$$

Define a trajectory $z : [0, 1 + \gamma] \to R^n$ as follows:

$$z(t) = x + 2\gamma^{-1}t(x_1 - x), \ t \in [0, \gamma/2],$$

$$z(t) = z_0(t - \gamma/2), \ t \in [\gamma/2, 1 + \gamma/2],$$

$$z(t) = y_1 + 2\gamma^{-1}(t - 1 - \gamma/2)(y - y_1), \ t \in [1 + \gamma/2, 1 + \gamma].$$

It is clear that the trajectory z is well defined and satisfies $z(0) = x$, $z(1 + \gamma) = y$. In view of (4.577) and (4.579), we have

$$|L(z(t), z'(t))| \le K, \ t \in (0, \gamma/2)$$

and

$$L(z(t), z'(t)) \le \Delta, \ t \in (1 + \gamma/2, 1 + \gamma).$$

When combined with (4.578) and the choice of z_0 these relations imply the validity of the lemma.   □

Proof of Theorem 4.48. Put

$$\mu(L) = \mu, \ M(L) = 5(a - b) + |d_L| + |\mu| + 1.$$

It should be mentioned that Theorem 4.43 is valid with v, μ, a, b. Let $z : [0, \infty) \to R^n$ be a trajectory. Theorem 4.43 imply that

$$\int_0^N [L(z(t), z'(t)) - \mu] dt \ge \sum_{i=0}^{N-1} [v(z(i), z(i+1)) - \mu]$$

$$\ge b - a \text{ for all natural numbers } N.$$

Let $T > 0$ be a real number. There exists a nonnegative integer N such that $N < T \le N + 1$. By (4.573), we have

$$\int_0^T [L(z(t), z'(t)) - \mu] dt \ge \int_N^T [L(z(t), z'(t)) - \mu] dt + b - a$$

$$\ge b - a - |d_L| - |\mu|. \tag{4.580}$$

Assertion 1 of Theorem 4.48 is proved.   □

Assertion 2 follows from assertion 1. We prove assertion 3. Let $z_0 \in R^n$ be given. In view of Theorem 4.43, there exists a sequence $\{x_i\}_{i=0}^\infty \subset R^n$ such that $x_0 = z_0$ and that

$$\left| \sum_{i=0}^N [v(x_i, x_{i+1}) - \mu] \right| \le 4(a - b) \text{ for all integers } N \ge 0.$$

Put

$$\epsilon_i = 2^{-i}(1 + |\mu|)^{-1}, \ i = 1, 2, \dots.$$

By induction using Lemma 4.50 we construct a sequence of real numbers $\gamma_i \in (0, \epsilon_i)$, $i = 1, 2 \ldots$ and a trajectory $z : [0, \infty) \to R^n$ such that for all nonnegative integers N, we have

$$z(\beta_N) = x_N, \quad \int_{\beta_N}^{\beta_{N+1}} L(z(t), z'(t))dt \leq v(x_N, x_{N+1}) + \epsilon_{N+1},$$

where $\beta_0 = 0$, $\beta_N = \sum_{i=1}^{N} \gamma_i + N$ for all integers $N \geq 1$. It follows from the relations above and the choice of $\{x_i\}_{i=0}^{\infty}$ that for all natural numbers $N = 1, 2, \ldots$,

$$\int_0^{\beta_N} [L(z(t), z'(t)) - \mu]dt \leq -\mu\beta_N + \sum_{i=0}^{N-1} [v(x_i, x_{i+1}) + \epsilon_{i+1}]$$

$$\leq \sum_{i=0}^{N-1} [v(x_i, x_{i+1}) - \mu] - \mu(\beta_N - N) + \sum_{i=1}^{N} \epsilon_i$$

$$\leq 4(a - b) + (1 + |\mu|) \sum_{i=1}^{N} \epsilon_i \leq 4(a - b) + 1.$$

Let $T > 0$ be a real number. Choose an integer $N \geq 1$ such that $\beta_N > T + 1$. Then in view of (4.580) which holds for any trajectory, we have

$$\int_0^T [L(z(t), z'(t)) - \mu]dt = \int_0^{\beta_N} [L(z(t), z'(t)) - \mu]dt$$

$$- \int_T^{\beta_N} [L(z(t), z'(t)) - \mu]dt$$

$$\leq 4(a - b) + 1 + (a - b + |d_L| + |\mu|) \leq M(L).$$

This completes the proof of the theorem.

For every point $x \in R^n$ put

$$\pi(x) = \inf \left\{ \liminf_{T \to \infty} \int_0^T [L(z(t), z'(t)) - \mu(L)]dt : \right.$$

$$\left. z : [0, \infty) \to R^n \text{ is a trajectory and } z(0) = x \right\}.$$

In view of Theorem 4.48 the function $\pi : R^n \to R^1$ is bounded:

$$|\pi(x)| \leq M(L) \text{ for each } x \in R^n,$$

$$\pi(x + m) = \pi(x) \text{ for each } x \in R^n \text{ and each } m \in \mathbf{Z}^n.$$

Let δ be a positive number. A trajectory $s : [0, \infty) \rightarrow R^n$ is called δ-weakly optimal if there exists a strictly increasing sequence of positive numbers $\{T_i\}_{i=1}^{\infty}$ such that $T_i \rightarrow \infty$ as $i \rightarrow \infty$ and that for any trajectory $z : [0, \infty) \rightarrow R^n$ satisfying $z(0) = s(0)$ the relation

$$\int_0^{T_i} [L(s(t), s'(t)) - L(z(t), z'(t))]dt \leq \delta$$

holds for all large natural numbers i.

Proposition 4.51. *For each point $x \in R^n$ and each positive number δ there exists a δ-weakly optimal trajectory $s : [0, \infty) \rightarrow R^n$ such that $s(0) = x$.*

Proof. There exists a trajectory $s : [0, \infty) \rightarrow R^n$ such that $s(0) = x$ and that

$$\liminf_{T \to \infty} \int_0^T [L(s(t), s'(t)) - \mu(L)]dt \leq \pi(x) + \delta/4.$$

To complete the proof it is sufficient to note that there exists a strictly increasing sequence of positive numbers $\{T_i\}_{i=1}^{\infty}$ such that $T_i \rightarrow \infty$ and

$$\lim_{i \to \infty} \int_0^{T_i} [L(s(t), s'(t)) - \mu(L)]dt \leq \pi(x) + \delta/2.$$

$\square$

Proposition 4.52. $\pi : R^n \rightarrow R^1$ *is a Lipschitzian function.*

Proof. Put

$$K = \sup\{|L(z, v)| : z, v \in R^n \text{ and } |v| \leq 16\}.$$

Let a pair of points $x, y \in R^n$ satisfy $0 < |x - y| \leq 1$ and let $z(\cdot) : [0, \infty) \rightarrow R^n$ be a trajectory such that $z(0) = y$. Define a trajectory $z_1 : [0, \infty) \rightarrow R^n$ as follows:

$$z_1(t) = x + t|x - y|^{-1}(y - x), \ t \in [0, |x - y|],$$

$$z_1(t + |x - y|) = z(t), \ t \in [0, \infty).$$

It is clear that the trajectory z_1 is well defined and

$$\pi(x) \leq \liminf_{T \to \infty} \int_0^T [L(z_1(t), z_1'(t)) - \mu(L)]dt$$

$$= \int_0^{|x-y|} [L(z_1(t), z_1'(t)) - \mu(L)]dt$$

$$+ \liminf_{T \to \infty} \int_0^T [L(z(t), z'(t)) - \mu(L)]dt$$

$$\leq \liminf_{T \to \infty} \int_0^T [L(z(t), z'(t)) - \mu(L)]dt + |x - y|(|\mu(L)| + K).$$

Clearly, this relation holds for any trajectory $z : [0, \infty) \to R^n$ satisfying $z(0) = y$. Therefore

$$\pi(x) \leq \pi(y) + |x - y|(|\mu(L)| + \delta).$$

This completes the proof of the proposition. $\qquad\square$

4.18 Nonautonomous Discrete-Time Periodic Problems

Let $v_i : R^n \times R^n \to R^1 \cup \{\infty\}$, $i = 0, 1, 2, \dots$ be a sequence of functions such that for each nonnegative integer i, the following conditions hold:

$$a_i = \sup\{v_i(x, y) : x = (x_1, \dots, x_n), \ y = (y_1, \dots, y_n) \in R^n, \qquad (4.581)$$

$$0 \leq x_j \leq 1, \ 0 \leq y_j - x_j \leq 1 \text{ for all } j = 1, \dots, n\} < \infty,$$

$$b_i = \inf\{v_i(x, y) : x, y \in R^n\} > -\infty, \qquad (4.582)$$

$$v_i(x + m, y + m) = v_i(x, y) \text{ for each } x, y \in R^n \text{ and each } m \in \mathbf{Z}^n, \qquad (4.583)$$

there exists a positive number Γ_i such that

$$\inf\{v_i(x, y) : x, y \in R^n \text{ and } |x - y| \geq \Gamma_i\} \geq a_i. \qquad (4.584)$$

We suppose that

$$a = \sup\{a_i : i = 0, 1, \dots\} < \infty, \qquad (4.585)$$

$$b = \inf\{b_i : i = 0, 1, \dots\} > -\infty. \qquad (4.586)$$

We may assume without loss of generality that

$$\Gamma_i \geq 2 \text{ for all nonnegative integers } i. \qquad (4.587)$$

For every point $x \in R^n$ and every integer $N \geq 1$, put

$$S(x, N) = \inf \left\{ \sum_{i=0}^{N-1} v_i(z_i, z_{i+1}) : \{z_i\}_{i=0}^N \subset R^n, \ z_0 = x \right\}.$$

Define

$$A = \{(x, y) \in R^n \times R^n : x = (x_1, \ldots, x_n), \ y = (y_1, \ldots, y_n), \quad (4.588)$$
$$0 \le y_i - x_i \le 1 \text{ for } i = 1, \ldots, n\}.$$

Relations (4.581), (4.583), and (4.584) imply the following lemma.

Lemma 4.53. *Let $\{z_i\}_{i=0}^{\infty} \subset R^n$ and let $q \ge 1$ be an integer such that $|z_q - z_{q-1}| \ge \Gamma_{q-1}$. Let*

$$y_i = z_i, \ i = 0, \ldots, q-1, \ y_q - z_q \in \mathbf{Z}^n, \ (y_{q-1}, y_q) \in A,$$

$$y_i = z_i + y_q - z_q \text{ for all integers } i \ge q.$$

Then $v_i(z_i, z_{i+1}) \ge v_i(y_i, y_{i+1}), \ i = 0, 1, \ldots.$

We prove the following result which was obtained in [31].

Theorem 4.54. *Let $v_i, \ i = 0, 1, \ldots$ be a sequence of lower semicontinuous functions. Then for every point $x \in R^n$ there exists a sequence $\{x_i\}_{i=0}^{\infty} \subset R^n$ such that*

$$x_0 = x, \ |x_i - x_{i+1}| \le \Gamma_i, \ i = 0, 1, \ldots.$$

$$\sum_{i=0}^{N-1} v_i(x_i, x_{i+1}) \le S(x, N) + a_N - b_N, \ N = 1, 2, \ldots.$$

Proof. Let $x \in K$ be given. By Lemma 4.53, for each integer $N \ge 1$, there exists a sequence $\{z_i^N\}_{i=0}^{N} \subset R^n$ such that

$$z_0^N = x, \ |z_{i+1}^N - z_i^N| \le \Gamma_i, \ i = 0, \ldots, N-1,$$

$$\sum_{i=0}^{N-1} v_i(z_i^N, z_{i+1}^N) = S(x, N).$$

Let m, N be integers such that $1 \le m < N$. It is clear that there exists a sequence $\{z_i\}_{i=0}^{N} \subset R^n$ such that

$$z_i = z_i^m, \ i = 0, \ldots, m, \ z_{m+1} - z_{m+1}^N \in \mathbf{Z}^n, \ (z_m, z_{m+1}) \in A,$$

$$z_i = z_i^N + z_{m+1} - z_{m+1}^N, \ i = m+1, \ldots, N.$$

By (4.581), (4.582), (4.588), and (4.583),

$$0 \le \sum_{i=0}^{N-1} [v_i(z_i, z_{i+1}) - v_i(z_i^N, z_{i+1}^N)]$$

$$= S(x, m) - \sum_{i=0}^{m-1} v_i(z_i^N, z_{i+1}^N) + v_m(z_m, z_{m+1}) - v_m(z_m^N, z_{m+1}^N),$$

$$\sum_{i=0}^{m-1} v_i(z_i^N, z_{i+1}^N) \le S(x, m) + a_m - b_m \qquad (4.589)$$

for each pair of integers m, N satisfying $1 \le m < N$. There exists a strictly increasing sequence of natural numbers $\{N_k\}_{k=1}^{\infty}$ such that $z_i^{N_k} \to x_i$ as $k \to \infty$ for every nonnegative integer i. It follows from (4.589) that

$$\sum_{i=0}^{m-1} v_i(x_i, x_{i+1}) \le S(x, m) + a_m - b_m, \ m = 1, 2, \dots.$$

The theorem is proved. $\qquad\qquad\qquad\qquad\qquad\qquad\qquad\qquad\qquad\qquad\qquad\quad$ $\square$

Theorem 4.54 implies the following result.

Theorem 4.55. *Let v_i, $i = 0, 1, \dots$ be a sequence of lower semicontinuous functions and $a_i - b_i \to 0$ as $i \to \infty$. Then for each point $x \in R^n$ there exists a sequence $\{x_i\}_{i=1}^{\infty} \subset R^n$ such that*

$$x_0 = x, \ |x_i - x_{i+1}| \le \Gamma_i, \ i = 0, 1, \dots,$$

$$S(x, N) - \sum_{i=0}^{N-1} v_i(x_i, x_{i+1}) \to 0 \text{ as } N \to \infty.$$

The following result was also obtained in [31].

Theorem 4.56. *Let $a_i - b_i \to 0$ as $i \to \infty$. Then for each point $x \in R^n$ and each real number $\epsilon > 0$ there exists a sequence $\{y_i\}_{i=0}^{\infty} \subset R^n$ such that*

$$y_0 = x, \ |y_i - y_{i+1}| \le \Gamma_i, \ i = 0, 1, \dots$$

and that

$$\sum_{i=0}^{N-1} v_i(y_i, y_{i+1}) \le S(x, N) + \epsilon$$

for all sufficient large natural numbers N.

Proof. Let $x \in R^n$ and $\epsilon > 0$ be given. Put $\epsilon_i = 2^{-i-3}\epsilon, i = 1, 2, \ldots$. By Lemma 4.53, for every integer $N \geq 1$ there exists a sequence $\{z_i^N\}_{i=0}^N \subset R^n$ such that $z_0^N = x$,

$$|z_i^N - z_{i+1}^N| \leq \Gamma_i, \ i = 0, \ldots, N-1, \tag{4.590}$$

and that

$$\sum_{i=0}^{N-1} v_i(z_i^N, z_{i+1}^N) \leq S(x, N) + \epsilon_N. \tag{4.591}$$

Let m, N be integers satisfying $1 \leq m < N$. There exists a sequence $\{z_i(m, N)\}_{i=0}^N$ such that

$$z_i(m, N) = z_i^m, \ i = 0, \ldots, m, \ (z_m(m, N), z_{m+1}(m, N)) \in A,$$

$$z_{m+1}(m, N) - z_{m+1}^N \in \mathbf{Z}^n, \ z_i(m, N) - z_i^N = z_{m+1}(m, N) - z_{m+1}^N,$$

$$i = m+1, \ldots, N.$$

By (4.591), (4.588), (4.582), (4.583), and (4.581),

$$\epsilon_N \geq \sum_{i=0}^{N-1} [v_i(z_i^N, z_{i+1}^N) - v_i(z_i(m, N), z_{i+1}(m, N))]$$

$$\geq \sum_{i=0}^{m-1} v_i(z_i^N, z_{i+1}^N) - S(x, m) - \epsilon_m + b_m - v_m(z_m(m, N), z_{m+1}(m, N))$$

$$\geq \sum_{i=0}^{m-1} v_i(z_i^N, z_{i+1}^N) - S(x, m) - \epsilon_m - a_m + b_m$$

$$\geq b_m - a_m - \epsilon_m, \tag{4.592}$$

$$\sum_{i=0}^{m-1} v_i(z_i^N, z_{i+1}^N) \leq S(x, m) + \epsilon_m + \epsilon_N + a_m - b_m \tag{4.593}$$

for each pair of integers m, N satisfying $1 \leq m < N$. Choose a strictly increasing sequence of nonnegative integers $\{N_i\}_{i=0}^\infty$ such that $N_0 = 0$, $N_{i+1} - N_i \geq 10$ for all nonnegative integers i and that

$$\sum_{i=1}^\infty (a_{N_i} - b_{N_i}) < \epsilon/8. \tag{4.594}$$

It is not difficult to see that there exists a sequence $\{y_i\}_{i=0}^{\infty} \subset R^n$ such that

$$y_0 = x, \ y_i = z_i^{N_1}, \ i = 1, \ldots, N_1$$

and that for all integers $k \geq 1$, we have

$$(y_{N_k}, y_{N_k+1}) \in A, \ y_{N_{k+1}} - z_{N_k+1}^{N_k+1} \in \mathbf{Z}^n,$$

$$y_i = z_i^{N_k+1} + y_{N_k+1} - z_{N_k+1}^{N_k+1}, \ i = N_k + 1, \ldots, N_{k+1}.$$

We claim that $\{y_i\}_{i=0}^{\infty}$ is the required sequence.

By induction we show that for all natural numbers $k \geq 2$,

$$\sum_{i=0}^{N_k-1} [v_i(y_i, y_{i+1}) - v_i(z_i^{N_k}, z_{i+1}^{N_k})] \leq \sum_{j=1}^{k-1} 2(\epsilon_{N_j} + a_{N_j} - b_{N_j}). \tag{4.595}$$

We show that inequality (4.595) is true for $k = 2$. It is not difficult to see that

$$\sum_{i=0}^{N_2-1} [v_i(y_i, y_{i+1}) - v_i(z_i(N_1, N_2), z_{i+1}(N_1, N_2))] \leq v_{N_1}(y_{N_1}, y_{N_1+1}) - b_{N_1}$$

$$\leq a_{N_1} - b_{N_1}$$

and when combined with (4.592) this relation implies (4.595) for $k = 2$. Assume now that inequality (4.595) is true for some natural number $k \geq 2$. In view of (4.592), we have

$$\sum_{i=0}^{N_{k+1}-1} [v_i(y_i, y_{i+1}) - v_i(z_i^{N_k+1}, z_{i+1}^{N_k+1})]$$

$$= \sum_{i=0}^{N_{k+1}-1} [v_i(y_i, y_{i+1}) - v_i(z_i(N_k, N_{k+1}), z_{i+1}(N_k, N_{k+1}))]$$

$$+ \sum_{i=0}^{N_{k+1}-1} [v_i(z_i(N_k, N_{k+1}), z_{i+1}(N_k, N_{k+1})) - v_i(z_i^{N_k+1}, z_{i+1}^{N_k+1})]$$

$$\leq \sum_{i=0}^{N_k-1} [v_i(y_i, y_{i+1}) - v_i(z_i^{N_k}, z_{i+1}^{N_k})]$$

$$+ v_{N_k}(y_{N_k}, y_{N_k+1}) - b_{N_k} + a_{N_k} - b_{N_k} + \epsilon_{N_k}$$

$$\leq \sum_{j=1}^{k-1} 2(\epsilon_{N_j} + a_{N_j} - b_{N_j}) + 2a_{N_k} - 2b_{N_k} + \epsilon_{N_k} \leq \sum_{j=1}^{k} 2(\epsilon_{N_j} + a_{N_j} - b_{N_j}).$$

Hence inequality (4.595) is true for all natural numbers $k \geq 2$. Let $j > N_2$ be an integer. There exists a natural number $k \geq 2$ such that $N_k < j \leq N_{k+1}$. By (4.593)–(4.595), we have

$$\sum_{i=0}^{j-1} v_i(z_i^{N_{k+1}}, z_{i+1}^{N_{k+1}}) \leq S(x, j) + \epsilon_j + \epsilon_{N_{k+1}} + a_j - b_j,$$

$$\sum_{i=0}^{j-1} [v_i(y_i, y_{i+1}) - v_i(z_i^{N_{k+1}}, z_{i+1}^{N_{k+1}})]$$

$$= \sum_{i=0}^{N_{k+1}-1} [v_i(y_{i+1}, y_{i+1}) - v_i(z_i^{N_{k+1}}, z_{i+1}^{N_{k+1}})] \leq \sum_{i=1}^{k} 2(\epsilon_{N_i} + a_{N_i} - b_{N_i}),$$

$$\sum_{i=0}^{j-1} v_i(y_i, y_{i+1}) \leq \sum_{i=1}^{k} 2(\epsilon_{N_i} + a_{N_i} - b_{N_i}) + S(x, j)$$

$$+ \epsilon_j + \epsilon_{N_{k+1}} + a_j - b_j \leq S(x, j) + 3\epsilon/4 + a_j - b_j.$$

This completes the proof of the theorem.   □

4.19 Periodic Continuous-Time Problems

We consider an optimal control system

$$C_T(u) = \int_0^T f_0(z(t), u(t), t) dt,$$

$$z' = f(z, u), \tag{4.596}$$

where $z(t) \in R^n$, $u(t) \in \Omega$ for all real numbers $t \in [0, T]$, $\Omega \subset R^m$ is a closed set, and $f_0 : R^n \times \Omega \times [0, \infty)$ and $f : R^n \times \Omega \to R^n$ are continuous functions. The admissible controls are all the measurable functions $u(t)$ for which the constraints $u(t) \in \Omega$ and $z(t) \in R^n$ are satisfied (where z and u are related as in (4.596). The results of this section were obtained in [31].

We assume the following:

1. $$f(z + q, u) = f(z, u)$$
   for all pairs of points $z \in R^n$, $u \in \Omega$ and each $q \in \mathbf{Z}^n$ and

$$f_0(z + q, u, t) = f_0(z, u, t)$$

for all $z \in R^n$, $u \in \Omega$, $q \in \mathbf{Z}^n$ and all $t \in [0, \infty)$.

2. For any bounded set $E \subset \Omega$ the function f_0 is bounded on the set $R^n \times E \times [0, \infty)$ and the function f is bounded on the set $R^n \times E$.

3. For any bounded set $E \subset \Omega$ the function $f_0(z, u, t) \to 0$ as $t \to \infty$ uniformly on $R^n \times E$ (for every positive number ϵ there exists a positive number t_ϵ such that

$$|f_0(z, u, t)| \leq \epsilon \text{ for each } z \in R^n, \ u \in E \text{ and each } t \in [t_\epsilon, \infty)).$$

4. There exist a number a positive d_0 and a bounded function

$$\phi_0 : [0, \infty) \to [0, \infty)$$

such that $\phi_0(t) \to 0$ as $t \to \infty$ and that

$$f_0(z, u, t) \geq -d_0 \phi_0(t) \text{ for each } z \in R^n, \ u \in \Omega \text{ and each } t \in [0, \infty).$$

5. There exists a positive number d_1 such that for each pair of points $x = (x_1, \ldots, x_n)$, $y = (y_1, \ldots, y_n) \in R^n$ satisfying

$$0 \leq x_i \leq 1, \ 0 \leq y_i - x_i \leq 1, \ i = 1, \ldots, n$$

there exists an admissible control $u(t), 0 \leq t \leq 1$ with a corresponding trajectory $z(t), t \in [0, 1]$ such that

$$z(0) = x, \ z(1) = y, \ |u(t)| \leq d_1, \ 0 \leq t \leq 1.$$

6. For every positive number T there exists a pair of positive numbers α_T, β_T such that

$$\alpha_T |f(z, u)| \leq f_0(z, u, t) \text{ for all } z \in R^n, \ t \in [0, T]$$

and each $u \in \Omega$ satisfying $|u| \geq \beta_T$.

For each point $x \in R^n$ and each positive number T, put

$$\sigma(x, T) = \inf \left\{ \int_0^T f_0(z(t), u(t), t) dt : \right.$$

$$\left. z' = f(z, u), \ z(t) \in R^n, \ u(t) \in \Omega, \ t \in [0, T], \ z(0) = x \right\}.$$

In view of assumptions 1, 2, 4, and 5 the number $\sigma(x, T)$ is well defined.

For each pair of points $x, y \in R^n$ and each real number $T \geq 0$ denote by $H(x, y, T)$ the set of all pairs of functions $(z(t), u(t))$, $t \in [T, T + 1]$ such that

$$z' = f(z, u), \ z(t) \in R^n, \ u(t) \in \Omega$$

for all $t \in [T, T + 1]$ and $z(T) = x$, $z(T + 1) = y$,

and put

$$v_T(x, y) = \inf\left\{\int_T^{T+1} f_0(z(t), u(t), t)dt : (z, u) \in H(x, y, T)\right\} \text{ if } H(x, y, T) \neq \emptyset;$$

otherwise $v_T(x, y) = \infty$.

For each nonnegative integer i, set

$$a_i = \sup\{v_i(x, y) : x = (x_1, \ldots, x_n), \ y = (y_1, \ldots, y_n) \in R^n,$$

$$0 \leq x_i \leq 1, \ 0 \leq y_i - x_i \leq 1 \text{ for each } i = 1, \ldots, n\},$$

$$b_i = \inf\{v_i(x, y) : x, y \in R^n\}.$$

In view of our assumptions, we have

$$b_i > -\infty, \ a_i < \infty, \ i = 0, 1, \ldots \tag{4.597}$$

$$v_i(x+q, y+q) = v_i(x, y) \text{ for each } x, y \in R^n \text{ each } q \in \mathbf{Z}^n \text{ and each } i = 0, 1, \ldots, \tag{4.598}$$

$$\sup\{a_i : i = 0, 1, \ldots\} < \infty, \tag{4.599}$$

$$\inf\{b_i : i = 0, 1, \ldots\} > -\infty, \tag{4.600}$$

$$a_i \to 0, \ b_i \to 0 \text{ as } i \to \infty. \tag{4.601}$$

Lemma 4.57. *For each nonnegative integer i there exists a real number $\Gamma_i \geq 0$ such that for each pair of points $x, y \in R^n$ satisfying $|x - y| \geq \Gamma_i$ the inequality $v_i(x, y) \geq a_i$ holds.*

Proof. Let $i \in \{0, 1, \ldots\}$ be given. Assumption 6 implies that

$$\alpha_{i+1}|f(z, u)| \leq f_0(z, u, t)$$

for each $z \in R^n$, each $t \in [0, i + 1]$ and each $u \in \Omega$ satisfying $|u| \geq \beta_{i+1}$, \tag{4.602}

where $\alpha_{i+1} > 0, \beta_{i+1} > 0$. It follows from Assumption 2 that there exists a positive number γ such that

$$|f(z, u)| \leq \gamma \text{ for each } z \in R^n \text{ and each } u \in \Omega \text{ satisfying } |u| \leq \beta_{i+1}. \tag{4.603}$$

Choose a positive number Γ_i such that

$$\Gamma_i > \gamma + (\alpha_{i+1})^{-1}[|a_i| + \sup\{|\phi_0(t)| : t \in [i, i + 1]\}](|d_0| + 1). \tag{4.604}$$

Let a pair of points $x, y \in R^n$ satisfy $|x - y| \geq \Gamma_i$ and put $(z(t), u(t))(i \leq t \leq i + 1) \in H(x, y, i)$. It is clear that

$$\Gamma_i \leq |x - y| \leq \int_i^{i+1} |z'(t)| dt = \int_i^{i+1} |f(z(t), u(t))| dt.$$

Put

$$E_1 = \{t \in [i, i + 1] : |u(t)| < \beta_{i+1}\}, \quad E_2 = [i, i + 1] \setminus E_1.$$

By (4.603), we have

$$\Gamma_i \leq \int_{E_1} |f(z(t), u(t))| dt + \int_{E_2} |f(z(t), u(t))| \leq \int_{E_2} |f(z(t), u(t))| + \gamma.$$

$$(4.605)$$

On the other hand it follows from (4.602), assumption 4, (4.605), and (4.604) that

$$\int_i^{i+1} f_0(z(t), u(t), t) dt = \int_{E_1} f_0(z(t), u(t), t) dt + \int_{E_2} f_0(z(t), u(t), t) dt$$

$$\geq \int_{E_2} \alpha_{i+1} |f(z(t), u(t))| dt - \sup\{|\phi_0(t)| : t \in [i, i + 1]\} d_0$$

$$\geq \alpha_{i+1}(\Gamma_i - \gamma) - \sup\{|\phi_0(t)| : t \in [i, i + 1]\} |d_0| \geq a_i.$$

This completes the proof of Lemma 4.57.   □

Lemma 4.58. *For each point $x \in R^n$,*

$$\sup\{|\sigma(x, T) - \sigma(x, i)| : T \in [i, i + 1]\} \to 0 \text{ as } i \to \infty$$

(here i is a nonnegative integer).

Proof. Let $x \in R^n$ be given, i be a nonnegative integer and let $T \in (i, i + 1]$. Let $u(t), t \in [0, T]$ be an admissible control with a corresponding trajectory $z(t)$, $t \in [0, T]$ such that $z(0) = x$. Assumption 4 implies that

$$\int_0^T f_0(z(t), u(t), t) dt \geq \int_0^i f_0(z(t), u(t), t) dt - \sup\{|\phi_0(t)| : t \in [i, i + 1]\} |d_0|$$

$$\geq \sigma(x, i) - \sup\{|\phi_0(t)| : t \in [i, i + 1]\} |d_0|.$$

This inequality implies that

$$\sigma(x, T) - \sigma(x, i) \geq -\sup\{|\phi_0(t)| : t \in [i, i + 1]\} |d_0|. \qquad (4.606)$$

Assume that $u(t)$, $t \in [0, i]$ is an admissible control with a corresponding trajectory $z(t)$, $t \in [0, i]$ such that $z(0) = x$. By assumptions 1 and 5, there exists an admissible control $u_1(t)$, $t \in [0, T]$ with a corresponding trajectory $z_1(t)$, $t \in [0, T]$ such that $u_1(t) = u(t)$, $z_1(t) = z(t)$ for all numbers $t \in [0, i]$ and that $|u_1(t)| \leq d_1$, for all $t \in [i, T]$. It is clear that

$$\int_0^i f_0(z(t), u(t), t) dt \geq \int_0^T f_0(z_1(t), u_1(t), t) dt$$

$$- \sup\{ f_0(y, h, \tau) : \ y \in R^n, \ h \in \Omega, \ |h| \leq d_1, \ \tau \in [i, \infty) \}$$

$$\geq \sigma(x, T) - \sup\{ |f_0(y, h, \tau)| : \ y \in R^n, \ h \in \Omega, \ |h| \leq d_1, \ \tau \in [i, \infty) \}.$$

By the relation above, we have

$$\sigma(x, i) - \sigma(x, T) \geq -\sup\{ |f_0(y, h, \tau)| : \ y \in R^n, \ h \in \Omega, \ |h| \leq d_1, \ \tau \in [i, \infty) \}.$$
$$(4.607)$$

Now the validity of the lemma follows from relations (4.606), (4.607) which hold for all integers $i \in \{0, 1, \ldots, \}$ and all numbers $T \in (i, i+1]$, and from assumptions 3 and 4. $\qquad \square$

It is not difficult to see that for each point $x \in R^n$ and each integer $N \in \{0, 1, \ldots\}$, we have

$$\sigma(x, N) = \inf \left\{ \sum_{i=0}^{N-1} v_i(x_i, x_{i+1}) : \ \{x_i\}_{i=0}^N \subset R^n, \ x_0 = x \right\}. \qquad (4.608)$$

Theorem 4.59. *For each point $x \in R^n$ and each positive number ϵ there exists an admissible control $u(t)$, $t \in [0, \infty)$ with a corresponding trajectory $z(t)$, $t \in [0, \infty)$ such that $z(0) = y$ and that*

$$\int_0^T f_0(z(t), u(t), t) dt \leq \sigma(x, T) + \epsilon$$

for all sufficient large positive numbers T.

Proof. It follows from (4.597)–(4.601) and Lemma 4.57 that Theorem 4.56 is valid for the functions v_i, $i = 0, 1, \ldots$. Let $x \in R^n$ and let $\epsilon > 0$ be given.

In view of Theorem 4.56 and (4.608), there exists a sequence $\{y_i\}_{i=0}^{\infty} \subset R^n$ such that $y_0 = x$ and that for large natural numbers N, we have

$$\sum_{i=0}^{N-1} v_i(y_i, y_{i+1}) \leq \sigma(x, N) + \epsilon/4.$$

It is clear that there exists an admissible control $u(t)$, $t \in [0, \infty)$ with a corresponding trajectory $z(t)$, $t \in [0, \infty)$ such that for each nonnegative integer i,

$$z(i) = y_i, \quad \int_i^{i+1} f_0(z(t), u(t), t)dt \leq v_i(y_i, y_{i+1}) + 2^{-i-4}\epsilon.$$

Clearly, for all sufficiently large natural numbers N,

$$\int_0^N f_0(z(t), u(t), t)dt \leq \sigma(x, N) + \epsilon/2. \tag{4.609}$$

Let $N \geq 0$ be an integer and let $T \in [N, N+1)$. By (4.609), assumption 4, and Lemma 4.58, for sufficiently large natural numbers N, we have

$$\int_0^T f_0(z(t), u(t), t)dt - \sigma(x, T) - \epsilon/2$$

$$= \int_0^{N+1} f_0(z(t), u(t), t)dt - \sigma(x, N+1) - \epsilon/2$$

$$+ \sigma(x, N+1) - \sigma(x, T) - \int_T^{N+1} f_0(z(t), u(t), t)dt$$

$$\leq \sigma(x, N+1) - \sigma(x, T) - \int_T^{N+1} f_0(z(t), u(t), t)dt \leq \mu_1 + \mu_2,$$

where

$$\mu_1 = 2\sup\{|\sigma(x, N) - \sigma(x, \tau)| : \tau \in [N, N+1]\}$$

and

$$\mu_2 = \sup\{|\phi_0(\tau)| : \tau \in [N, N+1]\}|d_0|.$$

Since $\mu_1, \mu_2 \to 0$ as $N \to \infty$, our theorem is proved. $\qquad\square$

The following result, in particular, asserts the existence of overtaking optimal solutions which we define as follows.

We say that a pair (z^*, u^*), where $u^*(\cdot)$ is an admissible control on $[0, \infty)$ with a corresponding trajectory $z^*(\cdot)$, is *overtaking optimal* if for each positive number ϵ there exists a positive number T_ϵ such that

$$\int_0^T f_0(z^*(t), u^*(t), t)dt < \int_0^T f_0(z(t), u(t), t)dt + \epsilon$$

for each real number $T > T_\epsilon$ and each admissible pair (z, u) on the interval $[0, T]$ satisfying $z(0) = z^*(0)$.

Clearly, in the definition above T_ϵ depends only on ϵ. In the usual definition of an overtaking optimal trajectory used in the literature the pair (z, u) is defined on the interval $[0, \infty)$ and T_ϵ depends on ϵ and (z, u). Here we can use the strong version of the overtaking optimality criterion because of Assumption 3.

Theorem 4.60. *Assume that for each nonnegative integer i the function $v_i(\cdot, \cdot)$ is well defined (namely the minimum is attained by a certain admissible control) and is lower semicontinuous on the space $R^n \times R^n$. Then for each point $x \in R^n$ there exists an admissible control $u(t)$ with a corresponding trajectory $z(t)$, $t \in [0, \infty)$ such that $z(0) = x$ and*

$$\lim_{T \to \infty} \left[\int_0^T f_0(z(t), u(t), t) dt - \sigma(x, T) \right] = 0.$$

In particular, this admissible pair (z, u) is overtaking optimal.

Proof. Let $x \in R^n$ be given. In view of (4.597)–(4.601) and Lemma 4.57, Theorem 4.55 is valid for the functions v_i, $i = 0, 1, \ldots,$. By Theorem 4.55 and (4.608), there exists a sequence $\{y_i\}_{i=0}^\infty \subset R^n$ such that $y_0 = x$ and that

$$\sum_{i=0}^{N-1} v_i(y_i, y_{i+1}) - \sigma(x, N) \to 0 \text{ as } N \to \infty.$$

It is clear that there exists an admissible control $u(t)$, $t \in [0, \infty)$ with a corresponding trajectory $z(t)$, $t \in [0, \infty)$ such that for each nonnegative integer i, we have $z(i) = y_i$ and

$$\int_i^{i+1} f_0(z(t), u(t), t) dt = v_i(y_i, y_{i+1}).$$

Then

$$\int_0^N f_0(z(t), u(t), t) dt - \sigma(x, N) \to 0 \text{ as } N \to \infty. \tag{4.610}$$

(here $N \geq 0$ is an integer).

Let $N \geq 0$ be an integer and $T \in [N, N+1)$ be given. It follows from assumption 4, Lemma 4.58, and (4.610) that

$$\int_0^T f_0(z(t), u(t), t) - \sigma(x, T)$$

$$= \int_0^{N+1} f_0(z(t), u(t), t) dt - \sigma(x, N + 1)$$

$$+ \sigma(x, N+1) - \sigma(x, T) - \int_T^{N+1} f_0(z(t), u(t), t)dt$$

$$\leq \int_0^{N+1} f_0(z(t), u(t), t)dt - \sigma(x, N+1) \leq \mu_1 + \mu_2,$$

where

$$\mu_1 = 2\sup\{|\sigma(x, \tau) - \sigma(x, N)| : \tau \in [N, N+1]\}$$

and

$$\mu_2 = \sup\{|\phi_0(t)| : t \in [N, N+1]\}|d_0| \rightarrow 0 \text{ as } N \rightarrow \infty$$

and we have $\mu_1 + \mu_2 \rightarrow 0$ as $N \rightarrow \infty$. This completes the proof of the theorem. $\square$

Chapter 5
Dynamic Discrete-Time Zero-Sum Games

In this chapter we study the existence and structure of solutions for dynamic discrete-time two-player zero-sum games and establish a turnpike result. This result describes the structure of approximate solutions which is independent of the length of the interval, for all sufficiently large intervals. We also show that for each initial state there exists a pair of overtaking equilibria strategies over an infinite horizon.

5.1 Preliminaries

Let M, N be nonempty sets and $h : M \times N \to R^1$. The triplet (f, M, N) describes the two-player zero-sum game, where M is the set of strategies of the first player, N is the set of strategies of the second player, and f and $-f$ are objective functions for the first and the second players, respectively. In this chapter we study the turnpike properties of solutions for a dynamic discrete-time two-player zero-sum game described below.

Denote by $|\cdot|$ the Euclidean norm in R^m. Let $X \subset R^{m_1}$ and $Y \subset R^{m_2}$ be nonempty convex compact sets. Denote by $\mathcal{M}$ the set of all continuous functions $f : X \times X \times Y \times Y \to R^1$ such that:

for each point $(y_1, y_2) \in Y \times Y$ the function $(x_1, x_2) \to f(x_1, x_2, y_1, y_2)$, $(x_1, x_2) \in X \times X$ is convex;

for each point $(x_1, x_2) \in X \times X$ the function $(y_1, y_2) \to f(x_1, x_2, y_1, y_2)$, $(y_1, y_2) \in Y \times Y$ is concave.

The set $\mathcal{M}$ is equipped with a metric $\rho : \mathcal{M} \times \mathcal{M} \to R^1$ defined by

$$\rho(f, g) = \sup\{|f(x_1, x_2, y_1, y_2) - g(x_1, x_2, y_1, y_2)| :$$
$$x_1, x_2 \in X, \quad y_1, y_2 \in Y\}, \quad f, g \in \mathcal{M}. \qquad (5.1)$$

It is clearly that $(\mathcal{M}, \rho)$ is a complete metric space.

© Springer International Publishing Switzerland 2014
A.J. Zaslavski, *Turnpike Phenomenon and Infinite Horizon Optimal Control*,
Springer Optimization and Its Applications 99, DOI 10.1007/978-3-319-08828-0_5

Given $f \in \mathcal{M}$ and a natural number n we consider a discrete-time two-player zero-sum game over the interval $[0, n]$. For this game $\{\{x_i\}_{i=0}^n : x_i \in X, i = 0, \ldots n\}$ is the set of strategies for the first player, $\{\{y_i\}_{i=0}^n : y_i \in Y, i = 0, \ldots n\}$ is the set of strategies for the second player, and the objective function for the first player associated with the strategies $\{x_i\}_{i=0}^n$, $\{y_i\}_{i=0}^n$ is given by $\sum_{i=0}^{n-1} f(x_i, x_{i+1}, y_i, y_{i+1})$.

Let $f \in \mathcal{M}$, n be a natural number and let $M \in [0, \infty)$. A pair of sequences $\{\bar{x}_i\}_{i=0}^n \subset X$, $\{\bar{y}_i\}_{i=0}^n \subset Y$ is called (f, M)-good if the following properties hold:

(i) for each sequence $\{x_i\}_{i=0}^n \subset X$ satisfying $x_0 = \bar{x}_0$, $x_n = \bar{x}_n$ the inequality

$$M + \sum_{i=0}^{n-1} f(x_i, x_{i+1}, \bar{y}_i, \bar{y}_{i+1}) \geq \sum_{i=0}^{n-1} f(\bar{x}_i, \bar{x}_{i+1}, \bar{y}_i, \bar{y}_{i+1}) \qquad (5.2)$$

holds;

(ii) for each sequence $\{y_i\}_{i=0}^n \subset Y$ satisfying $y_0 = \bar{y}_0$, $y_n = \bar{y}_n$ the inequality

$$M + \sum_{i=0}^{n-1} f(\bar{x}_i, \bar{x}_{i+1}, \bar{y}_i, \bar{y}_{i+1}) \geq \sum_{i=0}^{n-1} f(\bar{x}_i, \bar{x}_{i+1}, y_i, y_{i+1}) \qquad (5.3)$$

holds.

If a pair of sequences $\{x_i\}_{i=0}^n \subset X$, $\{y_i\}_{i=0}^n \subset Y$ is $(f, 0)$-good then it is called (f)-optimal.

In this chapter we study the turnpike property of good pairs of sequences.

Let $f \in \mathcal{M}$. We say that the function f possesses the *turnpike property* if there exists a unique pair $(x_f, y_f) \in X \times Y$ for which the following assertion holds:

For each positive number ϵ there exist an integer $n_0 \geq 2$ and a positive number δ such that for each integer $n \geq 2n_0$ and each (f, δ)-good pair of sequences $\{x_i\}_{i=0}^n \subset X$, $\{y_i\}_{i=0}^n \subset Y$ the inequalities $|x_i - x_f|$, $|y_i - y_f| \leq \epsilon$ holds for all integers $i \in [n_0, n - n_0]$.

In [53] we showed that the turnpike property holds for a generic $f \in \mathcal{M}$. Namely, in [53] we proved the existence of a set $\mathcal{F} \subset \mathcal{M}$ which is a countable intersection of open everywhere dense sets in $\mathcal{M}$ such that each $f \in \mathcal{F}$ has the turnpike property.

In [53,69] and in the present chapter we also study the existence of equilibria over an infinite horizon and employ the following version of the overtaking optimality criterion.

Let $f \in \mathcal{M}$. A pair of sequences $\{\bar{x}_i\}_{i=0}^\infty \subset X$, $\{\bar{y}_i\}_{i=0}^\infty \subset Y$ is called (f)-*overtaking optimal* if the following properties hold:

for each sequence $\{x_i\}_{i=0}^\infty \subset X$ satisfying $x_0 = \bar{x}_0$ the inequality

$$\lim_{T\to\infty}\sup[\sum_{i=0}^{T-1} f(\bar{x}_i, \bar{x}_{i+1}, \bar{y}_i, \bar{y}_{i+1}) - \sum_{i=0}^{T-1} f(x_i, x_{i+1}, \bar{y}_i, \bar{y}_{i+1})] \leq 0 \qquad (5.4)$$

holds;

for each sequence $\{y_i\}_{i=0}^{\infty} \subset Y$ satisfying $y_0 = \bar{y}_0$ the inequality

$$\lim_{T\to\infty}\sup[\sum_{i=0}^{T-1} f(\bar{x}_i, \bar{x}_{i+1}, y_i, y_{i+1}) - \sum_{i=0}^{T-1} f(\bar{x}_i, \bar{x}_{i+1}, \bar{y}_i, \bar{y}_{i+1})] \leq 0 \qquad (5.5)$$

holds.

In [53] we showed that for a generic $f \in \mathcal{M}$ and each point $(x, y) \in X \times Y$ there exists an (f)-overtaking optimal pair of sequences $\{x_i\}_{i=0}^{\infty} \subset X$, $\{y_i\}_{i=0}^{\infty} \subset Y$ such that $x_0 = x$, $y_0 = y$.

According to the results of [53] we know that for most functions $f \in \mathcal{M}$ the turnpike property holds and that (f)-overtaking optimal pairs of sequences exist. Nevertheless it is very important to have conditions on $f \in \mathcal{M}$ which imply the turnpike property and guarantee the existence of (f)-overtaking optimal pairs of sequences. These conditions were found in [69].

5.2   Minimal Pairs of Sequences

Let $f \in \mathcal{M}$. Define a function $\bar{f} : X \times Y \to R^1$ by

$$\bar{f}(x, y) = f(x, x, y, y), \quad x \in X, \ y \in Y. \qquad (5.6)$$

Then there exists a saddle point $(x_f, y_f) \in X \times Y$ for $\bar{f}$ [6] such that

$$\sup_{y\in Y} \bar{f}(x_f, y) = \bar{f}(x_f, y_f) = \inf_{x\in X} \bar{f}(x, y_f). \qquad (5.7)$$

Set

$$\mu(f) = \bar{f}(x_f, y_f). \qquad (5.8)$$

A pair of sequences $\{x_i\}_{i=0}^{\infty} \subset X$, $\{y_i\}_{i=0}^{\infty} \subset Y$ is called (f)-minimal if for each natural number $n \geq 2$ the pair of sequences $\{x_i\}_{i=0}^{n}$, $\{y_i\}_{i=0}^{n}$ is (f)-optimal.

The following three results were established in [53].

Proposition 5.1. *Let $n \geq 2$ be a natural number and*

$$\bar{x}_i = x_f, \quad \bar{y}_i = y_f, \quad i = 0, \ldots n. \qquad (5.9)$$

Then the pair of sequences $\{\bar{x}_i\}_{i=0}^{n}$, $\{\bar{y}_i\}_{i=0}^{n}$ is (f)-optimal.

Proof. Assume that $\{x_i\}_{i=0}^n \subset X, \{y_i\}_{i=0}^n \subset Y$ and that

$$x_0, x_n = x_f, \quad y_0, y_n = y_f. \tag{5.10}$$

It follows from (5.10), (5.9), and (5.7) that

$$\sum_{i=0}^{n-1} f(x_i, x_{i+1}, \bar{y}_i, \bar{y}_{i+1}) = \sum_{i=0}^{n-1} f(x_i, x_{i+1}, y_f, y_f)$$

$$\geq n f\left(n^{-1}\sum_{i=0}^{n-1} x_i, n^{-1}\sum_{i=0}^{n-1} x_{i+1}, y_f, y_f\right)$$

$$= n f\left(n^{-1}\sum_{i=0}^{n-1} x_i, n^{-1}\sum_{i=0}^{n-1} x_i, y_f, y_f\right)$$

$$\geq n f(x_f, x_f, y_f, y_f),$$

$$\sum_{i=0}^{n-1} f(\bar{x}_i, \bar{x}_{i+1}, y_i, y_{i+1}) = \sum_{i=0}^{n-1} f(x_f, x_f, y_i, y_{i+1})$$

$$\leq n f\left(x_f, x_f, n^{-1}\sum_{i=0}^{n-1} y_i, n^{-1}\sum_{i=0}^{n-1} y_{i+1}\right)$$

$$= n f\left(x_f, x_f, n^{-1}\sum_{i=0}^{n-1} y_i, n^{-1}\sum_{i=0}^{n-1} y_i\right)$$

$$\leq n f(x_f, x_f, y_f, y_f).$$

This completes the proof of the proposition. □

Proposition 5.2. *Let $n \geq 2$ be a natural number and let*

$$(\{x_i^{(k)}\}_{i=0}^n, \{y_i^{(k)}\}_{i=0}^n) \subset X \times Y, \quad k = 1, 2, \ldots$$

be a sequence of (f)-optimal pairs. Assume that

$$\lim_{k\to\infty} x_i^{(k)} = x_i, \quad \lim_{k\to\infty} y_i^{(k)} = y_i, \quad i = 0, 1, 2, \ldots, n. \tag{5.11}$$

Then the pair of sequences $(\{x_i\}_{i=0}^n, \{y_i\}_{i=0}^n)$ is (f)-optimal.

Proof. Assume that

$$\{u_i\}_{i=0}^n \subset X, \quad u_0 = x_0, u_n = x_n. \tag{5.12}$$

We claim that

$$\sum_{i=0}^{n-1} f(x_i, x_{i+1}, y_i, y_{i+1}) \le \sum_{i=0}^{n-1} f(u_i, u_{i+1}, y_i, y_{i+1}). \tag{5.13}$$

Assume the contrary. Then there exists a positive number ϵ such that

$$\sum_{i=0}^{n-1} f(x_i, x_{i+1}, y_i, y_{i+1}) > \sum_{i=0}^{n-1} f(u_i, u_{i+1}, y_i, y_{i+1}) + 8\epsilon. \tag{5.14}$$

There exists a positive number $\delta < \epsilon$ such that

$$|f(z_1, z_2, \xi_1, \xi_2) - f(\bar{z}_1, \bar{z}_2, \bar{\xi}_1, \bar{\xi}_2)| \le \epsilon(8n)^{-1} \tag{5.15}$$

for each $z_1, z_2, \bar{z}_1, \bar{z}_2 \in X$, $\xi_1, \xi_2, \bar{\xi}_1, \bar{\xi}_2 \in Y$ satisfying $|z_i - \bar{z}_i|$, $|\xi_i - \bar{\xi}_i| \le \delta$, $i = 1, 2$. There exists a natural number q such that

$$|x_i - x_i^{(q)}|, \ |y_i - y_i^{(q)}| \le \delta, \quad i = 0, \ldots n. \tag{5.16}$$

Define $\{u_i^{(q)}\}_{i=0}^n \subset X$ by

$$u_0^{(q)} = x_0^{(q)}, \ u_n^{(q)} = x_n^{(q)}, \quad u_i^{(q)} = u_i, \ i = 1, \ldots n - 1. \tag{5.17}$$

Since the pair of sequences $(\{x_i^{(q)}\}_{i=0}^n, \{y_i^{(q)}\}_{i=0}^n)$ is (f)-optimal relation (5.17) implies that

$$\sum_{i=0}^{n-1} f(x_i^{(q)}, x_{i+1}^{(q)}, y_i^{(q)}, y_{i+1}^{(q)}) \le \sum_{i=0}^{n-1} f(u_i^{(q)}, u_{i+1}^{(q)}, y_i^{(q)}, y_{i+1}^{(q)}). \tag{5.18}$$

In view of the definition of δ (see (5.15)), (5.16), (5.17), and (5.12), for $i = 0, \ldots, n - 1$, we have

$$|f(x_i^{(q)}, x_{i+1}^{(q)}, y_i^{(q)}, y_{i+1}^{(q)}) - f(x_i, x_{i+1}, y_i, y_{i+1})| \le (8n)^{-1}\epsilon,$$

$$|f(u_i^{(q)}, u_{i+1}^{(q)}, y_i^{(q)}, y_{i+1}^{(q)}) - f(u_i, u_{i+1}, y_i, y_{i+1})| \le (8n)^{-1}\epsilon.$$

By these relations and (5.14),

$$\sum_{i=0}^{n-1} f(x_i^{(q)}, x_{i+1}^{(q)}, y_i^{(q)}, y_{i+1}^{(q)}) - \sum_{i=0}^{n-1} f(u_i^{(q)}, u_{i+1}^{(q)}, y_i^{(q)}, y_{i+1}^{(q)}) > \epsilon.$$

This is contradictory to (5.18). The obtained contradiction proves that (5.13) is true. Analogously we can show that for each $\{u_i\}_{i=0}^n \subset Y$ satisfying $u_0 = y_0$, $u_n = y_n$, we have

$$\sum_{i=0}^{n-1} f(x_i, x_{i+1}, y_i, y_{i+1}) \geq \sum_{i=0}^{n-1} f(x_i, x_{i+1}, u_i, u_{i+1}).$$

This completes the proof of the proposition.   □

Proposition 5.3. *Let $f \in \mathcal{M}$ and let $x \in X$, $y \in Y$. Then there exists an (f)-minimal pair of sequences $\{x_i\}_{i=0}^\infty \subset X$, $\{y_i\}_{i=0}^\infty \subset Y$ such that $x_0 = x$, $y_0 = y$.*

Proof. In view of Theorem 8 of Sect. 2 of Chap. 6 of [6], for each natural number $n \geq 2$ there exists an (f)-optimal pair of sequences $\{x_i^{(n)}\}_{i=0}^n \subset X$, $\{y_i^{(n)}\}_{i=0}^n \subset Y$ such that $x_0^{(n)} = x$, $y_0^{(n)} = y$. There exist a pair of sequences $\{x_i\}_{i=0}^\infty \subset X$, $\{y_i\}_{i=0}^\infty \subset Y$ and a strictly increasing sequence of natural numbers $\{n_k\}_{k=1}^\infty$ such that for each nonnegative integer i, we have

$$x_i^{(n_k)} \to x_i, \quad y_i^{(n_k)} \to y_i \text{ as } k \to \infty.$$

By Proposition 5.2, the pair of sequences $\{x_i\}_{i=0}^\infty$, $\{y_i\}_{i=0}^\infty$ is (f)-minimal. The proposition is proved.   □

Let n be a natural number and let $\xi = (\xi_1, \xi_2, \xi_3, \xi_4) \in X \times X \times Y \times Y$. Set

$$\Lambda_X(\xi, n) = \{\{x_i\}_{i=0}^n \subset X : \quad x_0 = \xi_1, \ x_n = \xi_2\}, \tag{5.19}$$

$$\Lambda_Y(\xi, n) = \{\{y_i\}_{i=0}^n \subset Y : \quad y_0 = \xi_3, \ y_n = \xi_4\}, \tag{5.20}$$

$$f^{(\xi,n)}((x_0, \ldots, x_i, \ldots, x_n), (y_0, \ldots, y_i, \ldots y_n)) = \sum_{i=0}^{n-1} f(x_i, x_{i+1}, y_i, y_{i+1}),$$
$$\tag{5.21}$$

$$\{x_i\}_{i=0}^n \in \Lambda_X(\xi, n), \quad \{y_i\}_{i=0}^n \in \Lambda_Y(\xi, n).$$

5.3 Main Results

Let $f \in \mathcal{M}$. Then there exists $(x_f, y_f) \in X \times Y$ such that [6]

$$\sup_{y \in Y} f(x_f, x_f, y, y) = f(x_f, x_f, y_f, y_f) = \inf_{x \in X} f(x, x, y_f, y_f). \tag{5.22}$$

We suppose that the following assumptions hold:

(A1) for each point $x \in X \setminus \{x_f\}$ and each point $x' \in X$, the inequality

$$f(2^{-1}(x_f + x), 2^{-1}(x_f + x'), y_f, y_f) < 2^{-1} f(x_f, x_f, y_f, y_f)$$
$$+ 2^{-1} f(x, x', y_f, y_f)$$

holds;

(A2) for each point $y \in Y \setminus \{y_f\}$ and each point $y' \in Y$, the inequality

$$f(x_f, x_f, 2^{-1}(y_f + y), 2^{-1}(y' + y_f)) > 2^{-1} f(x_f, x_f, y_f, y_f)$$
$$+ 2^{-1} f(x_f, x_f, y, y')$$

holds.

Fix a real number

$$D_0 \geq \sup\{|f(x_1, x_2, y_1, y_2)| : x_1, x_2 \in X, \ y_1, y_2 \in Y\}. \tag{5.23}$$

In this chapter we prove the following results obtained in [69].

Theorem 5.4. *Let $\epsilon \in (0, 1)$. Then there exist a neighborhood $\mathcal{U}$ of the function f in the space $\mathcal{M}$, a natural number $n_1 \geq 4$, and a real number $\delta \in (0, \epsilon)$ such that for each function $g \in \mathcal{U}$, each natural number $n \geq 2n_1$, and each (g, δ)-good pair of sequences $\{x_i\}_{i=0}^n \subset X$, $\{y_i\}_{i=0}^n \subset Y$, the inequality*

$$|x_i - x_f|, \ |y_i - y_f| \leq \epsilon \tag{5.24}$$

holds for all integers $i \in [n_1, n - n_1]$. Moreover, if $|x_0 - x_f|, \ |y_0 - y_f| \leq \delta$, then (5.24) holds for all integers $i \in [0, n - n_1]$, and if $|x_n - x_f|, \ |y_n - y_f| \leq \delta$, then (5.24) is valid for all integers $i \in [n_1, n]$.

Theorem 5.5. *For each point $x \in X$ and each point $y \in Y$ there exists an (f)-overtaking optimal pair of sequences $\{x_i\}_{i=0}^\infty \subset X$, $\{y_i\}_{i=0}^\infty \subset Y$ such that $x_0 = x$, $y_0 = y$.*

5.4 Auxiliary Results for Theorem 5.4

Lemma 5.6. *Let $\epsilon \in (0, 1)$. Then there exists a positive number $\delta < \epsilon$ such that for each natural number $n \geq 2$ and each (f, δ)-good pair of sequences $\{x_i\}_{i=0}^n \subset X$, $\{y_i\}_{i=0}^n \subset Y$ satisfying*

$$x_n, x_0 = x_f, \quad y_n, y_0 = y_f, \tag{5.25}$$

the following inequality holds:

$$|x_i - x_f|, \ |y_i - y_f| \leq \epsilon, \quad i = 0, \ldots n. \tag{5.26}$$

Proof. In view of assumptions (A1), (A2) and continuity of the function f there exists a real number $\gamma > 0$ such that the following properties hold:

(P1) for each point $x \in X$ and each point $x' \in X$ satisfying $|x - x_f| \geq \epsilon$,

$$- f(2^{-1}(x_f + x), 2^{-1}(x_f + x'), y_f, y_f)$$
$$+ 2^{-1} f(x_f, x_f, y_f, y_f) + 2^{-1} f(x, x', y_f, y_f) \geq \gamma;$$

(P2) for each $y \in Y$ and each $y' \in Y$ satisfying $|y - y_f| \geq \epsilon$,

$$f(x_f, x_f, 2^{-1}(y_f + y), 2^{-1}(y' + y_f)) - 2^{-1} f(x_f, x_f, y_f, y_f)$$
$$- 2^{-1} f(x_f, x_f, y, y') \geq \gamma.$$

Fix a real number $\delta > 0$ such that

$$\delta < \gamma/4, \ \delta < 8^{-1}\epsilon. \tag{5.27}$$

Assume that a natural number $n \geq 2$, $\{x_i\}_{i=0}^n \subset X$, $\{y_i\}_{i=0}^n \subset Y$ is an (f, δ)-good pair of sequences and that (5.25) is true. Put

$$\xi_1, \xi_2 = x_f, \quad \xi_3, \xi_4 = y_f, \quad \xi = (\xi_1, \xi_2, \xi_3, \xi_4). \tag{5.28}$$

Consider the sets $\Lambda_X(\xi, n)$, $\Lambda_Y(\xi, n)$ and the functions $f^{(\xi, n)}$ (see (5.19)–(5.21)). By (5.25) and Proposition 5.1,

$$\sup \left\{ \sum_{i=0}^{n-1} f(x_f, x_f, u_i, u_{i+1}) : \{u_i\}_{i=0}^n \in \Lambda_Y(\xi, n) \right\}$$

$$= n f(x_f, x_f, y_f, y_f)$$

$$= \inf \left\{ \sum_{i=0}^{n-1} f(p_i, p_{i+1}, y_f, y_f) : \{p_i\}_{i=0}^n \in \Lambda_X(\xi, n) \right\}. \tag{5.29}$$

By (5.25) and (5.28),

$$\{x_i\}_{i=0}^n \in \Lambda_X(\xi, n)\}, \ \{y_i\}_{i=0}^n \in \Lambda_Y(\xi, n). \tag{5.30}$$

Since $(\{x_i\}_{i=0}^n, \{y_i\}_{i=0}^n)$ is an (f, δ)-good pair of sequences we have

$$\sup \left\{ \sum_{i=0}^{n-1} f(x_i, x_{i+1}, u_i, u_{i+1}) : \{u_i\}_{i=0}^n \in \Lambda_Y(\xi, n) \right\} - \delta$$

$$\leq \sum_{i=0}^{n-1} f(x_i, x_{i+1}, y_i, y_{i+1})$$

$$\leq \inf\left\{\sum_{i=0}^{n-1} f(p_i, p_{i+1}, y_i, y_{i+1}) : \{p_i\}_{i=0}^{n} \in \Lambda_X(\xi, n)\right\} + \delta. \tag{5.31}$$

By (5.28)–(5.31),

$$nf(x_f, x_f, y_f, y_f) \leq \sum_{i=0}^{n-1} f(x_i, x_{i+1}, y_f, y_f) \leq \sum_{i=0}^{n-1} f(x_i, x_{i+1}, y_i, y_{i+1}) + \delta$$

$$\leq \sum_{i=0}^{n-1} f(x_f, x_f, y_i, y_{i+1}) + 2\delta \leq nf(x_f, x_f, y_f, y_f) + 2\delta. \tag{5.32}$$

In view of (5.32),

$$\left|\sum_{i=0}^{n-1} f(x_i, x_{i+1}, y_i, y_{i+1}) - nf(x_f, x_f, y_f, y_f)\right| \leq \delta, \tag{5.33}$$

$$\sum_{i=0}^{n-1} f(x_i, x_{i+1}, y_f, y_f) - nf(x_f, x_f, y_f, y_f) \in [0, 2\delta], \tag{5.34}$$

$$\sum_{i=0}^{n-1} f(x_f, x_f, y_i, y_{i+1}) - nf(x_f, x_f, y_f, y_f) \in [-2\delta, 0]. \tag{5.35}$$

Put

$$\tilde{x}_i = 2^{-1}(x_i + x_f), \quad \tilde{y}_i = 2^{-1}(y_i + y_f), \quad i = 0, \ldots, n. \tag{5.36}$$

It follows from (5.29), (5.30), and (5.36) that

$$\sum_{i=0}^{n-1} f(\tilde{x}_i, \tilde{x}_{i+1}, y_f, y_f) \geq nf(x_f, x_f, y_f, y_f) \geq \sum_{i=0}^{n-1} f(x_f, x_f, \tilde{y}_i, \tilde{y}_{i+1}). \tag{5.37}$$

In order to complete the proof of the lemma it is sufficient to show that (5.26) holds.
Assume the contrary. Then relation (5.25) implies that there exists a natural number $k \in [1, n-1]$ such that

$$\max\{|x_k - x_f|, |y_k - y_f|\} > \epsilon. \tag{5.38}$$

In view of (5.36), for all integers $i = 0, \ldots, n-1$,

$$f(\tilde{x}_i, \tilde{x}_{i+1}, y_f, y_f) \leq 2^{-1} f(x_i, x_{i+1}, y_f, y_f) + 2^{-1} f(x_f, x_f, y_f, y_f),$$
(5.39)

$$f(x_f, x_f, \tilde{y}_i, \tilde{y}_{i+1}) \geq 2^{-1} f(x_f, x_f, y_i, y_{i+1}) + 2^{-1} f(x_f, x_f, y_f, y_f).$$
(5.40)

It follows from (5.36), (5.38)–(5.40) and properties (P1) and (P2) that

$$2^{-1} f(x_k, x_{k+1}, y_f, y_f) + 2^{-1} f(x_f, x_f, y_f, y_f) - f(\tilde{x}_k, \tilde{x}_{k+1}, y_f, y_f)$$

$$+ f(x_f, x_f, \tilde{y}_k, \tilde{y}_{k+1}) - 2^{-1} f(x_f, x_f, y_k, y_{k+1})$$

$$- 2^{-1} f(x_f, x_f, y_f, y_f) \geq \gamma.$$
(5.41)

Relations (5.34), (5.35), (5.37), and (5.39)–(5.41) imply that

$$\gamma \leq \sum_{i=0}^{n-1} [2^{-1} f(x_i, x_{i+1}, y_i, y_{i+1}) + 2^{-1} f(x_f, x_f, y_f, y_f) - f(\tilde{x}_i, \tilde{x}_{i+1}, y_f, y_f)]$$

$$+ \sum_{i=0}^{n-1} [f(x_f, x_f, \tilde{y}_i, \tilde{y}_{i+1}) - 2^{-1} f(x_f, x_f, y_i, y_{i+1}) - 2^{-1} f(x_f, x_f, y_f, y_f)]$$

$$= \sum_{i=0}^{n-1} [f(x_f, x_f, \tilde{y}_i, \tilde{y}_{i+1}) - f(\tilde{x}_i, \tilde{x}_{i+1}, y_f, y_f)$$

$$+ 2^{-1} f(x_i, x_{i+1}, y_f, y_f) - 2^{-1} f(x_f, x_f, y_i, y_{i+1})] \leq \delta.$$

This contradicts (5.27). The contradiction we have reached proves (5.26). This completes the proof of Lemma. 5.6.   □

We can easily prove the following result.

Lemma 5.7. *Let* $n \geq 2$ *be a natural number,* M *be a positive number, and* $\{x_i\}_{i=0}^n \subset X$, $\{y_i\}_{i=0}^n \subset Y$ *be an* (f, M)-*good pair of sequences. Then the pair of sequences* $\{\tilde{x}_i\}_{i=0}^n \subset X$, $\{\tilde{y}_i\}_{i=0}^n \subset Y$ *defined by*

$$\bar{x}_i = x_i, \quad \bar{y}_i = y_i, \ i = 1, \ldots n-1, \quad \bar{x}_0, \bar{x}_n = x_f, \quad \bar{y}_0, \bar{y}_n = y_f$$

is $(f, M + 8D_0)$-*good.*

By using the uniform continuity of the function $f : X \times X \times Y \times Y \to R^1$ we can easily prove the following lemma.

Lemma 5.8. *Let* $\epsilon > 0$. *Then there exists a positive number* δ *such that for each natural number* $n \geq 2$ *and each*

$$\{x_i\}_{i=0}^n, \ \{\bar{x}_i\}_{i=0}^n \subset X, \ \{y_i\}_{i=0}^n, \ \{\bar{y}_i\}_{i=0}^n \subset Y$$

which satisfy

$$|\bar{x}_j - x_j|, \ |\bar{y}_j - y_j| \le \delta, \ j = 0, n, \quad x_j = \bar{x}_j, \ y_j = \bar{y}_j, \ j = 1, \dots n - 1,$$
$$(5.42)$$

the inequality

$$\left| \sum_{i=0}^{n-1} [f(x_i, x_{i+1}, y_i, y_{i+1}) - f(\bar{x}_i, \bar{x}_{i+1}, \bar{y}_i, \bar{y}_{i+1})] \right| \le \epsilon$$

holds.

Lemma 5.8 implies the following result.

Lemma 5.9. *Assume that ϵ is a positive number. Then there exists a positive number δ such that for each natural number $n \ge 2$, each (f, ϵ)-good pair of sequences $\{x_i\}_{i=0}^n \subset X$, $\{y_i\}_{i=0}^n \subset Y$, and each pair of sequences $\{\bar{x}_i\}_{i=0}^n \subset X$, $\{\bar{y}_i\}_{i=0}^n \subset Y$, the following assertion holds:*
if (5.42) is valid then the pair of sequences $(\{\bar{x}_i\}_{i=0}^n, \{\bar{y}_i\}_{i=0}^n)$ is $(f, 2\epsilon)$-good.

Lemmas 5.9 and 5.6 imply the following auxiliary result.

Lemma 5.10. *Let $\epsilon \in (0, 1)$. Then there exists a positive number $\delta < \epsilon$ such that for each natural number $n \ge 2$ and each (f, δ)-good pair of sequences $\{x_i\}_{i=0}^n \subset X$, $\{y_i\}_{i=0}^n \subset Y$ which satisfies $|x_j - x_f|, |y_j - y_f| \le \delta, \ j = 0, n$, the following inequalities hold: $|x_i - x_f|, |y_i - y_f| \le \epsilon, \ i = 0, \dots n.$*

Denote by $\mathrm{Card}(E)$ the cardinality of a set E.

Lemma 5.11. *Let M be a positive number and let $\epsilon \in (0, 1)$. Then there exists a natural number $n_0 \ge 4$ such that for each (f, M)-good pair of sequences $\{x_i\}_{i=0}^{n_0} \subset X$, $\{y_i\}_{i=0}^{n_0} \subset Y$ which satisfies*

$$x_0, x_{n_0} = x_f, \quad y_0, y_{n_0} = y_f \tag{5.43}$$

there exists an integer $j \in \{1, \dots n_0 - 1\}$ such that

$$|x_j - x_f|, \ |y_j - y_f| \le \epsilon. \tag{5.44}$$

Proof. In view of assumptions (A1), (A2) and continuity of the function f, there exists a real number $\gamma > 0$ such that:
    for each point $x \in X$ and each point $x' \in X$ satisfying $|x - x_f| \ge \epsilon$, we have

$$- f(2^{-1}(x_f + x), 2^{-1}(x_f + x'), y_f, y_f)$$
$$+ 2^{-1} f(x_f, x_f, y_f, y_f) + 2^{-1} f(x, x', y_f, y_f) \ge \gamma; \tag{5.45}$$

for each point $y \in Y$ and each point $y' \in Y$ satisfying $|y - y_f| \geq \epsilon$, we have

$$f(x_f, x_f, 2^{-1}(y_f + y), 2^{-1}(y' + y_f))$$
$$- 2^{-1} f(x_f, x_f, y_f, y_f) - 2^{-1} f(x_f, x_f, y, y') \geq \gamma. \qquad (5.46)$$

Choose a natural number:

$$n_0 > 8 + 2(\gamma)^{-1} M. \qquad (5.47)$$

Put

$$\xi_1, \xi_2 = x_f, \quad \xi_3, \xi_4 = y_f, \quad \xi = \{\xi_i\}_{i=1}^{4}. \qquad (5.48)$$

Assume that $\{x_i\}_{i=0}^{n_0} \subset X$, $\{y_i\}_{i=0}^{n_0} \subset Y$ is an (f, M)-good pair of sequences and that (5.43) holds. We claim that there exists $j \in \{1, \ldots, n_0 - 1\}$ such that (5.44) holds. Assume the contrary. Then

$$\max\{|x_j - x_f|, \; |y_j - y_f|\} > \epsilon, \quad j = 1, \ldots, n_0 - 1. \qquad (5.49)$$

By (5.43), (5.48) and Proposition 5.1,

$$\sup\Big\{ \sum_{i=0}^{n_0-1} f(x_f, x_f, u_i, u_{i+1}) : \{u_i\}_{i=0}^{n_0} \in \Lambda_Y(\xi, n_0) \Big\} = n_0 f(x_f, x_f, y_f, y_f)$$

$$= \inf\Big\{ \sum_{i=0}^{n_0-1} f(p_i, p_{i+1}, y_f, y_f) : \{p_i\}_{i=0}^{n_0} \in \Lambda_X(\xi, n_0) \Big\}. \qquad (5.50)$$

Relations (5.43) and (5.48) imply that

$$\{x_i\}_{i=0}^{n_0} \in \Lambda_X(\xi, n_0), \; \{y_i\}_{i=0}^{n_0} \in \Lambda_Y(\xi, n_0). \qquad (5.51)$$

Since $(\{x_i\}_{i=0}^{n_0}, \{y_i\}_{i=0}^{n_0})$ is an (f, M)-good pair of sequences we have

$$\sup\Big\{ \sum_{i=0}^{n_0-1} f(x_i, x_{i+1}, u_i, u_{i+1}) : \{u_i\}_{i=0}^{n_0} \in \Lambda_Y(\xi, n_0) \Big\} - M$$

$$\leq \sum_{i=0}^{n_0-1} f(x_i, x_{i+1}, y_i, y_{i+1})$$

$$\leq \inf\Big\{ \sum_{i=0}^{n_0-1} f(p_i, p_{i+1}, y_i, y_{i+1}) : \{p_i\}_{i=0}^{n_0} \in \Lambda_X(\xi, n_0) \Big\} + M. \qquad (5.52)$$

By (5.48), (5.50)–(5.52),

$$n_0 f(x_f, x_f, y_f, y_f) \le \sum_{i=0}^{n_0-1} f(x_i, x_{i+1}, y_f, y_f) \le \sum_{i=0}^{n_0-1} f(x_i, x_{i+1}, y_i, y_{i+1}) + M$$

$$\le \sum_{i=0}^{n_0-1} f(x_f, x_f, y_i, y_{i+1}) + 2M$$

$$\le n_0 f(x_f, x_f, y_f, y_f) + 2M. \tag{5.53}$$

In view of (5.53),

$$\left| n_0 f(x_f, x_f, y_f, y_f) - \sum_{i=0}^{n_0-1} f(x_i, x_{i+1}, y_i, y_{i+1}) \right| \le M, \tag{5.54}$$

$$\sum_{i=0}^{n_0-1} f(x_i, x_{i+1}, y_f, y_f) - n_0 f(x_f, x_f, y_f, y_f) \in [0, 2M], \tag{5.55}$$

$$\sum_{i=0}^{n_0-1} f(x_f, x_f, y_i, y_{i+1}) - n_0 f(x_f, x_f, y_f, y_f) \le [-2M, 0]. \tag{5.56}$$

Put

$$\tilde{x}_i = 2^{-1}(x_i + x_f), \; \tilde{y}_i = 2^{-1}(y_i + y_f), \; i = 0, 1, \ldots, n_0. \tag{5.57}$$

Relations (5.48), (5.50), (5.51), and (5.57) imply that

$$\sum_{i=0}^{n_0-1} f(\tilde{x}_i, \tilde{x}_{i+1}, y_f, y_f) \ge n_0 f(x_f, x_f, y_f, y_f) \ge \sum_{i=0}^{n_0-1} f(x_f, x_f, \tilde{y}_i, \tilde{y}_{i+1}).$$
$$\tag{5.58}$$

In view of (5.57) for $i = 0, \ldots, n_0 - 1$,

$$f(\tilde{x}_i, \tilde{x}_{i+1}, y_f, y_f) \le 2^{-1} f(x_i, x_{i+1}, y_f, y_f) + 2^{-1} f(x_f, x_f, y_f, y_f), \tag{5.59}$$

$$f(x_f, x_f, \tilde{y}_i, \tilde{y}_{i+1}) \ge 2^{-1} f(x_f, x_f, y_i, y_{i+1}) + 2^{-1} f(x_f, x_f, y_f, y_f). \tag{5.60}$$

It follows from (5.59), (5.49), the choice of γ (see (5.45) and (5.46)), and (5.57) that for each natural number $i = 1, \ldots, n_0 - 1$, at least one of the following inequalities holds:

$$2^{-1} f(x_i, x_{i+1}, y_f, y_f) + 2^{-1} f(x_f, x_f, y_f, y_f) - f(\tilde{x}_i, \tilde{x}_{i+1}, y_f, y_f) \ge \gamma,$$

$$f(x_f, x_f, \tilde{y}_i, \tilde{y}_{i+1}) - 2^{-1} f(x_f, x_f, y_i, y_{i+1}) - 2^{-1} f(x_f, x_f, y_f, y_f) \geq \gamma.$$

Together with (5.59) and (5.60) this implies that for each natural number $i = 1, \ldots, n_0 - 1$, we have

$$2^{-1} f(x_i, x_{i+1}, y_f, y_f) + 2^{-1} f(x_f, x_f, y_f, y_f) - f(\tilde{x}_i, \tilde{x}_{i+1}, y_f, y_f)$$
$$+ f(x_f, x_f, \tilde{y}_i, \tilde{y}_{i+1})$$
$$- 2^{-1} f(x_f, x_f, y_i, y_{i+1}) - 2^{-1} f(x_f, x_f, y_f, y_f) \geq \gamma.$$

When combined with (5.59), (5.60), and (5.58) this implies that

$$\gamma(n_0 - 1) \leq \sum_{i=0}^{n_0 - 1} [2^{-1} f(x_i, x_{i+1}, y_f, y_f)$$
$$+ 2^{-1} f(x_f, x_f, y_f, y_f) - f(\tilde{x}_i, \tilde{x}_{i+1}, y_f, y_f)$$
$$+ f(x_f, x_f, \tilde{y}_i, \tilde{y}_{i+1}) - 2^{-1} f(x_f, x_f, y_i, y_{i+1})$$
$$- 2^{-1} f(x_f, x_f, y_f, y_f)]$$
$$= \sum_{i=0}^{n_0 - 1} [f(x_f, x_f, \tilde{y}_i, \tilde{y}_{i+1}) - f(\tilde{x}_i, \tilde{x}_{i+1}, y_f, y_f)$$
$$+ 2^{-1} f(x_i, x_{i+1}, y_f, y_f) - 2^{-1} f(x_f, x_f, y_i, y_{i+1})])$$
$$\leq n_0 f(x_f, x_f, y_f, y_f) - n_0 f(x_f, x_f, y_f, y_f)$$
$$+ 2^{-1} n_0 f(x_f, x_f, y_f, y_f) + M$$
$$- 2^{-1} n_0 f(x_f, x_f, y_f, y_f) + M \leq 2M, \gamma(n_0 - 1) \leq 2M.$$

This contradicts (5.47). The contradiction we have reached proves that there exists a natural number $j \in \{1, \ldots, n_0 - 1\}$ such that (5.44) holds. Lemma 5.11 is proved. $\qquad\square$

Lemmas 5.11 and 5.7 imply the following auxiliary result.

Lemma 5.12. *Let* $\epsilon \in (0, 1)$ *and* $M \in (0, \infty)$. *Then there exists a natural number* $n_0 \geq 4$ *such that for each* (f, M)-*good pair of sequences* $\{x_i\}_{i=0}^{n_0} \subset X$, $\{y_i\}_{i=0}^{n_0} \subset Y$ *there exists a natural number* $j \in \{1, \ldots, n_0 - 1\}$ *such that* $|x_f - x_j|, |y_f - y_j| \leq \epsilon.$

Lemma 5.13. *Let* $\epsilon \in (0, 1)$ *and* $M \in (0, \infty)$. *Then there exist a natural number* $n_0 \geq 4$ *and a neighborhood* $\mathcal{U}$ *of the function* f *in the space* $\mathcal{M}$ *such that for each function* $g \in \mathcal{U}$ *and each* (g, M)-*good pair of sequences* $\{x_i\}_{i=0}^{n_0} \subset X$, $\{y_i\}_{i=0}^{n_0} \subset Y$ *there exists a natural number* $j \in \{1, \ldots n_0 - 1\}$ *such that*

$$|x_f - x_j|, \ |y_f - y_j| \le \epsilon. \tag{5.61}$$

Proof. In view of Lemma 5.12 there exists a natural number $n_0 \ge 4$ such that for each $(f, M + 8)$-good pair of sequences $\{x_i\}_{i=0}^{n_0} \subset X$, $\{y_i\}_{i=0}^{n_0} \subset Y$ there exists a natural number $j \in \{1, \ldots n_0 - 1\}$ such that (5.61) is valid. Put

$$\mathcal{U} = \{g \in \mathcal{M} : \rho(f, g) \le (16 n_0)^{-1}\}. \tag{5.62}$$

Assume that $g \in \mathcal{U}$ and $\{x_i\}_{i=0}^{n_0} \subset X$, $\{y_i\}_{i=0}^{n_0} \subset Y$ is a (g, M)-good pair of sequences. It follows from (5.62) that the pair of sequences $\{x_i\}_{i=0}^{n_0}$, $\{y_i\}_{i=0}^{n_0}$ is $(f, M + 8)$-good. By the definition of n_0, there exists an integer $j \in \{1, \ldots n_0 - 1\}$ such that (5.61) is valid. The lemma is proved. □

5.5 Proof of Theorem 5.4

In view of Lemma 5.10 there exists a real number $\delta_0 \in (0, \epsilon)$ such that the following property holds:

(P3) for each natural number $n \ge 2$ and each (f, δ_0)-good pair of sequences $\{x_i\}_{i=0}^{n} \subset X$, $\{y_i\}_{i=0}^{n} \subset Y$ satisfying

$$|x_j - x_f|, \ |y_j - y_f| \le \delta_0, \quad j = 0, n$$

the inequality

$$|x_j - x_f|, \ |y_j - y_f| \le \epsilon \tag{5.63}$$

holds for all integers $i = 0, \ldots, n$.

Lemma 5.13 implies that there exists a natural number $n_0 \ge 4$ and a neighborhood $\mathcal{U}_0$ of the function f in the space $\mathcal{M}$ such that the following property holds:

(P4) for each function $g \in \mathcal{U}_0$ and each $(g, 8)$-good pair of sequences $\{x_i\}_{i=0}^{n_0} \subset X$, $\{y_i\}_{i=0}^{n_0} \subset Y$ there exists a natural number $j \in \{1, \ldots n_0 - 1\}$ such that

$$|x_j - x_f|, \ |y_j - y_f| \le \delta_0.$$

Choose a natural number

$$n_1 \ge 4 n_0 \tag{5.64}$$

and a positive number

$$\delta < 4^{-1} \delta_0. \tag{5.65}$$

Set

$$\mathcal{U} = \mathcal{U}_0 \cap \{g \in \mathcal{M} : \quad \rho(g, f) \leq 16^{-1} \delta n_1^{-1}\}. \tag{5.66}$$

Assume that a function $g \in \mathcal{U}$, a natural number $n \geq 2n_1$, and $\{x_i\}_{i=0}^n \subset X$, $\{y_i\}_{i=0}^n \subset Y$ is a (g, δ)-good pair of sequences. By (5.64)–(5.66) and property (P4), there exists a sequence of integers $\{t_i\}_{i=1}^k \subset [0, n]$ such that

$$t_1 \leq n_0, \ t_{i+1} - t_i \in [n_0, 3n_0], \ i = 1, \ldots k - 1, \tag{5.67}$$

$$n - t_k \leq n_0, \quad |x_{t_i} - x_f|, \ |y_{t_i} - y_f| \leq \delta_0, \ i = 1, \ldots k \tag{5.68}$$

and, moreover, if $|x_0 - x_f|, \ |y_0 - y_f| \leq \delta$, then $t_1 = 0$, and if $|x_n - x_f|, \ |y_n - y_f| \leq \delta$ then $t_k = n$. It is clear that $k \geq 2$. Fix a natural number $q \in \{1, \ldots k - 1\}$. To complete the proof of the theorem it is sufficient to show that for each natural number $i \in [t_q, t_{q+1}]$ the relation (5.63) holds.

Define sequences $\{x_i^{(q)}\}_{i=0}^{t_{q+1} - t_q} \subset X, \{y_i^{(q)}\}_{i=0}^{t_{q+1} - t_q} \subset Y$ by

$$x_i^{(q)} = x_{i+t_q}, \ y_i^{(q)} = y_{i+t_q}, \quad i \in [0, t_{q+1} - t_q]. \tag{5.69}$$

It is not difficult to see that

$$\{x_i^{(q)}\}_{i=0}^{t_{q+1} - t_q}, \ \{y_i^{(q)}\}_{i=0}^{t_{q+1} - t_q}$$

is a (g, δ)-good pair of sequences. When combined with (5.64)–(5.68) this implies that the pair $\{x_i^{(q)}\}_{i=0}^{t_{q+1} - t_q}, \ \{y_i^{(q)}\}_{i=0}^{t_{q+1} - t_q}$ is (f, δ_0)-good. By (5.67), (5.68), and property (P3),

$$|x_i^{(q)} - x_f|, \ |y_i^{(q)} - y_f| \leq \epsilon, \quad i = 0, \ldots t_{q+1} - t_q. \tag{5.70}$$

When combined with (5.69) relation (5.70) implies that $|x_i - x_f|, \ |y_i - y_f| \leq \epsilon, \ i = t_q, \ldots t_{q+1}$. This completes the proof of Theorem 5.4.

5.6   Preliminary Lemmas for Theorem 5.5

Let $f \in \mathcal{M}$. $x_f \in X, y_f \in Y$ satisfy (5.7).

For each metric space K denote by $C(K)$ the space of all continuous functions on K with the topology of uniform convergence ($\|\phi\| = \sup\{|\phi(z)| : z \in K\}$, $\phi \in C(K)$).

Define functions $f^{(X)} : X \times X \to R^1, f^{(Y)} : Y \times Y \to R^1$ as follows:

$$f^{(X)}(x_1, x_2) = f(x_1, x_2, y_f, y_f), \quad x_1, x_2 \in X, \tag{5.71}$$

$$f^{(Y)}(y_1, y_2) = f(x_f, x_f, y_1, y_2), \quad y_1, y_2 \in Y. \tag{5.72}$$

Lemma 5.14. *Let $\epsilon \in (0, 1)$. Then there exists a positive number $\delta < \epsilon$ for which the following assertion holds:*
Assume that $n \geq 2$ is a natural number,

$$\{x_i\}_{i=0}^n \subset X, \quad x_0, x_n = x_f, \tag{5.73}$$

and that for each sequence $\{z_i\}_{i=0}^n \subset X$ satisfying

$$z_0 = x_0, \quad z_n = x_n \tag{5.74}$$

the inequality

$$\sum_{i=0}^{n-1} f^{(X)}(x_i, x_{i+1}) \leq \sum_{i=0}^{n-1} f^{(X)}(z_i, z_{i+1}) + \delta \tag{5.75}$$

holds. Then

$$|x_i - x_f| \leq \epsilon, \quad i = 0, \ldots n. \tag{5.76}$$

Proof. In view of assumption (A1) and continuity of the function f there exists a number $\gamma > 0$ such that the following property holds:

(P5) for each point $x \in X$ and each point $x' \in X$ satisfying $|x - x_f| \geq \epsilon$,

$$- f(2^{-1}(x_f + x), 2^{-1}(x_f + x'), y_f, y_f)$$
$$+ 2^{-1} f(x_f, x_f, y_f, y_f) + 2^{-1} f(x, x', y_f, y_f) \geq \gamma.$$

Fix a real number $\delta > 0$ such that

$$\delta < \min\{8^{-1}\epsilon, \ \gamma/4\}. \tag{5.77}$$

Assume that $n \geq 2$ is a natural number, $\{x_i\}_{i=0}^n \subset X$, (5.73) is valid and for each sequence $\{z_i\}_{i=0}^n \subset X$ satisfying (5.74), relation (5.75) is true. We claim that (5.76) holds. Assume the contrary. Then there is a natural number $j \in \{1, \ldots, n-1\}$ such that

$$|x_j - x_f| > \epsilon. \tag{5.78}$$

Since (5.75) holds with $z_i = x_f, i = 0, \ldots, n$, Proposition 5.1 implies that

$$\sum_{i=0}^{n-1} f(x_i, x_{i+1}, y_f, y_f) \leq nf(x_f, x_f, y_f, y_f) + \delta \leq \sum_{i=0}^{n-1} f(x_i, x_{i+1}, y_f, y_f) + \delta. \tag{5.79}$$

Set

$$\tilde{x}_i = 2^{-1}(x_i + x_f), \quad i = 0, \ldots, n - 1. \tag{5.80}$$

In view of (5.71) and (5.80), for all integers $i = 0, \ldots, n - 1$,

$$f^{(X)}(\tilde{x}_i, \tilde{x}_{i+1}) \leq 2^{-1} f^{(X)}(x_i, x_{i+1}) + 2^{-1} f^{(X)}(x_f, x_f). \tag{5.81}$$

It follows from (5.78), (5.79) and property (P5) that

$$f(\tilde{x}_j, \tilde{x}_{j+1}, y_f, y_f) \leq 2^{-1} f(x_j, x_{j+1}, y_f, y_f) + 2^{-1} f(x_f, x_f, y_f, y_f) - \gamma. \tag{5.82}$$

By (5.73), (5.79)–(5.82) and Proposition 5.1, we have

$$nf(x_f, x_f, y_f, y_f) \leq \sum_{i=0}^{n-1} f(\tilde{x}_i, \tilde{x}_{i+1}, y_f, y_f)$$

$$\leq 2^{-1} \sum_{i=0}^{n-1} f(x_i, x_{i+1}, y_f, y_f) + 2^{-1} nf(x_f, x_f, y_f, y_f) - \gamma$$

$$\leq nf(x_f, x_f, y_f, y_f) + \delta - \gamma.$$

This contradicts (5.77). The contradiction we have reached proves that (5.76) holds. Lemma 5.14 is proved.   □

Analogously to Lemma 5.14 we can establish the following auxiliary result.

Lemma 5.15. *Let $\epsilon \in (0, 1)$. Then there exists a positive number $\delta < \epsilon$ for which the following assertion holds:*
Assume that $n \geq 2$ is a natural number,

$$\{y_i\}_{i=0}^n \subset Y, \quad y_0, y_n = y_f$$

and for that each $\{z_i\}_{i=0}^n \subset Y$ satisfying

$$z_0 = y_0, \quad z_n = y_n \tag{5.83}$$

the inequality

$$\sum_{i=0}^{n-1} f^{(Y)}(y_i, y_{i+1}) \geq \sum_{i=0}^{n-1} f^{(Y)}(z_i, z_{i+1}) - \delta \tag{5.84}$$

holds. Then

$$|y_i - y_f| \leq \epsilon, \quad i = 0, \ldots n. \tag{5.85}$$

Let $g \in C(X \times X)$, $n \geq 1$ be a natural number and let $M \in [0, \infty)$. A sequence $\{\bar{x}_i\}_{i=0}^{n} \subset X$ is called (g, X, M)-good if for each sequence $\{x_i\}_{i=0}^{n} \subset X$ satisfying $x_0 = \bar{x}_0$, $x_n = \bar{x}_n$ the inequality $M + \sum_{i=0}^{n-1} g(x_i, x_{i+1}) \geq \sum_{i=0}^{n-1} g(\bar{x}_i, \bar{x}_{i+1})$ is true.

Let $g \in C(Y \times Y)$, $n \geq 1$ be a natural number and let $M \in [0, \infty)$. A sequence $\{\bar{y}_i\}_{i=0}^{n} \subset Y$ is called (g, Y, M)-good if for each sequence $\{y_i\}_{i=0}^{n} \subset Y$ satisfying $y_0 = \bar{y}_0$, $y_n = \bar{y}_n$ the inequality

$$\sum_{i=0}^{n-1} g(y_i, y_{i+1}) \leq M + \sum_{i=0}^{n-1} g(\bar{y}_i, \bar{y}_{i+1})$$

is true.

Let $n_1 \geq 0$, $n_2 > n_1$ be a pair of integers, and let $\{g_i\}_{i=n_1}^{n_2-1} \subset C(X \times X)$, $M \in [0, \infty)$. A sequence $\{\bar{x}_i\}_{i=n_1}^{n_2} \subset X$ is called $(\{g_i\}_{i=n_1}^{n_2-1}, X, M)$-good if for each sequence $\{x_i\}_{i=n_1}^{n_2} \subset X$ satisfying $x_{n_1} = \bar{x}_{n_1}$, $x_{n_2} = \bar{x}_{n_2}$, we have

$$M + \sum_{i=n_1}^{n_2-1} g_i(x_i, x_{i+1}) \geq \sum_{i=n_1}^{n_2-1} g_i(\bar{x}_i, \bar{x}_{i+1}).$$

Let $n_1 \geq 0$, $n_2 > n_1$ be a pair of integers, and let $\{g_i\}_{i=n_1}^{n_2-1} \subset C(Y \times Y)$, $M \in [0, \infty)$. A sequence $\{\bar{y}_i\}_{i=n_1}^{n_2} \subset Y$ is called $(\{g_i\}_{i=n_1}^{n_2-1}, Y, M)$-good if for each sequence $\{y_i\}_{i=n_1}^{n_2} \subset Y$ satisfying $y_{n_1} = \bar{y}_{n_1}$, $y_{n_2} = \bar{y}_{n_2}$, we have

$$\sum_{i=n_1}^{n_2-1} g_i(y_i, y_{i+1}) \leq \sum_{i=n_1}^{n_2-1} g_i(\bar{y}_i, \bar{y}_{i+1}) + M.$$

By using Lemmas 5.14 and 5.8 we can easily deduce the following auxiliary result.

Lemma 5.16. *Let $\epsilon \in (0, 1)$. Then there exists a positive number δ such that for each natural number $n \geq 2$ and each $(f^{(X)}, X, \delta)$-good sequence $\{x_i\}_{i=0}^{n} \subset X$ satisfying*

$$|x_0 - x_f|, \ |x_n - x_f| \leq \delta \qquad .$$

the following inequality holds: $|x_i - x_f| \leq \epsilon$, $i = 0, \ldots n$.

By using Lemmas 5.15 and 5.8 we can easily deduce the following auxiliary result.

Lemma 5.17. *Let $\epsilon \in (0, 1)$. Then there exists a positive number δ such that for each natural number $n \geq 2$ and each $(f^{(Y)}, Y, \delta)$-good sequence $\{y_i\}_{i=0}^{n} \subset Y$ satisfying*

$$|y_0 - y_f|, \ |y_n - y_f| \leq \delta$$

the following inequality holds:

$$|y_i - y_f| \leq \epsilon, \ i = 0, \ldots n.$$

Lemma 5.18. *Let $\epsilon \in (0, 1)$ and M be a positive number. Then there exists a natural number $n_0 \geq 4$ such that for each $(f^{(X)}, X, M)$-good sequence $\{x_i\}_{i=0}^{n_0} \subset X$ satisfying*

$$x_0 = x_f, \ x_{n_0} = x_f \tag{5.86}$$

there exists a natural number $j \in \{1, \ldots n_0 - 1\}$ such that

$$|x_j - x_f| \leq \epsilon. \tag{5.87}$$

Proof. Assumption (A1) implies that there exists a real number $\gamma > 0$ such that the following property holds:

(P6) for each point $x \in X$ and each point $x' \in X$ satisfying $|x - x_f| \geq \epsilon$, we have

$$- f(2^{-1}(x_f + x), 2^{-1}(x_f + x'), y_f, y_f)$$
$$+ 2^{-1} f(x_f, x_f, y_f, y_f) + 2^{-1} f(x, x', y_f, y_f) \geq \gamma.$$

Fix an integer

$$n_0 > 8 + M\gamma^{-1}. \tag{5.88}$$

Assume that an $(f^{(X)}, X, M)$-good sequence $\{x_i\}_{i=0}^{n_0} \subset X$ satisfies (5.86). We claim that there exists a natural number $j \in \{1, \ldots, n_0 - 1\}$ such that (5.87) holds. Assume the contrary. Then

$$|x_i - x_f| > \epsilon, \ i = 1, \ldots, n_0 - 1. \tag{5.89}$$

Put

$$\tilde{x}_i = 2^{-1} x_i + 2^{-1} x_f, \ i = 0, \ldots, n_0. \tag{5.90}$$

In view of (5.90), for $i = 0, \ldots, n_0 - 1$,

$$f(\tilde{x}_i, \tilde{x}_{i+1}, y_f, y_f) \leq 2^{-1} f(x_i, x_{i+1}, y_f, y_f) + 2^{-1} f(x_f, x_f, y_f, y_f). \tag{5.91}$$

By (5.89), (5.90) and property (P6), for all natural numbers $i = 1, \ldots, n_0 - 1$,

$$f(\tilde{x}_i, \tilde{x}_{i+1}, y_f, y_f) \leq 2^{-1} f(x_i, x_{i+1}, y_f, y_f) + 2^{-1} f(x_f, x_f, y_f, y_f) - \gamma. \tag{5.92}$$

It follows from (5.86), (5.90), (5.92), (5.94) and Proposition 5.1 that

$$
n_0 f(x_f, x_f, y_f, y_f) \le \sum_{i=0}^{n_0-1} f(\tilde{x}_i, \tilde{x}_{i+1}, y_f, y_f)
$$

$$
\le 2^{-1} \sum_{i=0}^{n_0-1} f(x_i, x_{i+1}, y_f, y_f)
$$

$$
+ 2^{-1} n_0 f(x_f, x_f, y_f, y_f) - (n_0 - 1)\gamma
$$

$$
\le 2^{-1}(M + n_0 f(x_f, x_f, y_f, y_f))
$$

$$
+ 2^{-1} n_0 f(x_f, x_f, y_f, y_f) - (n_0 - 1)\gamma,
$$

$$
(n_0 - 1)\gamma \le 2^{-1} M.
$$

This contradicts (5.88). The contradiction we have reached proves that there exists a natural number $j \in \{1, \dots, n_0 - 1\}$ such that (5.87) holds. Lemma 5.18 is proved.

□

Analogously to Lemma 5.18 we can establish the following auxiliary result.

Lemma 5.19. *Let $\epsilon \in (0, 1)$ and M be a positive number. Then there exists a natural number $n_0 \ge 4$ such that for each $(f^{(Y)}, Y, M)$-good sequence $\{y_i\}_{i=0}^{n_0} \subset Y$ satisfying $y_0 = y_f$, $y_{n_0} = y_f$ there exists a natural number $j \in \{1, \dots n_0 - 1\}$ such that $|y_j - y_f| \le \epsilon$.*

We can easily prove the following result.

Lemma 5.20. *1. Assume that $n \ge 2$ is a natural number, M is a positive number, a sequence $\{x_i\}_{i=0}^n \subset X$ is $(f^{(X)}, X, M)$-good, and $\bar{x}_0 = x_f$, $\bar{x}_n = x_f$, $\bar{x}_i = x_i$, $i = 1, \dots n - 1$. Then the sequence $\{\bar{x}_i\}_{i=0}^n$ is $(f^{(X)}, X, M + 8D_0)$-good.*
2. Assume that $n \ge 2$ is a natural number, M is a positive number, a sequence $\{y_i\}_{i=0}^n \subset Y$ is $(f^{(Y)}, Y, M)$-good and $\bar{y}_0 = y_f$, $\bar{y}_n = y_f$, $\bar{y}_i = y_i$, $i = 1, \dots n - 1$. Then the sequence $\{\bar{y}_i\}_{i=0}^n$ is $(f^{(Y)}, Y, M + 8D_0)$-good.

Lemmas 5.18, 5.19 and 5.20 imply the following two results.

Lemma 5.21. *Let $\epsilon \in (0, 1)$ and M be a positive number. Then there exists a natural number $n_0 \ge 4$ such that for each $(f^{(X)}, X, M)$-good sequence $\{x_i\}_{i=0}^{n_0} \subset X$ there exists a natural number $j \in \{1, \dots n_0 - 1\}$ for which $|x_j - x_f| \le \epsilon$.*

Lemma 5.22. *Let $\epsilon \in (0, 1)$ and M be a positive number. Then there exists a natural number $n_0 \ge 4$ such that for each $(f^{(Y)}, Y, M)$-good sequence $\{y_i\}_{i=0}^{n_0} \subset Y$ there exists a natural number $j \in \{1, \dots n_0 - 1\}$ for which $|y_j - y_f| \le \epsilon$.*

By using Lemmas 5.21 and 5.22, analogously to the proof of Lemma 5.13, we can establish the following two results.

Lemma 5.23. *Let $\epsilon \in (0, 1)$, $M \in (0, \infty)$. Then there exist a natural number $n_0 \geq 4$ and a neighborhood $\mathcal{U}$ of the function $f^{(X)}$ in the space $C(X \times X)$ such that for each $\{g_i\}_{i=0}^{n_0-1} \subset \mathcal{U}$ and each $(\{g_i\}_{i=0}^{n_0-1}, X, M)$-good sequence $\{x_i\}_{i=0}^{n_0} \subset X$ there exists a natural number $j \in \{1, \ldots n_0 - 1\}$ for which $|x_f - x_j| \leq \epsilon$.*

Lemma 5.24. *Let $\epsilon \in (0, 1)$, $M \in (0, \infty)$. Then there exist a natural number $n_0 \geq 4$ and a neighborhood $\mathcal{U}$ of the function $f^{(Y)}$ in the space $C(Y \times Y)$ such that for each $\{g_i\}_{i=0}^{n_0-1} \subset \mathcal{U}$ and each $(\{g_i\}_{i=0}^{n_0-1}, Y, M)$-good sequence $\{y_i\}_{i=0}^{n_0} \subset Y$ there exists a natural number $j \in \{1, \ldots n_0 - 1\}$ for which $|y_f - y_j| \leq \epsilon$.*

Lemma 5.25. *Let $\epsilon \in (0, 1)$. Then there exist a neighborhood $\mathcal{U}$ of the function $f^{(X)}$ in the space $C(X \times X)$, a positive number $\delta < \epsilon$, and a natural number $n_1 \geq 4$ such that for each integer $n \geq 2n_1$, each $\{g_i\}_{i=0}^{n-1} \subset \mathcal{U}$, and each $(\{g_i\}_{i=0}^{n-1}, X, \delta)$-good sequence $\{x_i\}_{i=0}^{n} \subset X$ the inequality*

$$|x_i - x_f| \leq \epsilon \tag{5.93}$$

holds for all integers $i \in [n_1, n - n_1]$. Moreover if $|x_0 - x_f| \leq \delta$, then (5.93) holds for all integers $i \in [0, n - n_1]$, and if $|x_n - x_f| \leq \delta$, then (5.93) is valid for all integers $i \in [n_1, n]$.

Proof. In view of Lemma 5.16 there exists a positive $\delta_0 < \epsilon$ such that the following property holds:

(P7) for each natural number $n \geq 2$ and each $(f^{(X)}, X, \delta_0)$-good sequence $\{x_i\}_{i=0}^{n} \subset X$ satisfying $|x_0 - x_f|$, $|x_n - x_f| \leq \delta_0$ relation (5.93) is valid for integers $i = 0, \ldots n$.

Lemma 5.23 implies that there exist a natural number integer $n_0 \geq 4$ and a neighborhood $\mathcal{U}_0$ of the function $f^{(X)}$ in the space $C(X \times X)$ such that the following property holds:

(P8) for each $\{g_i\}_{i=0}^{n_0-1} \subset \mathcal{U}_0$ and each $(\{g_i\}_{i=0}^{n_0-1}, X, 8)$-good sequence $\{x_i\}_{i=0}^{n} \subset X$ there exists a natural number $j \in \{1, \ldots n_0 - 1\}$ for which $|x_j - x_f| \leq \delta_0$.

Fix a natural number $n_1 \geq 4n_0$ and a real number $\delta \in (0, 4^{-1}\delta_0)$. Set

$$\mathcal{U} = \mathcal{U}_0 \cap \{g \in C(X \times X) : \|g - f^{(X)}\| \leq (16n_1)^{-1}\delta\}.$$

Assume that $n \geq 2n_1$ is a natural number, $\{g_i\}_{i=0}^{n-1} \subset \mathcal{U}$, and a sequence $\{x_i\}_{i=0}^{n} \subset X$ is $(\{g_i\}_{i=0}^{n-1}, X, \delta)$-good. Arguing as in the proof of Theorem 5.4 we complete the proof of Lemma 5.25. □

Analogously to Lemma 5.25 we can prove the following.

Lemma 5.26. *Let $\epsilon \in (0, 1)$. Then there exist a neighborhood $\mathcal{U}$ of the function $f^{(Y)}$ in $C(Y \times Y)$, a real number $\delta \in (0, \epsilon)$, and a natural number $n_1 \geq 4$ such that for each natural number $n \geq 2n_1$, each $\{g_i\}_{i=0}^{n-1} \subset \mathcal{U}$, and each $(\{g_i\}_{i=0}^{n-1}, Y, \delta)$-good sequence $\{y_i\}_{i=0}^{n} \subset Y$ the inequality*

$$|y_i - y_f| \leq \epsilon \tag{5.94}$$

holds for all integers $i \in [n_1, n - n_1]$. Moreover if $|y_0 - y_f| \le \delta$, then (5.94) holds for all integers $i \in [0, n - n_1]$, and if $|y_n - y_f| \le \delta$, then (5.94) is valid for all integers $i \in [n_1, n]$.

5.7 Proof of Theorems 5.5

Let $x \in X$ and $y \in Y$. In view of Proposition 5.3 there exists an (f)-minimal pair of sequences $\{\bar{x}_j\}_{j=0}^\infty \subset X$, $\{\bar{y}_j\}_{j=0}^\infty \subset Y$ such that

$$\bar{x}_0 = x, \quad \bar{y}_0 = y. \tag{5.95}$$

We claim that the pair of sequences $(\{\bar{x}_j\}_{j=0}^\infty, \{\bar{y}_j\}_{j=0}^\infty)$ is (f)-overtaking optimal. By Theorem 5.4,

$$\bar{x}_j \to x_f, \quad \bar{y}_j \to y_f \text{ as } j \to \infty. \tag{5.96}$$

Let $\{x_i\}_{i=0}^\infty \subset X$ and $x_0 = x$. We show that

$$\limsup_{T \to \infty} \left[\sum_{j=0}^{T-1} f(\bar{x}_j, \bar{x}_{j+1}, \bar{y}_j, \bar{y}_{j+1}) - \sum_{j=0}^{T-1} f(x_j, x_{j+1}, \bar{y}_j, \bar{y}_{j+1}) \right] \le 0. \tag{5.97}$$

Assume the contrary. Then there exists a positive number Γ_0 and a strictly increasing sequence of natural numbers $\{T_k\}_{k=1}^\infty$ such that for all natural numbers k, we have

$$\sum_{j=0}^{T_k-1} f(\bar{x}_j, \bar{x}_{j+1}, \bar{y}_j, \bar{y}_{j+1}) - \sum_{j=0}^{T_k-1} f(x_j, x_{j+1}, \bar{y}_j, \bar{y}_{j+1}) \ge \Gamma_0. \tag{5.98}$$

We claim that

$$x_j \to x_f \text{ as } j \to \infty.$$

For $j = 0, 1, \ldots$ define a function $g_j : X \times X \to R^1$ by

$$g_j(u_1, u_2) = f(u_1, u_2, \bar{y}_j, \bar{y}_{j+1}), \quad u_1, u_1 \in X. \tag{5.99}$$

In view of (5.96),

$$\lim_{j \to \infty} \|g_j - f^{(X)}\| = 0. \tag{5.100}$$

Since the pair of sequences $(\{\bar{x}_j\}_{j=0}^{\infty}, \{\bar{y}_j\}_{j=0}^{\infty})$ is (f)-minimal there exists a positive constant c_0 such that for each natural number T, we have

$$\sum_{i=0}^{T-1} f(\bar{x}_j, \bar{x}_{j+1}, \bar{y}_j, \bar{y}_{j+1}) \le \inf\{\sum_{j=0}^{T-1} f(u_j, u_{j+1}, \bar{y}_j, \tilde{u}_{j+1}) :$$

$$\{u_j\}_{j=0}^{T} \subset X, \ u_0 = z\} + c_0. \tag{5.101}$$

It follows from (5.98), (5.99), and (5.101) that the following property holds:

(P9) For each positive number Δ there exists an integer $j(\Delta) \ge 1$ such that for each pair of integers $n_1 \ge j(\Delta)$, $n_2 > n_1$ the sequence $\{x_j\}_{j=n_1}^{n_2}$ is $(\{g_j\}_{j=n_1}^{n_2-1}, X, \Delta)$-good.

Property (P9) and Theorem 5.4 imply that

$$\lim_{j\to\infty} x_j = x_f. \tag{5.102}$$

There exists a positive number ϵ_0 such that for each z_1, z_2, $\bar{z}_1$, $\bar{z}_2 \in X$ and each ξ_1, ξ_2, $\bar{\xi}_1$, $\bar{\xi}_2 \in Y$ which satisfy

$$|z_j - \bar{z}_j|, \ |\xi_j - \bar{\xi}_j| \le 2\epsilon_0, \quad j = 1, 2, \tag{5.103}$$

we have

$$|f(z_1, z_2, \xi_1, \xi_2) - f(\bar{z}_1, \bar{z}_2, \bar{\xi}_1, \bar{\xi}_2)| \le 8^{-1}\Gamma_0. \tag{5.104}$$

In view of (5.102) and (5.96), there exists an integer $j_0 \ge 8$ such that for all integers $j \ge j_0$,

$$|x_j - x_f| \le 2^{-1}\epsilon_0, \quad |\bar{x}_j - x_f| \le 2^{-1}\epsilon_0. \tag{5.105}$$

There exists a natural number s such that

$$T_s > j_0. \tag{5.106}$$

Define a sequence $\{x_j^*\}_{j=0}^{T_s} \subset X$ by

$$x_j^* = x_j, \ j = 0, \ldots T_s - 1, \quad x_{T_s}^* = \bar{x}_{T_s}. \tag{5.107}$$

Since the pair of sequences $(\{\bar{x}_j\}_{j=0}^{\infty}, \{\bar{y}_j\}_{j=0}^{\infty})$ is (f)-minimal we have

$$\sum_{j=0}^{T_s-1} f(\bar{x}_j, \bar{x}_{j+1}, \bar{y}_j, \bar{y}_{j+1}) - \sum_{j=0}^{T_s-1} f(x_j^*, x_{j+1}^*, \bar{y}_j, \bar{y}_{j+1}) \le 0. \tag{5.108}$$

On the other hand by (5.98), (5.105)–(5.107) and the definition of ϵ_0 (see (5.103), (5.104)),

$$
\sum_{j=0}^{T_s-1} f(\bar{x}_j, \bar{x}_{j+1}, \bar{y}_j, \bar{y}_{j+1}) - \sum_{j=0}^{T_s-1} f(x_j^*, x_{j+1}^*, \bar{y}_j, \bar{y}_{j+1})
$$

$$
= \sum_{j=0}^{T_s-1} f(\bar{x}_j, \bar{x}_{j+1}, \bar{y}_j, \bar{y}_{j+1}) - \sum_{j=0}^{T_s-1} f(x_j, x_{j+1}, \bar{y}_j, \bar{y}_{j+1})
$$

$$
+ f(x_{T_s-1}, x_{T_s}, \bar{y}_{T_s-1}, \bar{y}_{T_s}) - f(x_{T_s-1}^*, x_{T_s}^*, \bar{y}_{T_s-1}, \bar{y}_{T_s})
$$

$$
\geq \Gamma_0 + f(x_{T_s-1}, x_{T_s}, \bar{y}_{T_s-1}, \bar{y}_{T_s}) - f(x_{T_s-1}, \bar{x}_{T_s}, \bar{y}_{T_s-1}, \bar{y}_{T_s})
$$

$$
\geq \Gamma_0 - 8^{-1}\Gamma_0.
$$

This is contradictory to (5.108). The obtained contradiction proves that (5.97) holds.

Analogously we can show that for each sequence $\{y_j\}_{j=0}^{\infty} \subset Y$ satisfying $y_0 = y$

$$
\limsup_{T \to \infty} [\sum_{j=0}^{T-1} f(\bar{x}_j, \bar{x}_{j+1}, y_j, y_{j+1}) - \sum_{j=0}^{T-1} f(\bar{x}_j, \bar{x}_{j+1}, \bar{y}_j, \bar{y}_{j+1})] \leq 0.
$$

This implies that the pair of sequences $(\{\bar{x}_j\}_{j=0}^{\infty}, \{\bar{y}_j\}_{j=0}^{\infty})$ is (f)-overtaking optimal. This completes the proof of Theorem 5.5.

References

1. Anderson, B. D. O., & Moore, J. B. (1971). *Linear optimal control*. Englewood Cliffs: Prentice-Hall.
2. Arkin, V. I., & Evstigneev, I. V. (1987). *Stochastic models of control and economic dynamics*. London: Academic.
3. Aseev, S. M., & Kryazhimskiy, A. V. (2004). The Pontryagin maximum principle and transversality conditions for a class of optimal control problems with infinite time horizons. *SIAM Journal on Control and Optimization, 43*, 1094–1119.
4. Aseev, S. M., & Veliov, V. M. (2012). Maximum principle for infinite-horizon optimal control problems with dominating discount. *Dynamics of Continuous, Discrete and Impulsive Systems Series B, 19*, 43–63.
5. Atsumi, H. (1965). Neoclassical growth and the efficient program of capital accumulation. *Review of Economic Studies, 32*, 127–136.
6. Aubin, J. P., & Ekeland, I. (1984). *Applied nonlinear analysis*. New York: Wiley Interscience.
7. Aubry, S., & Le Daeron, P. Y. (1983). The discrete Frenkel-Kontorova model and its extensions I. *Physica D, 8*, 381–422.
8. Baumeister, J., Leitao, A., & Silva, G. N. (2007). On the value function for nonautonomous optimal control problem with infinite horizon. *Systems and Control Letters, 56*, 188–196.
9. Berkovitz, L. D. (1974). *Optimal control theory*. New York: Springer.
10. Berkovitz, L. D. (1974). Lower semicontinuity of integral functionals. *Transactions of the American Mathematical Society, 192*, 51–57.
11. Blot, J. (2009). Infinite-horizon Pontryagin principles without invertibility. *Journal of Nonlinear and Convex Analysis, 10*, 177–189.
12. Blot, J., & Cartigny, P. (2000). Optimality in infinite-horizon variational problems under sign conditions. *The Journal of Optimization Theory and Applications, 106*, 411–419.
13. Blot, J., & Hayek, N. (2000). Sufficient conditions for infinite-horizon calculus of variations problems. *ESAIM: Control, Optimisation and Calculus of Variations, 5*, 279–292.
14. Brock, W. A. (1970). On existence of weakly maximal programmes in a multi-sector economy. *Review of Economic Studies, 37*, 275–280.
15. Carlson, D. A. (1990). The existence of catching-up optimal solutions for a class of infinite horizon optimal control problems with time delay. *SIAM Journal on Control and Optimization, 28*, 402–422.
16. Carlson, D. A., Haurie, A., & Leizarowitz, A. (1991). *Infinite horizon optimal control*. Berlin: Springer.

© Springer International Publishing Switzerland 2014
A.J. Zaslavski, *Turnpike Phenomenon and Infinite Horizon Optimal Control*,
Springer Optimization and Its Applications 99, DOI 10.1007/978-3-319-08828-0

17. Carlson, D. A., Jabrane, A., & Haurie, A. (1987). Existence of overtaking solutions to infinite dimensional control problems on unbounded time intervals. *SIAM Journal on Control and Optimizaton, 25,* 517–1541.
18. Cartigny, P., & Michel, P. (2003). On a sufficient transversality condition for infinite horizon optimal control problems. *Automatica Journal IFAC, 39,* 1007–1010.
19. Cesari, L. (1983). *Optimization: Theory and applications.* New York: Springer.
20. Coleman, B. D., Marcus, M., & Mizel, V. J. (1992). On the thermodynamics of periodic phases. *Archive for Rational Mechanics and Analysis, 117,* 321–347.
21. Evstigneev, I. V., & Flam, S. D. (1998). Rapid growth paths in multivalued dynamical systems generated by homogeneous convex stochastic operators. *Set-Valued Analysis, 6,* 61–81.
22. Gaitsgory, V., Rossomakhine, S., & Thatcher, N. (2012). Approximate solution of the HJB inequality related to the infinite horizon optimal control problem with discounting. *Dynamics of Continuous, Discrete and Impulsive Systems Series B, 19,* 65–92.
23. Gale, D. (1967). On optimal development in a multi-sector economy. *Review of Economic Studies, 34,* 1–18.
24. Guo, X., & Hernandez-Lerma, O. (2005). Zero-sum continuous-time Markov games with unbounded transition and discounted payoff rates. *Bernoulli, 11,* 1009–1029.
25. Hayek, N. (2011). Infinite horizon multiobjective optimal control problems in the discrete time case. *Optimization, 60,* 509–529.
26. Jasso-Fuentes, H., & Hernandez-Lerma, O. (2008). Characterizations of overtaking optimality for controlled diffusion processes. *Applied Mathematics and Optimization, 57,* 349–369.
27. Kolokoltsov, V., & Yang, W. (2012). The turnpike theorems for Markov games. *Dynamic Games and Applications, 2,* 294–312.
28. Leizarowitz, A. (1985). Infinite horizon autonomous systems with unbounded cost. *Applied Mathematics and Optimization, 13,* 19–43.
29. Leizarowitz, A. (1986). Tracking nonperiodic trajectories with the overtaking criterion. *Applied Mathematics and Optimization, 14,* 155–171.
30. Leizarowitz, A., & Mizel, V. J. (1989). One dimensional infinite horizon variational problems arising in continuum mechanics. *Archive for Rational Mechanics and Analysis, 106,* 161–194.
31. Leizarowitz, A., & Zaslavski. A. J. (2005). On a class of infinite horizon optimal control problems with periodic cost functions. *Journal of Nonlinear and Convex Analysis 6,* 71–91.
32. Lykina, V., Pickenhain, S., & Wagner, M. (2008). Different interpretations of the improper integral objective in an infinite horizon control problem. *Journal of Mathematical Analysis and Applications, 340,* 498–510.
33. Makarov, V. L., & Rubinov, A. M. (1977) *Mathematical theory of economic dynamics and equilibria.* New York: Springer.
34. Malinowska, A. B., Martins, N., & Torres, D. F. M. (2011). Transversality conditions for infinite horizon variational problems on time scales. *Optimization Letters, 5,* 41–53.
35. Marcus, M., & Zaslavski, A. J. (1999). On a class of second order variational problems with constraints. *Israel Journal of Mathematics, 111,* 1–28.
36. Marcus. M., & Zaslavski, A. J. (1999). The structure of extremals of a class of second order variational problems. *Annales de l'Institut Henri Poincaré, Analyse Non Linéaire, 16,* 593–629.
37. Marcus, M., & Zaslavski, A. J. (2002). The structure and limiting behavior of locally optimal minimizers. *Annales de l'Institut Henri Poincaré, Analyse Non Linéaire, 19,* 343–370.
38. McKenzie, L. W. (1976). Turnpike theory. *Econometrica, 44,* 841–866.
39. Mordukhovich, B. S. (1990). Minimax design for a class of distributed parameter systems. *Automation and Remote Control, 50,* 1333–1340.
40. Mordukhovich, B. S. (2011). Optimal control and feedback design of state-constrained parabolic systems in uncertainly conditions. *Applied Analysis, 90,* 1075–1109.
41. Mordukhovich, B. S., Shvartsman, I. (2004). Optimization and feedback control of constrained parabolic systems under uncertain perturbations. In *Optimal control, stabilization and non-smooth analysis* (pp. 121–132). Lecture Notes in Control and Information Science. Berlin: Springer.

42. Moser, J. (1986). Minimal solutions of variational problems on a torus. *Annales de l'Institut Henri Poincaré, Analyse Non Linéaire, 3,* 229–272.
43. Ocana Anaya, E., Cartigny, P., & Loisel, P. (2009). Singular infinite horizon calculus of variations. Applications to fisheries management. *Journal of Nonlinear and Convex Analysis, 10,* 157–176.
44. Pickenhain, S., Lykina, V., & Wagner, M. (2008). On the lower semicontinuity of functionals involving Lebesgue or improper Riemann integrals in infinite horizon optimal control problems. *Control and Cybernetics, 37,* 451–468.
45. Rockafellar, R. T. (1970). *Convex analysis.* Princeton: Princeton University Press.
46. Rubinov, A. M. (1984). Economic dynamics. *Journal of Soviet Mathematics, 26,* 1975–2012.
47. Samuelson, P. A. (1965). A catenary turnpike theorem involving consumption and the golden rule. *American Economic Review, 55,* 486–496.
48. von Weizsacker, C. C. (1965). Existence of optimal programs of accumulation for an infinite horizon. *Review of Economic Studies, 32,* 85–104.
49. Zaslavski, A. J. (1987). Ground states in Frenkel-Kontorova model. *Mathematics of the USSR-Izvestiya, 29,* 323–354.
50. Zaslavski, A. J. (1995). Optimal programs on infinite horizon 1. *SIAM Journal on Control and Optimization, 33,* 1643–1660.
51. Zaslavski, A. J. (1995). Optimal programs on infinite horizon 2. *SIAM Journal on Control and Optimization, 33,* 1661–1686.
52. Zaslavski, A. J. (1998). Turnpike theorem for convex infinite dimensional discrete-time control systems. *Convex Analysis, 5,* 237–248.
53. Zaslavski, A. J. (1999). Turnpike property for dynamic discrete time zero-sum games. *Abstract and Applied Analysis, 4,* 21–48.
54. Zaslavski, A. J. (2000). Turnpike theorem for nonautonomous infinite dimensional discrete-time control systems. *Optimization, 48,* 69–92.
55. Zaslavski, A. J. (2005). The turnpike property of discrete-time control problems arising in economic dynamics. *Discrete and Continuous Dynamical Systems, B, 5,* 861–880.
56. Zaslavski, A. J. (2006). *Turnpike properties in the calculus of variations and optimal control.* New York: Springer.
57. Zaslavski, A. J. (2006). A nonintersection property for extremals of variational problems with vector-valued functions. *Annales de l'Institut Henri Poincaré, Analyse Non Linéaire, 23,* 929–948.
58. Zaslavski, A. J. (2007). Turnpike results for a discrete-time optimal control systems arising in economic dynamics. *Nonlinear Analysis, 67,* 2024–2049.
59. Zaslavski, A. J. (2008). A turnpike result for a class of problems of the calculus of variations with extended-valued integrands. *Journal of Convex Analysis, 15,* 869–890.
60. Zaslavski, A. J. (2009). Two turnpike results for a discrete-time optimal control systems. *Nonlinear Analysis, 71,* 902–909.
61. Zaslavski, A. J. (2009). Structure of approximate solutions of variational problems with extended-valued convex integrands. *ESAIM: Control, Optimization and the Calculus of Variations 15,* 872–894.
62. Zaslavski, A. J. (2010). Stability of a turnpike phenomenon for a discrete-time optimal control systems. *Journal of Optimization theory and Applications, 145,* 597–612.
63. Zaslavski, A. J. (2010). Optimal solutions for a class of infinite horizon variational problems with extended-valued integrands. *Optimization, 59,* 181–197.
64. Zaslavski, A. J. (2011). Turnpike properties of approximate solutions for discrete-time control systems. *Communications in Mathematical Analysis, 11,* 36–45.
65. Zaslavski, A. J. (2011). Structure of approximate solutions for a class of optimal control systems. *Journal of Mathematics and Applications, 34,* 00–14.
66. Zaslavski, A. J. (2011). Stability of a turnpike phenomenon for a class of optimal control systems in metric spaces. *Numerical Algebra, Control and Optimization, 1,* 245–260.

67. Zaslavski, A. J. (2011). Structure of approximate solutions for a class of optimal control systems. *Journal of Mathematics and Applications*, 34, 00–14.
68. Zaslavski, A. J. (2011). One dimensional infinite horizon nonconcave optimal control problems arising in economic dynamics. *Applied Mathematics and Optimization*, 64, 417–440.
69. Zaslavski, A. J. (2011). The existence and structure of approximate solutions of dynamic discrete time zero-sum games. *Journal of Nonlinear and Convex Analysis*, 12, 49–68.
70. Zaslavski, A. J. (2012). A generic turnpike result for a class of discrete-time optimal control systems. *Dynamics of Continuous, Discrete and Impulsive Systems Series B*, 19, 225–265.
71. Zaslavski, A. J. (2012). Existence and structure of solutions for a class of optimal control systems with discounting arising in economic dynamics. *Nonlinear Analysis: Real World Applications*, 13, 1749–1760.
72. Zaslavski, A. J. (2012). Existence of solutions for a class of infinite horizon optimal control problems with discounting. *Journal of Nonlinear and Convex Analysis*, 13, 637–655.
73. Zaslavski, A. J. (2013) *Structure of solutions of variational problems*. New York: Springer-Briefs in Optimization.
74. Zaslavski, A. J. (2013). Existence of solutions for a class of nonconcave infinite horizon optimal control problems. *Optimization*, 62, 115–130.
75. Zaslavski, A. J. (2013). Existence of solutions for a class of infinite horizon optimal control problems without discounting arising in economic dynamics. In *Proceedings of an International Conference, Complex Analysis and Dynamical Systems V. Contemporary Mathematics* (Vol. 591, pp. 291–314). Providence: American Mathematical Society.
76. Zaslavski, A. J., Leizarowitz, A. (1997). Optimal solutions of linear control systems with nonperiodic integrands. *Mathematical Methods of Operations Research*, 22, 726–746.

Index

A
Absolutely continuous function, 16, 147
Admissible control, 330
Admissible trajectory, 15
Approximate solution, 1, 5, 15, 21
Asymptotic turnpike property, 16, 18, 25, 149
Autonomous discrete-time control system, 15, 23
Autonomous variational problem, 147

B
Baire category approach, 16
Balanced equilibrium path, 2
Borel function, 318

C
Cardinality of a set, 5
Compact metric space, 15, 17, 23
Complete metric space, 339
Constrained problems, 16
Control system, 15
Convex discrete-time problems, 1
Convex function, 15, 148
Convex set, 16

D
Differentiable function, 2, 3, 15
Discrete-time problem, 15, 313

E
Euclidean norm, 147
Euclidean space, 1, 16
Extended-valued integrand, 147

G
Good function, 149
Good pair of sequences, 340
Good program, 16, 18
Good sequence, 10, 11, 14

I
Increasing function, 20, 148
Infinite horizon, 15
Infinite horizon optimal control problem, 19
Inner product, 1, 147
Interior point, 17, 24, 148

L
Lebesgue measurable set, 147
Lebesgue measure, 147
Lower semicontinuous function, 148
Lower semicontinuous integrand, 147

M
Minimal function, 207

N
Neumann path, 2

O
Objective function, 2, 15
One-sector model, 20, 221
Optimal control problem, 2
Optimal pair of sequences, 341
Optimal trajectory, 2
Optimality criterion, 13, 15

© Springer International Publishing Switzerland 2014
A.J. Zaslavski, *Turnpike Phenomenon and Infinite Horizon Optimal Control*,
Springer Optimization and Its Applications 99, DOI 10.1007/978-3-319-08828-0

Overtaking optimal function, 150
Overtaking optimal pair, 341, 345
Overtaking optimal program, 19, 25
Overtaking optimal sequence, 13, 14
Overtaking optimal solution, 19

P
Periodic integrand, 318
Program, 15, 17, 20, 23, 222, 223, 226

S
Strategy, 339
Strictly convex function, 2, 3, 53

T
Trajectory, 16, 319
Turnpike, 15, 147
Turnpike phenomenon, 15
Turnpike property, 2, 5, 16, 142, 340
Turnpike result, 15
Two-sector model, 20, 260

U
Upper semicontinuous function, 15, 17, 23

Z
Zero-sum game, 339

Printed in the United States
By Bookmasters